Nanoparticle-Mediated Targeted Drug Delivery Systems

Nanoparticle-Mediated Targeted Drug Delivery Systems

Guest Editors

Nicolas Tournier
Toshihiko Tashima

MDPI

Basel • Beijing • Wuhan • Barcelona • Belgrade • Novi Sad • Cluj • Manchester

Guest Editors
Nicolas Tournier
Laboratoire d'Imagerie
Biomédicale Multimodale
Université Paris-Saclay
Orsay
France

Toshihiko Tashima
Research and Development
Office
Tashima Laboratories of Arts
and Sciences
Yokohama
Japan

Editorial Office
MDPI AG
Grosspeteranlage 5
4052 Basel, Switzerland

This is a reprint of the Special Issue, published open access by the journal *Pharmaceutics* (ISSN 1999-4923), freely accessible at: https://www.mdpi.com/journal/pharmaceutics/special_issues/YR2L99MIJM.

For citation purposes, cite each article independently as indicated on the article page online and as indicated below:

Lastname, A.A.; Lastname, B.B. Article Title. *Journal Name* **Year**, *Volume Number*, Page Range.

ISBN 978-3-7258-6470-6 (Hbk)
ISBN 978-3-7258-6471-3 (PDF)
https://doi.org/10.3390/books978-3-7258-6471-3

Contents

About the Editors

Nicolas Tournier

Nicolas Tournier (PhD, PharmD) is a hospital pharmacist and pharmacologist at the Centre Hospitalier Frédéric Joliot (CEA/SHFJ) in Orsay (France). He leads the Clinical Applications group of the IMIV research unit (Imagerie Moléculaire in Vivo—UMR 1023 Inserm/CEA/Université Paris Sud/Université Paris Saclay—ERL 9218 CNRS). The main aim of this group is to promote the clinical translation of original imaging probes and systems and develop their clinical applications. Their work builds upon the concept of pharmacological and pharmacokinetic imaging. They notably use radiolabeled analogues of drugs to study the impact of membrane transporters on tissue distribution and elimination and estimate the corresponding risk of drug–drug interactions. To that end, the team benefits from a translational and multidisciplinary environment, making the best with recent developments in radiochemistry, medical physics, molecular imaging, nuclear medicine, and pharmacokinetics.

Toshihiko Tashima

Toshihiko Tashima has been involved in synthetic research and the discovery of physiologically active substances for many years and is currently engaged in the pharmaceutical department. His research interests encompass pharmaceuticals, with a focus on drug development and drug delivery systems. His research keywords include medicinal chemistry, drug designing, and bio-organics. Dr. Tashima serves as an Editorial Member and reviewer for several internationally renowned journals and has fulfilled various administrative responsibilities. He has authored numerous research articles and books related to medicine. He believes that science should be enjoyable. With advancements in internet technology and open-access journals, science has become infinitely familiar to us. He encourages many people to take advantage of the opportunities to enjoy science.

Preface

Nanoparticles are increasingly utilized as drug delivery carriers in pharmaceutical research and development. Nanoparticle-based drug delivery systems offer multiple advantages, including enhanced drug bioavailability and absorption, as well as the reduction in drug aggregation, enzymatic degradation, renal clearance, and undesired drug interactions. Naked drugs, such as RNAs and proteins, are susceptible to degradation by enzymes present in serum and the gastrointestinal tract. The tight junctions of capillary endothelial cells pose a significant barrier, preventing both low- and high-molecular-weight drugs from crossing the endothelium from the systemic circulation into target tissues such as the brain, eyes, and heart. Furthermore, unmodified nanoparticles are rapidly cleared by cell-dependent mechanisms—including the reticuloendothelial system (RES) and the mononuclear phagocyte system (MPS)—as well as by cell-independent mechanisms, such as glomerular filtration, hepatic sinusoidal uptake, and splenic sinusoidal filtration. These biological barriers contribute to the unmet medical needs of patients with diseases affecting these critical tissues. In addition to these challenges, there are various technical hurdles. Nanoparticles must be biocompatible, biodegradable, and capable of controlled drug release. Appropriate materials and designs are required to match the characteristics of the drug cargo and the target site, in order to enable active and non-invasive delivery to diseased tissues.

To address these issues, the Special Issue titled 'Nanoparticle-Mediated Targeted Drug Delivery Systems' has been curated. This Issue includes outstanding contributions that explore the use of nanocarriers for drug delivery. The reprint features one Editorial, six original Articles, and five Review papers. We hope this compilation provides valuable insights into recent scientific advancements and supports researchers in their ongoing work.

Nicolas Tournier and Toshihiko Tashima
Guest Editors

pharmaceutics

MDPI

Editorial

Nanoparticle-Mediated Targeted Drug Delivery Systems

Toshihiko Tashima [1,*,†] and Nicolas Tournier [2,*,†]

1 Tashima Laboratories of Arts and Sciences, 1239-5 Toriyama-cho, Kohoku-ku, Yokohama 222-0035, Japan
2 Laboratoire d'Imagerie Biomédicale Multimodale, BIOMAPS, Université Paris-Saclay, CEA, CNRS, Inserm, Service Hospitalier Frédéric Joliot, 4 Place du Général Leclerc, 91401 Orsay, France
* Correspondence: tashima_lab@yahoo.co.jp (T.T.); nicolas.tournier@cea.fr (N.T.)
† These authors contributed equally to this work.

We will be serving as the Guest Editors for this very interesting Special Issue on "Nanoparticle-Mediated Targeted Drug Delivery Systems". In drug discovery and development, selective and efficient transmembrane delivery of therapeutic substances into cells remains a major challenge. This difficulty arises from several factors, including molecular size, hydrophilic/hydrophobic balance, and the substrate specificity of membrane transporters—such as influx transporters from the solute carrier (SLC) family [1,2] and efflux transporters like multidrug resistance protein 1 (MDR1, also known as P-glycoprotein) [3,4]—as well as limited enzymatic stability. These barriers are rooted in complex transport and metabolic systems within the cell, based on the machinery systems regulated by structuralism, wherein the structure of biological components fundamentally determines their function and interaction within cellular pathways [5,6]. As a result, numerous pharmaceutical challenges arise. Substance impermeability significantly contributes to low bioavailability, particularly in the case of central nervous system (CNS) drugs, whose entry from the bloodstream into the brain is restricted by the blood–brain barrier (BBB) [7,8]. This barrier is maintained by tight junctions and efflux transporters such as MDR1. In addition, off-target side effects may occur due to improper tissue distribution or unintended drug–protein interactions during systemic distribution. Substances such as RNAs and peptides are often enzymatically degraded in the serum or small intestine before reaching their target sites. Therefore, innovative drug delivery systems are needed to overcome these challenges. Nanodelivery systems have shown promise in addressing these issues [9]. Notably, microRNAs (miRNAs) encapsulated in exosomes released from donor cells can be delivered to recipient cells through intercellular communication, without degradation by RNases [10,11]. Exosomes are small, lipid-based extracellular vesicles. Interestingly, cancer cells have exploited this system by utilizing exosomal miRNAs to modulate the tumor microenvironment, particularly during metastasis. Similarly, nanoparticles can encapsulate therapeutic agents as cargo to protect them from degradation and unwanted interactions. Comirnaty® (tozinameran, BNT162b2), approved by the FDA on 23 August 2021, and Spikevax® (Moderna COVID-19 vaccine), approved on 31 January 2022, are examples of lipid nanoparticle-based mRNA vaccines developed for COVID-19 treatment [12]. Furthermore, nanoparticles can be functionalized with targeting vectors that enhance cellular uptake, such as monoclonal antibodies (mAbs) [13], cell-penetrating peptides (CPPs) [14], or tumor-homing peptides (THPs) [15]. Although the mechanisms of CPP internalization are not fully understood, both endocytosis and direct translocation are generally accepted pathways. It is thought that negatively charged heparan sulfate chains, which extend from proteoglycans (HSPGs) on the cell surface, facilitate receptor-mediated endocytosis by serving as receptors for cationic CPPs, such as the trans-activator of transcription (TAT) peptide (GRKKRRQRRRPQ) derived from HIV-1 [16]. Among representative THPs, RGD peptides

(Arg-Gly-Asp) specifically bind to αvβ3 and αvβ5 integrins on the surface of cancer cells [17], while NGR peptides (Asn-Gly-Arg) target the receptor aminopeptidase N [18]. Although the binding of THPs to their receptors can induce various biological responses, endocytosis is one of the primary mechanisms involved. Additionally, the molecular Trojan horse strategy utilizes monoclonal antibodies (mAbs) targeting specific receptors to initiate receptor-mediated endocytosis [19]. For example, Izcargo® (pabinafusp alfa), a conjugate of an anti-transferrin receptor (TfR) mAb and human iduronate-2-sulfatase, was approved in Japan in May 2021 for the treatment of all forms of mucopolysaccharidosis type II (MPS II). This therapeutic crosses the BBB via TfR-mediated transcytosis [20]. Numerous types of nanoparticles functionalized with such targeting vectors have since been developed [21,22]. This Special Issue aims to highlight recent advances and emerging trends in this rapidly evolving field.

Currently, nanoparticle-based drug delivery systems are widely utilized due to their exceptional versatility and adaptability, which can be achieved through appropriate material selection and practical surface modifications to suit various therapeutic needs and conditions. Nanoparticles are primarily composed of one or more of the following constituents: (i) biodegradable synthetic polymers, (ii) natural polymers (e.g., chitosan, PLGA, PGA), (iii) lipids (e.g., liposomes, micelles, exosomes), (iv) inorganic materials (e.g., gold, silica, Fe_3O_4), (v) organic materials (e.g., albumin, monoclonal antibodies, virosomes), (vi) emulsions, and (vii) other specialized components. Biodegradability and biocompatibility are critical factors in minimizing adverse side effects. In practice, functionalized or engineered nanoparticles [23] have been developed with various specialized features, including: (i) passive targeting based on the enhanced permeability and retention (EPR) effect, particularly in solid tumors [24,25]; (ii) active targeting through ligand–receptor interactions that facilitate receptor-mediated endocytosis, using the aforementioned vectors; (iii) magnetic responsiveness; (iv) pH sensitivity; (v) thermosensitivity; (vi) enteric coating for oral administration [26]; and (vii) other application-specific functionalities. Nanoparticles are employed in drug therapy for the treatment of a wide range of diseases, including (i) cancers; (ii) central nervous system (CNS) disorders such as Alzheimer's disease, Parkinson's disease, ischemic stroke, and glioma; (iii) infectious diseases, including influenza, cytomegalovirus, and COVID-19; (iv) cardiovascular diseases; (v) pulmonary diseases; (vi) ocular diseases; and (vii) other conditions. As drug carriers, nanoparticles can load or encapsulate a variety of therapeutic agents, including (i) low-molecular-weight compounds; (ii) peptides; (iii) nucleic acids (e.g., RNAs and DNAs); and (iv) other bioactive molecules.

Active targeting based on ligand–receptor interactions, which facilitate receptor-mediated endocytosis through specific vectors, has been utilized to deliver therapeutic agents to selected tissues such as (1) the heart, (2) (3) the respiratory system, and (4) the brain. While drug delivery to the brain has been extensively studied [23], research on targeted delivery to other tissues remains limited.

(1) Toshihiko Tashima has focused on nanoparticle-based delivery to the heart using nanodelivery systems. In this approach, nanoparticles actively traverse the cardiac capillary endothelium via receptor-mediated transcytosis. Conversely, during myocardial infarction (MI) and chronic heart failure (CHF), the tight junctions of the cardiac capillary endothelium become disrupted, allowing nanoparticles to passively diffuse through these gaps into the heart [27]. This passive transport mechanism resembles the EPR effect observed in solid tumors [24,25].

(2) Devesh U. Kapoor et al. [28] reviewed the use of polymeric nanoparticles for targeted lung cancer therapy. Reported delivery routes include intravenous, pulmonary (inhalation), oral, and intratumoral administration. Multifunctional nanoparticles have been shown to enhance tumor-specific accumulation and cellular uptake, thereby improv-

ing bioavailability while reducing systemic toxicity. (3) Moreover, the coronavirus disease 2019 (COVID-19) pandemic, caused by severe acute respiratory syndrome coronavirus 2 (SARS-CoV-2) [29], had a devastating impact on global public health and triggered the most significant economic crisis in over a century. Comirnaty® and Spikevax®, both mRNA-encapsulated lipid nanoparticles, were administered via intramuscular injection; however, injection site reactions such as pain, redness, and swelling were commonly reported [30]. Nicolette Frank et al. investigated novel nasal mucoadhesive nanoformulations containing lipid-soluble epigallocatechin-3-gallate (EGCG) for the treatment of long COVID. EGCG-monopalmitate (EC16m), a derivative of green tea, is known to possess multiple antiviral mechanisms in addition to anti-inflammatory, antioxidant, and neuroprotective properties. It is hypothesized that persistent SARS-CoV-2 infection in the nasal neuroepithelium may invade support and stem cells in the olfactory mucosa, thereby contributing to neurological symptoms, including CNS inflammation and oxidative stress. A 5 min exposure to a saline-based EC16m mucoadhesive nanoformulation—specifically, formulation D (FD), with a particle size of 257 ± 134 nm and a Zeta potential of -51.31 ± 1.22 mV—resulted in 99.9% inactivation of β-coronavirus OC43 in TCID50 assays using MRC-5 cells. In contrast, when HCT-8 human intestinal epithelial cells were exposed to the same formulation for one hour, no significant difference in cell viability was observed compared to the untreated control. Typically, a full nasal spray volume is approximately 0.07 mL; each spray of the FD formulation delivers about 455 million EC16m nanoparticles, an amount considered sufficient to inactivate coronavirus present in the olfactory mucosa. Additionally, a small portion of EC16m nanoparticles is expected to enter the CNS before clearance by nasal cilia, where they subsequently release free EGCG [31].

(4) Nasal administration of nanoparticles is considered a promising strategy for the treatment of CNS diseases, offering an alternative route for brain delivery that bypasses the BBB. Despite this, non-invasive delivery across the BBB remains a major pharmacokinetic challenge in CNS drug development. Alberta De Capua et al. reported the peptide functionalization of emulsion-based nanocarriers to enhance BBB penetration. The CRT peptide (CRTIGPSVC), a bioactive iron-mimetic peptide, selectively targets bEnd.3 cells—a widely used in vitro model of the BBB—via Tf–TfR interactions. Paclitaxel-loaded nanoemulsions coated with two functional layers—chitosan and hyaluronic acid conjugated to the CRT peptide—showed a 41.5% increase in cellular uptake compared to the negative control in bEnd.3 cells. Additionally, these functionalized nanoemulsions resulted in a 33.05% increase in cytotoxicity compared to undecorated polyethylene glycol (PEG)-based nanoemulsions. Paclitaxel, a plant-derived alkaloid, inhibits mitotic progression. The CRT–PEG–streptavidin–hyaluronic acid–biotin–chitosan nanoemulsions exhibited an average diameter of 137 ± 7 nm and a Zeta potential of -30 ± 1 mV [32]. The use of TfR-mediated transcytosis represents a promising strategy for nanoparticle delivery to the brain. Streptavidin, a biotin-binding protein, enables facile surface modification of nanoparticles through strong non-covalent interactions [33,34]. Other examples of non-covalent, transient linkages include electrostatic interactions, such as those formed between glutamic acid (Glu) and arginine (Arg) residues [35].

Nanodelivery systems are suitable to deliver certain types of substances, such as polyphenols including (5) resveratrol, cytotoxic agents including (6) cabazitaxel and RAS-selective lethal 3, and enzymatically unstable materials including (7) peptides, (8) nucleic acids, and (9) antibiotics.

(5) Resveratrol (Res) is a type of polyphenol found in the skin of red grapes and other botanicals, and it possesses strong antioxidant potential. Kengo Banshoya et al. developed a water-soluble nanomicellar formulation loaded with trans-resveratrol using polyethylene glycol monostearate (stPEG) for the treatment of intracerebral hemorrhage (ICH). They also successfully formulated poorly water-soluble compounds such as α-tocopherol [36] and

coenzyme Q10 [37] into water-soluble stPEG micelles. Trans-resveratrol reduces oxidative stress, which plays an important role in brain injury following ICH. In an in vivo assay using a collagenase-induced ICH mouse model, stPEG40/Res micelles administered via the tail vein exhibited neuroprotective effects by suppressing motor dysfunction, compared to stPEG40 micelles alone as a control [38]. Nanoparticles with diameters smaller than 100 nm can evade rapid sequestration by the reticuloendothelial system in the liver, spleen, and lungs [39]. The diameter of stPEG/Res micelles suitable for intravenous administration was adjusted within this range—85.4 nm for stPEG10/Res (Res content: 10.3% w/w) and 11.8 nm for stPEG40/Res (Res content: 8.6% w/w). Therefore, poorly water-soluble, low-molecular-weight compounds such as α-tocopherol, coenzyme Q10, and (4) Res can be effectively delivered using this system.

(6) Remya Valsalakumari et al. investigated the potential of cabazitaxel (CBZ) and RAS-selective lethal 3 (RSL3) in combinatorial nanomedicine. It is well established that combination therapies in cancer treatment often exhibit greater efficacy than monotherapies. In this context, the combination of CBZ and RSL3, co-encapsulated within poly(2-ethyl butyl cyanoacrylate) (PEBCA) nanoparticles, produced synergistic effects against various breast cancer cell lines [40].

(7) Nanodelivery systems are effective for delivering enzymatically unstable substances such as nucleic acids and peptides. Angiotensin-converting enzyme (ACE) inhibitors, such as captopril, were originally discovered from snake venom toxins [41]. Thus, venom-derived toxins have considerable potential as drug leads [42,43], despite the historical anecdote that Cleopatra VII, Queen of the Ptolemaic Kingdom of Egypt, is believed to have committed suicide by snakebite. As an example of venom-based substance delivery, Álisson E. F. Alves et al. reported the development of snake venom-loaded nanobiosystems for advanced medical applications [44]. Snake venoms, especially in their structurally unmodified form, exhibit diverse biological activities, including cytotoxic, neurotoxic, and haemotoxic effects. Encapsulation of these toxins in nanocarriers can not only protect them from enzymatic degradation but also help mitigate off-target side effects by controlling their biodistribution. Interestingly, certain snake venom components—such as bradykinin-potentiating peptides (BPPs) derived from *Bothrops jararaca* venom [45]—function as cell-penetrating peptides (CPPs) and can thus serve as vectors when incorporated into nanoparticles. Therefore, snake venoms possess both pharmacodynamic and pharmacokinetic utility in nanoparticle-based drug delivery systems.

(8) Furthermore, the clustered regularly interspaced short palindromic repeats (CRISPR)/CRISPR-associated (Cas) system has emerged as a revolutionary and straightforward tool for genome editing [46,47]. However, in vivo gene editing remains significantly more challenging than in vitro or ex vivo approaches, due to factors such as the enzymatic instability of RNAs, difficulty in achieving selective distribution to target tissues, and barriers to transmembrane transport into cells. In contrast, in vitro and ex vivo genome editing—such as gene delivery into fertilized eggs, chimeric antigen receptor (CAR)-T cells, or mesenchymal stem cells (MSCs)—can be achieved using established techniques including lipofection (cationic lipid-mediated transfection), electroporation (electric pulse-mediated delivery), viral vector systems, or microinjection [48]. To address the limitations of in vivo gene editing, CRISPR/Cas systems delivered via nanocarriers offer a promising solution. As a strategy for delivering enzymatically unstable substances such as nucleic acids and peptides, Toshihiko Tashima reviewed non-invasive transmembrane delivery methods for CRISPR/Cas9 ribonucleoproteins (Cas9 RNPs) into cells [49].

(9) Javiera Carrasco-Rojas et al. developed a nanostructured lipid carrier (NLC)-based delivery system for rifampicin. The NLCs exhibited sustained drug release, best described by the Korsmeyer–Peppas model, and significantly enhanced antibacterial activity against

Staphylococcus aureus, with an IC_{50} value of 0.46 ng/mL—approximately three times lower than that of rifampicin alone [50]. The Korsmeyer–Peppas model is commonly used to analyze the drug release kinetics from pharmaceutical formulations [51–53].

Among various nanocarriers, nanogels are attracting increasing attention as a highly promising drug delivery platform. They are composed of hydrophilic cross-linked polymers capable of absorbing large amounts of water or biological fluids, along with encapsulated cargo materials. This structure imparts several advantageous properties, including high drug loading capacity, tunable permeability, controllable particle size, minimal immunogenicity, colloidal stability, biocompatibility, and biodegradability [54].

(10) Tapinarof is an aryl hydrocarbon receptor (AhR) agonist that exerts therapeutic effects by downregulating proinflammatory cytokines, and is used in the treatment of psoriasis [55]. Clinical studies have shown that tapinarof cream (1%) is both effective and generally well-tolerated [56]. Barbara Balogh et al. explored emerging therapeutic strategies involving nanogel-based formulations of tapinarof as a novel alternative for topical psoriasis treatment. Nanogels offer several key advantages in transdermal drug delivery, including enhanced skin hydration, targeted delivery to skin layers, and controlled drug release. In their study, nanogel formulations containing tapinarof were prepared using Carbopol 940 and 936 polymers, along with excipients such as Tween 80, Kolliphor, and oleic acid, which facilitated the release of tapinarof [57].

(11) Moreover, Shaul D. Cemal et al. demonstrated that voriconazole-loaded nanohydrogels exhibited potent antifungal activity against clinical fungal isolates obtained from the respiratory secretions of cystic fibrosis (CF) patients. Drug release was assessed by detecting the fluorescence of the released voriconazole. These findings provide a foundation for future in vivo investigations [58].

Overall, the articles in this Special Issue highlight recent advances in nanoparticle-mediated targeted drug delivery systems and are expected to contribute significantly to the development of this field. We extend our sincere gratitude to all the authors for their outstanding contributions, as well as to the Assistant Editors for their invaluable support.

Funding: This research received no external funding.

Institutional Review Board Statement: Not applicable.

Conflicts of Interest: The authors declare no conflicts of interest.

References

1. Wang, W.W.; Gallo, L.; Jadhav, A.; Hawkins, R.; Parker, C.G. The Druggability of Solute Carriers. *J. Med. Chem.* **2020**, *63*, 3834–3867. [CrossRef]
2. Dvorak, V.; Superti-Furga, G. Structural and functional annotation of solute carrier transporters: Implication for drug discovery. *Expert Opin. Drug Discov.* **2023**, *18*, 1099–1115. [CrossRef] [PubMed]
3. Löscher, W.; Potschka, H. Drug resistance in brain diseases and the role of drug efflux transporters. *Nat. Rev. Neurosci.* **2005**, *6*, 591–602. [CrossRef] [PubMed]
4. Pote, M.S.; Gacche, R.N. ATP-binding cassette efflux transporters and MDR in cancer. *Drug Discov. Today* **2023**, *28*, 103537. [CrossRef] [PubMed]
5. Laughlin, C.D.; D'Aquili, E.G. *Biogenetic Structuralism*; Columbia University Press: New York, NY, USA, 1974.
6. Leavy, S.A. Biogenetic Structuralism. *Yale J. Biol. Med.* **1976**, *49*, 420–421.
7. Wu, D.; Chen, Q.; Chen, X.; Han, F.; Chen, Z.; Wang, Y. The blood–brain barrier: Structure, regulation and drug delivery. *Signal Transduct. Target. Ther.* **2023**, *8*, 217. [CrossRef]
8. Terstappen, G.C.; Meyer, A.H.; Bell, R.D.; Zhang, W. Strategies for delivering therapeutics across the blood–brain barrier. *Nat. Rev. Drug Discov.* **2021**, *20*, 362–383. [CrossRef]
9. Mitchell, M.J.; Billingsley, M.M.; Haley, R.M.; Wechsler, M.E.; Peppas, N.A.; Langer, R. Engineering precision nanoparticles for drug delivery. *Nat. Rev. Drug Discov.* **2021**, *20*, 101–124. [CrossRef]

10. Shang, R.; Lee, S.; Senavirathne, G.; Lai, E.C. microRNAs in action: Biogenesis, function and regulation. *Nat. Rev. Genet.* **2023**, *24*, 816–833. [CrossRef]
11. Saadh, M.J.; Khalifehsoltani, A.; Hussein, A.H.A.; Allela, O.Q.B.; Sameer, H.N.; Rizaev, J.; Hameed, H.G.; Idan, A.H.; Alsaikhan, F. Exosomal microRNAs in cancer metastasis: A bridge between tumor micro and macroenvironment. *Pathol. Res. Pract.* **2024**, *263*, 155666. [CrossRef]
12. Szebeni, J.; Kiss, B.; Bozó, T.; Turjeman, K.; Levi-Kalisman, Y.; Barenholz, Y.; Kellermayer, M. Insights into the Structure of Comirnaty Covid-19 Vaccine: A Theory on Soft, Partially Bilayer-Covered Nanoparticles with Hydrogen Bond-Stabilized mRNA–Lipid Complexes. *ACS Nano* **2023**, *17*, 13147–13157. [CrossRef] [PubMed]
13. Pardridge, W.M. Kinetics of Blood-Brain Barrier Transport of Monoclonal Antibodies Targeting the Insulin Receptor and the Transferrin Receptor. *Pharmaceuticals* **2022**, *15*, 3. [CrossRef] [PubMed]
14. Varnamkhasti, B.S.; Jafari, S.; Taghavi, F.; Alaei, L.; Izadi, Z.; Lotfabadi, A.; Dehghanian, M.; Jaymand, M.; Derakhshankhah, H.; Saboury, A.A. Cell-Penetrating Peptides: As a Promising Theranostics Strategy to Circumvent the Blood-Brain Barrier for CNS Diseases. *Curr. Drug Deliv.* **2020**, *17*, 375–386. [CrossRef] [PubMed]
15. Cho, C.-F.; Farquhar, C.E.; Fadzen, C.M.; Scott, B.; Zhuang, P.; von Spreckelsen, N.; Loas, A.; Hartrampf, N.; Pentelute, B.L.; Lawler, S.E. A Tumor-Homing Peptide Platform Enhances Drug Solubility, Improves Blood-Brain Barrier Permeability and Targets Glioblastoma. *Cancers* **2022**, *14*, 2207. [CrossRef]
16. Christianson, H.C.; Belting, M. Heparan sulfate proteoglycan as a cell-surface endocytosis receptor. *Matrix Biol.* **2014**, *35*, 51–55. [CrossRef]
17. Javid, H.; Oryani, M.A.; Rezagholinejad, N.; Esparham, A.; Tajaldini, M.; Karimi-Shahri, M. RGD peptide in cancer targeting: Benefits, challenges, solutions, and possible integrin–RGD interactions. *Cancer Med.* **2024**, *13*, e6800. [CrossRef]
18. Li, X.; Fu, H.; Wang, J.; Liu, W.; Deng, H.; Zhao, P.; Liao, W.; Yang, Y.; Wei, H.; Yang, X.; et al. Multimodality labeling of NGR-functionalized hyaluronan for tumor targeting and radiotherapy. *Eur. J. Pharm. Sci.* **2021**, *161*, 105775. [CrossRef]
19. Pardridge, W.M. Molecular Trojan horses for blood-brain barrier drug delivery. *Discov. Med.* **2006**, *6*, 139–143. [CrossRef]
20. Sonoda, H.; Minami, K. IZCARGO®: The world's first biological drug applied with brain drug delivery technology. *Drug Deliv. Syst.* **2023**, *38*, 68–74. [CrossRef]
21. Chen, L.; Hong, W.; Ren, W.; Xu, T.; Qian, Z.; He, Z. Recent progress in targeted delivery vectors based on biomimetic nanoparticles. *Signal Transduct. Target. Ther.* **2021**, *6*, 225. [CrossRef]
22. Diez-Pascual, A.M.; Rahdar, A. Functional Nanomaterials in Biomedicine: Current Uses and Potential Applications. *ChemMedChem* **2022**, *17*, e202200142. [CrossRef] [PubMed]
23. Tashima, T. Smart Strategies for Therapeutic Agent Delivery into Brain across the Blood-Brain Barrier Using Receptor-Mediated Transcytosis. *Chem. Pharm. Bull.* **2020**, *68*, 316–325. [CrossRef] [PubMed]
24. Torrice, M. Does nanomedicine have a delivery problem? *ACS Cent. Sci.* **2016**, *2*, 434–437. [CrossRef] [PubMed]
25. Matsumura, Y.; Maeda, H. A new concept for macromolecular therapeutics in cancer chemotherapy: Mechanism of tumor-itropic accumulation of proteins and the antitumor agent smancs. *Cancer Res.* **1986**, *46*, 6387–6392.
26. Tashima, T. Delivery of Orally Administered Digestible Antibodies Using Nanoparticles. *Int. J. Mol. Sci.* **2021**, *22*, 3349. [CrossRef]
27. Tashima, T. Nanoparticle Based TargetedDrugDelivery MethodsforHeart-Specific Distribution in Cardiovascular Therapy. *Pharmaceutics* **2025**, *17*, 1365. [CrossRef]
28. Kapoor, D.U.; Gandhi, S.M.; Swarn, S.; Lal, B.; Prajapati, B.G.; Khondee, S.; Mangmool, S.; Singh, S.; Chittasupho, C. Polymeric Nanoparticles for Targeted Lung Cancer Treatment: Review and Perspectives. *Pharmaceutics* **2025**, *17*, 1091. [CrossRef]
29. The World Health Organization. Coronavirus Disease (COVID-19) Pandemic. Available online: https://www.who.int/europe/emergencies/situations/covid-19 (accessed on 1 January 2025).
30. Tomita, K.; Okada, S.; Sugihara, S.; Ikeuchi, T.; Touge, H.; Hasegawa, J.; Yamasaki, A. Physical Characteristics of Injection Site Pain After COVID-19 mRNA BNT162b2 Vaccination. *Yonago Acta Medica* **2021**, *64*, 339–344. [CrossRef]
31. Frank, N.; Dickinson, D.; Lovett, G.; Liu, Y.; Yu, H.; Cai, J.; Yao, B.; Jiang, X.; Hsu, S. Evaluation of Novel Nasal Mucoadhesive Nanoformulations Containing Lipid-Soluble EGCG for Long COVID Treatment. *Pharmaceutics* **2024**, *16*, 791. [CrossRef]
32. De Capua, A.; Vecchione, R.; Sgambato, C.; Chino, M.; Lagreca, E.; Lombardi, A.; Netti, P.A. Peptide Functionalization of Emulsion-Based Nanocarrier to Improve Uptake across Blood–Brain Barrier. *Pharmaceutics* **2024**, *16*, 1010. [CrossRef]
33. Jain, A.; Cheng, K. The Principles and Applications of Avidin-Based Nanoparticles in Drug Delivery and Diagnosis. *J. Control. Release* **2017**, *245*, 27–40. [CrossRef]
34. Fathi-Karkan, S.; Sargazi, S.; Shojaei, S.; Farasati Far, B.; Mirinejad, S.; Cordani, M.; Khosravi, A.; Zarrabi, A.; Ghavami, S. Biotin-functionalized nanoparticles: An overview of recent trends in cancer detection. *Nanoscale* **2024**, *16*, 12750–12792. [CrossRef]
35. Lee, Y.W.; Mout, R.; Luther, D.C.; Liu, Y.; Castellanos-García, L.; Burnside, A.S.; Ray, M.; Tonga, G.Y.; Hardie, J.; Nagaraj, H.; et al. In Vivo Editing of Macrophages through Systemic Delivery of CRISPR Cas9-Ribonucleoprotein-Nanoparticle Nanoassemblies. *Adv. Ther.* **2019**, *2*, 1900041. [CrossRef] [PubMed]

36. Banshoya, K.; Nakamura, T.; Tanaka, T.; Kaneo, Y. Development of α-Tocopherol Nanomicellar Formulation Using Polyethylene Glycol Monostearate for the Oxidative Stress-Related Disease. *J. Drug Deliv. Sci. Technol.* **2021**, *61*, 102310. [CrossRef]
37. Banshoya, K.; Nakamura, T.; Tanaka, T.; Kaneo, Y. Coenzyme Q10-Polyethylene Glycol Monostearate Nanoparticles: An Injectable Water-Soluble Formulation. *Antioxidants* **2020**, *9*, 86. [CrossRef] [PubMed]
38. Banshoya, K.; Machida, A.; Kawamura, S.; Yamada, T.; Okada, R.; Kawamoto, Y.; Kimura, H.; Shibata, S.; Hieda, Y.; Kaneo, Y.; et al. Development of a Water-Soluble Nanomicellar Formulation Loaded with Trans-Resveratrol Using Polyethylene Glycol Monostearate for the Treatment of Intracerebral Hemorrhage. *Pharmaceutics* **2024**, *16*, 1462. [CrossRef]
39. Minchin, R.F.; Martin, D.J. Minireview: Nanoparticles for Molecular Imaging—An Overview. *Endocrinology* **2010**, *151*, 474–481. [CrossRef]
40. Valsalakumari, R.; Feith, M.; Pettersen, S.; Åslund, A.K.O.; Mørch, Ý.; Skotland, T.; Sandvig, K.; Mælandsmo, G.M.; Iversen, T.-G. Combinatorial Effects of Free and Nanoencapsulated Forms of Cabazitaxel and RAS-Selective Lethal 3 in Breast Cancer Cells. *Pharmaceutics* **2025**, *17*, 657. [CrossRef]
41. Cushman, D.W.; Ondetti, M.A. History of the Design of Captopril and Related Inhibitors of Angiotensin Converting Enzyme. *Hypertension* **1991**, *17*, 589–592. [CrossRef]
42. Waheed, H.; Moin, S.F.; Choudhary, M.I. Snake Venom: From Deadly Toxins to Life-saving Therapeutics. *Curr. Med. Chem.* **2017**, *24*, 1874–1891. [CrossRef]
43. Oliveira, A.L.; Viegas, M.F.; da Silva, S.L.; Soares, A.M.; Ramos, M.J.; Fernandes, P.A. The chemistry of snake venom and its medicinal potential. *Nat. Rev. Chem.* **2022**, *6*, 451–469. [CrossRef] [PubMed]
44. Alves, Á.E.F.; Barros, A.B.C.; Silva, L.C.F.; Carvalho, L.M.M.; Pereira, G.M.A.; Uchôa, A.F.C.; Barbosa-Filho, J.M.; Silva, M.S.; Luna, K.P.O.; Soares, K.S.R.; et al. Emerging Trends in Snake Venom-Loaded Nanobiosystems for Advanced Medical Applications: A Comprehensive Overview. *Pharmaceutics* **2025**, *17*, 204. [CrossRef] [PubMed]
45. Sciani, J.M.; Vigerelli, H.; Costa, A.S.; Câmara, D.A.D.; Junior, P.L.-S.; Pimenta, D.C. An unexpected cell-penetrating peptide from Bothrops jararaca venom identified through a novel size exclusion chromatography screening. *J. Pept. Sci.* **2017**, *23*, 68–76. [CrossRef]
46. Ran, F.A.; Hsu, P.D.; Wright, J.; Agarwala, V.; Scott, D.A.; Zhang, F. Genome engineering using the CRISPR-Cas9 system. *Nat. Protoc.* **2013**, *8*, 2281–2308. [CrossRef]
47. Arora, L.; Narula, A. Gene Editing and Crop Improvement Using CRISPR-Cas9 System. *Front. Plant Sci.* **2017**, *8*, 1932. [CrossRef]
48. Waddington, S.N.; Privolizzi, R.; Karda, R.; O'Neill, H.C. A broad overview and review of CRISPR-Cas technology and stem cells. *Curr. Stem Cell Rep.* **2016**, *2*, 9–20. [CrossRef]
49. Tashima, T. Non-Invasive Delivery of CRISPR/Cas9 Ribonucleoproteins (Cas9 RNPs) into Cells via Nanoparticles for Membrane Transport. *Pharmaceutics* **2025**, *17*, 201. [CrossRef]
50. Carrasco-Rojas, J.; Sandoval, F.I.; Schuh, C.M.A.P.; Lagos, C.F.; Morales, J.O.; Arriagada, F.; Ortiz, A.C. NLC-Based Rifampicin Delivery System: Development and Characterization for Improved Drug Performance Against Staphylococcus aureus. *Pharmaceutics* **2025**, *17*, 799. [CrossRef]
51. Korsmeyer, R.W.; Lustig, S.R.; Peppas, N.A. Solute and penetrant diffusion in swellable polymers. I. Mathe matical modeling. *J. Polym. Sci. Part B Polym. Phys.* **1986**, *24*, 395–408. [CrossRef]
52. Ritger, P.L.; Peppas, N.A. A Simple Equation for Description of Solute Release I. Fickian and Non-Fickian Release from Non-Swellable Devices in the Form of Slabs, Spheres, Cylinders or Discs. Journal of Controlled Release. *J. Control. Release* **1987**, *5*, 23–36. [CrossRef]
53. Ritger, P.L.; Peppas, N.A. A simple equation for description of solute release II. Fickian and anomalous release from swellable devices. *J. Control. Release* **1987**, *5*, 37–42. [CrossRef]
54. Zhang, Y.; Zou, Z.; Liu, S.; Miao, S.; Liu, H. Nanogels as Novel Nanocarrier Systems for Efficient Delivery of CNS Therapeutics. *Front. Bioeng. Biotechnol.* **2022**, *10*, 954470. [CrossRef]
55. Bissonnette, R.; Stein Gold, L.; Rubenstein, D.S.; Tallman, A.M.; Armstrong, A. Tapinarof in the treatment of psoriasis: A review of the unique mechanism of action of a novel therapeutic aryl hydrocarbon receptor–modulating agent. *J. Am. Acad. Dermatol.* **2021**, *84*, 1059–1067. [CrossRef]
56. Tapinarof cream for the treatment of plaque psoriasis: Efficacy and safety results from 2 Japanese phase 3 trials. *J. Dermatol.* **2024**, *51*, 1269 1278. [CrossRef]
57. Balogh, B.; Pető, Á.; Fehér, P.; Ujhelyi, Z.; Bácskay, I. Tapinarof Nanogels as a Promising Therapeutic Approach. *Pharmaceutics* **2025**, *17*, 731. [CrossRef]
58. Cemal, S.D.; Ladetto, M.F.; Alava, K.H.; Kazimirsky, G.; Cucher, M.; Glisoni, R.J.; Cuestas, M.L.; Byk, G. Voriconazole-Loaded Nanohydrogels Towards Optimized Antifungal Therapy for Cystic Fibrosis Patients. *Pharmaceutics* **2025**, *17*, 725. [CrossRef]

pharmaceutics

MDPI

Review

Nanoparticle-Based Targeted Drug Delivery Methods for Heart-Specific Distribution in Cardiovascular Therapy

Toshihiko Tashima

Tashima Laboratories of Arts and Sciences, 1239-5 Toriyama-cho, Kohoku-ku, Yokohama 222-0035, Kanagawa, Japan; tashima_lab@yahoo.co.jp

Abstract: Cardiovascular diseases remain the leading cause of death worldwide and are often managed through invasive surgical procedures such as heart transplantation, ventricular assist device implantation, coronary artery bypass grafting, and stent placement. However, significant unmet medical needs persist in this field. The development of pharmaceutical agents using non-invasive delivery strategies is therefore of critical importance. Current treatments often target peripheral tissues or organs—such as capillary endothelial cells, vascular smooth muscle, and renal tubules—to reduce cardiac workload by lowering blood pressure. However, effective drug delivery directly to the myocardium continues to pose a significant challenge. For conditions such as congestive heart failure (CHF) and myocardial infarction (MI), targeted delivery of therapeutic agents to the heart is essential. In this perspective review, I discuss the potential and emerging strategies for non-invasive cardiac drug delivery, focusing on receptor-mediated endocytosis and transcytosis using nanoparticle-based delivery systems that have frequently been employed for targeting the brain or cancer cells although their use for cardiac delivery remains largely unexplored.

Keywords: cardiovascular diseases; heart-targeted drug delivery; nanoparticle-mediated delivery; receptor-mediated endocytosis/transcytosis

1. Introduction

Cardiovascular diseases—including coronary heart disease, cerebrovascular disease, peripheral arterial disease, rheumatic heart disease, congenital heart disease, deep vein thrombosis, pulmonary embolism, and congestive heart failure (CHF)—are the leading cause of death worldwide, accounting for an estimated 17.9 million deaths in 2019, of which 85% were due to heart attacks and strokes [1,2]. Therefore, in drug research and development, it is crucial to design pharmaceutical agents for the treatment of cardiovascular diseases using non-invasive approaches that do not rely on commonly performed surgical interventions such as heart transplantation, ventricular assist devices, bypass surgeries, or stent implantation. Despite advances in treatment, significant unmet medical needs remain in this therapeutic area. Clinically, medications such as angiotensin-converting enzyme inhibitors (e.g., enalapril), diuretics (e.g., furosemide), and others (Figure 1) are widely used to reduce cardiac workload by lowering blood pressure. These drugs target tissues or organs—such as capillary endothelial cells, vascular smooth muscle, and renal tubules—that are readily accessible through systemic circulation, even with oral administration. As such, they are well-suited for managing chronic conditions and preventing acute cardiovascular events. However, the primary target organ of inotropes such as digoxin (Figure 1), which increase heart rate, and β-blockers such as atenolol (Figure 1), which decrease heart rate, is the heart itself. These agents must be delivered to the heart via systemic circulation in most cases, although β-adrenergic receptors are also present in various non-cardiac tissues. Moreover, therapeutic drug monitoring is required to control blood concentrations of digitoxin, which can negatively impact patients' quality of life. The shift toward nanoparticle-based delivery could be further driven by clinical needs and existing therapeutic challenges. Therefore, it is essential to develop drug delivery systems capable of selectively and efficiently transporting appropriate amounts of pharmaceutical agents to the heart, while minimizing off-target side effects and excessive pharmacological activity. In this perspective review, I discuss the potential and practical applications of non-invasive cardiac drug delivery via receptor-mediated endocytosis and carrier-mediated transport using transporters (Figure 2), based on nanodelivery systems grounded in structuralist principles [3,4].

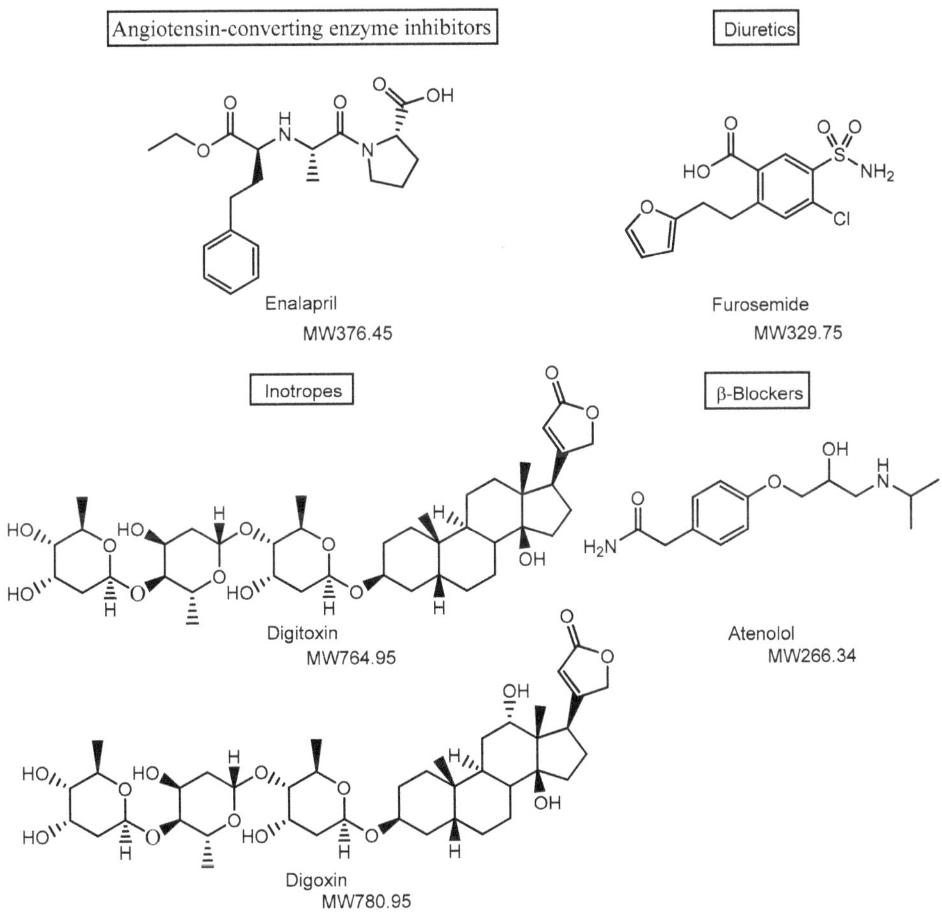

Figure 1. Structures of drugs used in the treatment of cardiovascular diseases. Medications such as angiotensin-converting enzyme inhibitors (e.g., enalapril) and diuretics (e.g., furosemide) are widely used to reduce cardiac workload by lowering blood pressure, primarily targeting tissues and organs such as capillary endothelial cells, vascular smooth muscle, and renal tubules. In contrast, digitoxin and digoxin act on the plasma membrane Na^+/K^+-ATPase of cardiac myocytes, directly targeting the heart. MW denotes molecular weight.

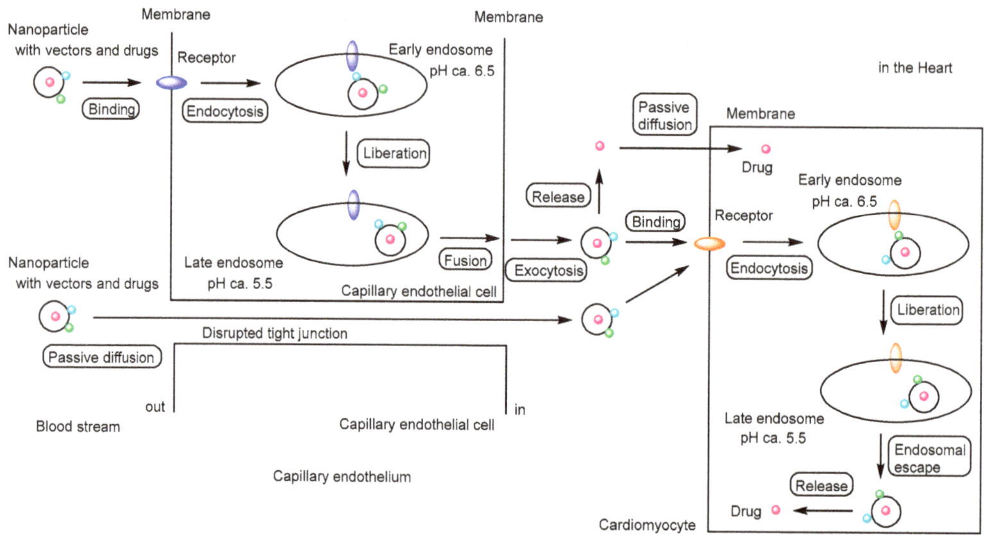

Figure 2. Schematic representation of the pathway by which drug-loaded nanoparticles, equipped with targeting vectors, exert their effects on cardiomyocytes, illustrating the concept of drug delivery specificity described in this review. Primarily, vector-functionalized nanoparticles cross the capillary endothelium via receptor-mediated transcytosis along the transcellular route. Subsequently, these nanoparticles enter cardiomyocytes through receptor-mediated endocytosis and release their cargo following endosomal escape. The released drugs then exert their biological activity. During myocardial infarction (MI) and chronic heart failure (CHF), disruption of tight junctions allows moderately sized nanoparticles to penetrate the capillary endothelium via the paracellular route. Additionally, drugs released into the cardiac interstitium may enter cardiomyocytes by passive diffusion. Red circles indicate drugs; light blue and green circles represent vectors; blue and orange ellipses indicate receptors.

2. Discussion

2.1. Strategies for Targeted Drug Delivery to the Heart

2.1.1. Overview of Cardiac Anatomy

The heart is a hollow, muscular organ that plays a vital role in the cardiovascular system. It is located behind the sternum and between the lungs. Structurally, the heart consists of four chambers: the right atrium, left atrium, right ventricle, and left ventricle [5]. The right atrium receives deoxygenated blood from the venae cavae and the coronary sinus. This blood is then passed to the right ventricle, which pumps it into the pulmonary trunk. The left atrium receives oxygenated blood from the pulmonary veins, which is subsequently transferred to the left ventricle. The left ventricle then pumps the arterial blood into the aorta for systemic circulation. The right atrium receives deoxygenated blood from the venae cavae and the coronary sinus. This blood is then transferred to the right ventricle, which pumps it into the pulmonary trunk. The left atrium receives oxygenated blood from the pulmonary veins, which is subsequently delivered to the left ventricle. The left ventricle then pumps this arterial blood into the aorta for systemic circulation. In general, vertebrate muscle tissue is classified into three types: skeletal muscle, smooth muscle, and cardiac muscle (myocardium). Among these, the myocardium, composed of cardiomyocytes, functions as a contractile syncytium responsible for generating the heart's pumping force. It forms the middle layer of the heart wall, situated between the inner endocardium—a single-cell layer—and the outer epicardium (Figure 3) [6]. Coronary arteries, which originate from the aortic root and branch into smaller vessels, are formed by endocardial cells and supply nutrients and oxygen to the myocardium via the capillary endothelium. The endocardial endothelium and the myocardial capillary endothelium share common structural and functional features [7]. The myocardial capillary endothelium is continuous and characterized by tight junctions with an approximate diameter of 4 nm [8,9] Therefore, the transepithelial transport of most compounds occurs primarily via the transcellular route rather than the paracellular pathway. Transporters such as sodium/glucose co-transporter 1 (SGLT1) are expressed in cardiomyocytes and capillary endothelial cells, facilitating the uptake of hydrophilic, low-molecular-weight nutrients into the heart through their pore [10]. Nevertheless, transporter-focused studies on cardiac biology involving SGLT1 are not likely to be actively pursued. Endocytosis also takes place in the myocardial capillary endothelium to internalize macromolecules; for instance, insulin is likely internalized via receptor-mediated endocytosis involving the insulin receptor [11]. Consequently, intravenously administered substances intended for delivery to the heart must cross the capillary endothelium branching from the coronary arteries within the heart walls, rather than the inner endocardial layer.

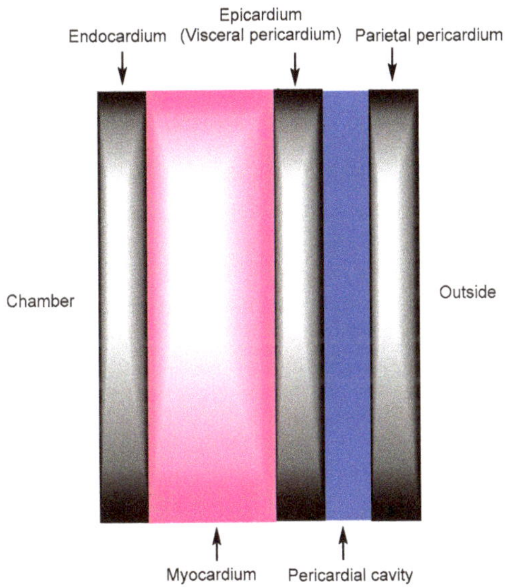

Figure 3. The heart wall is composed of three layers: the middle myocardium, the inner endocardium, and the outer epicardium. The pericardial cavity is the fluid-filled space located between the epicardium and the parietal pericardium. Drug-loaded nanoparticles administered intravenously should enter the heart through the myocardial capillary endothelium located in the mid-myocardium, rather than through the endocardial endothelium in the innermost layer of the endocardium, whose process likely occurs via receptor-mediated endocytosis or transcytosis through specific receptors expressed on the surface of mesh-patterned myocardial capillary endothelial cells.

2.1.2. Non-Nanoparticle Drug Therapy Approaches

In general, orally administered cardiac drugs are transported from the systemic circulation into the heart across continuous capillary endothelial cells [12,13]. Therefore, digitoxin and digoxin, which are representative inotropes, must cross the capillary endothelium to reach cardiac tissue. Additionally, oral administration requires their absorption in the small intestine. While digitoxin undergoes hepatic metabolism, digoxin is primarily eliminated via renal excretion. Digoxin is known to be transported from systemic circulation into urine by the kidney through the human organic anion transporter OATP4C1 located on the basolateral membrane of proximal tubule cells (Km = 7.8 μM), followed by efflux mediated by multiple drug resistance 1 (MDR1, P-glycoprotein) at the apical membrane of these cells [14]. The uptake of digitoxin in the small intestine has been shown to be enhanced by inhibition of MDR1 [15]. In general, compounds are filtered from the glomerular capillaries into Bowman's capsules, reabsorbed from Bowman's capsules or renal tubules into the peritubular capillaries via active transport, secreted from the peritubular capillaries into the renal tubules, and ultimately excreted in the urine. The sequence of reabsorption and secretion may be reversed for certain compounds. Furthermore, intestinal absorption of digoxin in rats has been reported to be facilitated by organic anion transporting polypeptides (OATPs) and enhanced by GF120918, a specific MDR1 inhibitor [16]. These findings suggest that digitoxin and digoxin may cross cellular membranes not only via carrier-mediated transport involving OATP4C1 but also through passive diffusion, as MDR1 functions to actively efflux hydrophobic compounds that have passively diffused across the membrane [17].

Digitoxin and digoxin inhibit the Na^+/K^+-ATPase on the plasma membrane of cardiac myocytes, leading to intracellular Ca^{2+} accumulation through the secondary suppression of Na^+/Ca^{2+} exchangers. As a result, calcium-induced calcium release via ryanodine receptors on the sarcoplasmic reticulum enhances myocardial contractile force [18]. The Na^+/K^+-ATPase is ubiquitously expressed in the plasma membranes of all animal cells [19]. In addition, although digitoxin and digoxin are substrates of MDR1, they are capable of crossing the capillary endothelium via passive diffusion. Consequently, these drugs may cause off-target side effects and severe toxicities due to overdose, necessitating routine monitoring of serum digoxin levels. To address this, the targeted delivery of cardiovascular drugs—including low-molecular-weight compounds—to the heart should be established using nanoparticle-based systems. Such systems, designed according to structuralist principles, can modulate material behavior within biological systems, including mechanisms such as receptor-mediated endocytosis and transcytosis.

2.1.3. Nanoparticle Drug Therapy Approaches

Nanoparticle-based drug delivery systems have been developed to offer multiple benefits, including improved drug bioavailability and absorption, as well as the reduction of drug aggregation, enzymatic degradation, renal clearance, and undesired drug interactions [20]. A variety of materials with unique properties can be used to construct biocompatible, biodegradable, and controlled-release nanoparticles. These nanoparticles may be composed of: (i) synthetic biodegradable polymers, (ii) natural polymers such as chitosan, poly(lactic-co-glycolic acid) (PLGA), and poly(glycolic acid) (PGA), (iii) lipids including liposomes, micelles, and exosomes, (iv) inorganic materials such as gold (Au), silicon (Si), and magnetite (Fe_3O_4), (v) organic materials including albumin, monoclonal antibodies, and virosomes, (vi) emulsions, or (vii) other components (Table 1). Functionalized or engineered nanoparticles have been developed to exhibit various features, including: (i) passive targeting via the enhanced permeability and retention (EPR) effect, particularly in solid tumors [21,22]; (ii) active targeting through ligand–receptor interactions that induce receptor-mediated endocytosis; (iii) magnetic responsiveness; (iv) pH sensitivity; (v) thermosensitivity; (vi) enteric protection for oral administration; and (vii) other beneficial properties. Nanoparticles can be readily surface-modified with targeting vectors such as antibodies, tumor-homing peptides, and cell-penetrating peptides (CPPs). Currently, active targeting via receptor-mediated endocytosis or transcytosis, enabled by nanodelivery systems, is receiving considerable attention [23,24].

Table 1. Types of Nanoparticles Based on Material Composition.

#	Nanoparticle Formulations	Material Components
i	Synthetic biodegradable polymers	Poly(α-hydroxy esters), polyethylene glycol (PEG), polyurethane
ii	Natural polymers	Chitosan, poly(lactic-co-glycolic acid) (PLGA), poly(glycolic acid) (PGA)
iii	Vesicles (liposomes, micelles, exosomes)	Lipids
iv	Inorganic materials	Gold (Au), silicon (Si), magnetite (Fe_3O_4)
v	Organic materials	Albumin, monoclonal antibodies, virosomes
vi	Emulsions	Surfactants
vii	Other components	Others

Doxorubicin (Figure 4), an anthracycline antibiotic widely used in cancer treatment—along with paclitaxel—has been associated with the development of cardiomyopathy [25], MI, and CHF [26,27]. Although the precise mechanisms underlying doxorubicin-induced cardiomyopathy remain unclear, it is proposed that oxidative stress, enhanced by free radical generation, contributes to the progressive loss of myofibrils and vacuolization of myocardial cells. Notably, doxorubicin-loaded nanoparticles have been shown to reduce cell viability more effectively than free doxorubicin [28]. As a result, surgical intervention is required in some cases. Currently, no established drug therapy exists for doxorubicin-induced cardiomyopathy. To minimize doxorubicin distribution to the heart as a precautionary measure, doxorubicin-loaded nanoparticles based on lactosylated bovine serum albumin (BSA-Lac) were designed to target human liver cancer cells (HepG2) via receptor-mediated endocytosis through the asialoglycoprotein receptor (ASGPR), which is almost exclusively expressed and highly abundant in liver cells. In cytotoxicity assays, BSA-Lac-based nanoparticles carrying doxorubicin reduced cell viability more effectively than free doxorubicin, potentially reducing off-target cardiotoxicity in the treatment of hepatocellular carcinoma (HCC) [29]. ASGPR specifically interacts with galactose residues or lactose moieties [30]. Nonetheless, various pharmaceutical agents are under investigation as potential novel treatments for doxorubicin-induced cardiomyopathy, particularly those targeting left ventricular dysfunction or mitochondrial function. These agents include cannabidiol, carvedilol, hydrogen sulfide, coenzyme Q10, aldose reductase inhibitors, statins, angiotensin receptor blockers, α1-adrenergic receptor agonists, phosphodiesterase III (PDE-3) inhibitors, vitamin B1, monounsaturated fats (e.g., olive oil), phenolic antioxidants (e.g., oleuropein), and a range of bioactive compounds with hemodynamic, histological, and biochemical effects, such as flavonoids, hesperidin (a flavanone glycoside), pycnogenol, chrysin, acacetin, berberine, and cardamonin [31]. If antioxidant agents can be effectively and selectively delivered to the heart, non-invasive drug therapy may serve as an alternative to surgical intervention.

Nanoparticle-mediated drug delivery into the heart through the following examples (i)–(viii) is examined.

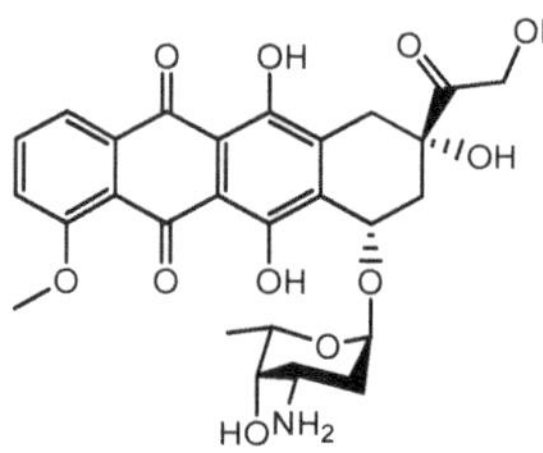

Doxorubicin

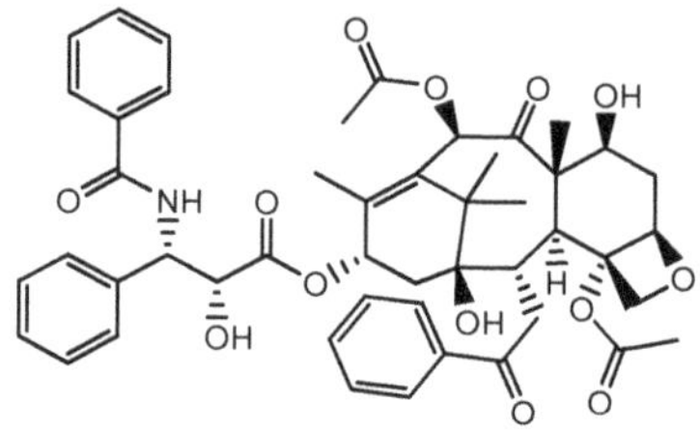

Paclitaxel

Figure 4. The structures of anti-cancer drugs.

(i) Milrinone (MRN) (Figure 5) is a low-molecular-weight phosphodiesterase 3 (PDE3) inhibitor used in the treatment of CHF. Among the PDE3 isoforms, PDE3A—rather than PDE3B—regulates the inotropic response by hydrolyzing cyclic adenosine monophosphate (cAMP) and modulating protein kinase A (PKA) activity. Calcium uptake into the sarcoplasmic reticulum (SR) via SR Ca^{2+}-ATPase (SERCA2a) is dependent on PKA-mediated phosphorylation of phospholamban (PLN) at Ser-16. However, excessive SR Ca^{2+} release, or "sparks", can occur following the opening of ryanodine receptor 2 (RyR2), especially when hyperphosphorylated at sites such as Ser-2808 and Ser-2030 by PKA, or Ser-2814 by Ca^{2+}/calmodulin-dependent protein kinase II (CaMKII). This dysregulation may contribute to cardiac arrhythmias or sudden cardiomyocyte death. Nevertheless, PDE3 inhibitors are also considered to have cytoprotective effects on the heart [32]. The angiotensin II type 1 (AT1) receptor, which specifically binds the angiotensin II peptide (an 8-amino acid sequence: DRVYIHPFHL), is overexpressed in the myocardium under conditions of MI and CHF [33,34]. MRN-loaded nanoparticles based on human serum albumin (HSA), surface-modified with angiotensin II through chemical conjugation using PA-$(PEG)_4$-SPA, EDC, and sulfo-NHS, had an average diameter of 190.2 $\pm$ 5.7 nm. These nanoparticles were internalized into cardiac tissue via AT1 receptor-mediated endocytosis, resulting in a 3.92-fold higher concentration of MRN in the heart at 2 h post-injection, compared to free MRN. In a rat model of CHF, intravenous administration of these targeted nanoparticles significantly improved cardiac function, as evidenced by enhanced left ventricular fractional shortening (LVFS), relative to treatment with free MRN [35] (Figure 6). Clustering on the plasma membrane surface can induce endocytosis [36–38]. In the renin–angiotensin system, binding of angiotensin II to the AT1 receptor triggers intracellular signal transduction. In cardiac myocytes, AT1 receptor activation plays a role in processes such as cardiac fibroblast activation [39]. Clustering of AT1 receptors with nanoparticles may also promote receptor-mediated endocytosis. Additionally, it is known that HSA binds to the Gp60 receptor and to secreted protein acidic and rich in cysteine (SPARC). The Gp60 receptor is specifically expressed on the surface of continuous capillary endothelium in organs such as the heart and lungs, but not in the cortical brain [40]. SPARC, on the other hand, is expressed by various cell types including fibroblasts and endothelial cells, particularly in cancerous tissues [41]. Thus, HSA-based nanoparticles may be internalized into cardiac tissue via receptor-mediated, caveolae-dependent endocytosis through interactions with the Gp60 receptor or SPARC. However, no studies have yet demonstrated the delivery of HSA-based nanoparticles into the heart specifically via Gp60 receptor-mediated caveolae-dependent endocytosis. In contrast, angiotensin II has been shown to undergo AT1 receptor-mediated transcytosis across the BBB, resulting in its delivery to the brain [42]. This suggests that AT1 receptor mediated transcytosis might also occur in the endothelial cells of the heart, although it remains unclear whether AT1 receptors are highly expressed in cardiac endothelial cells. Therefore, elucidating the expression status of AT1 receptors in cardiac endothelial cells is essential for establishing drug delivery through AT1 receptor-mediated transcytosis.

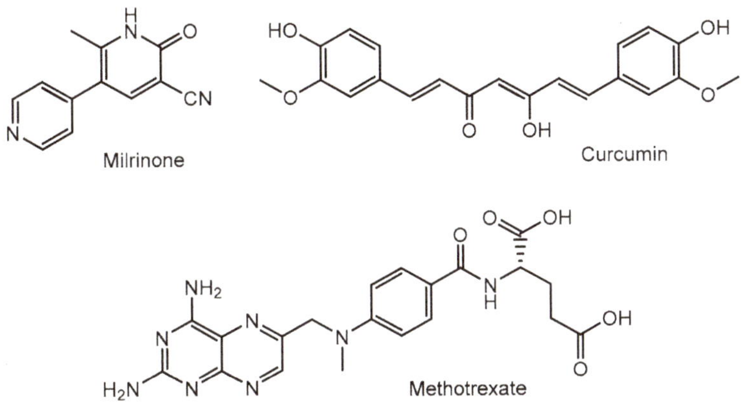

Figure 5. Structures of nanoparticle-delivered low-molecular-weight compounds such as milrinone, curcumin, and methotrexate.

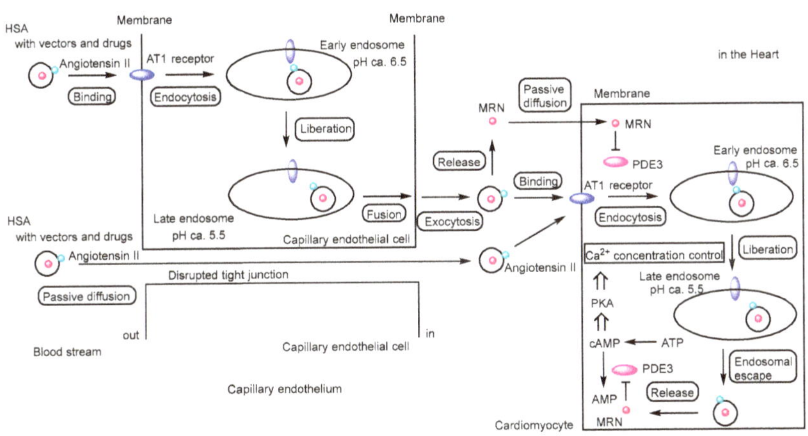

Figure 6. Schematic representation of the pathway by which milrinone (MRN)-loaded human serum albumin (HSA) nanoparticles exert their effects on cardiomyocytes. MRN-loaded HSA nanoparticles conjugated with angiotensin II (indicated by light blue circles) can enter the heart either via receptor-mediated transcellular transcytosis through angiotensin II type 1 (AT1) receptors (indicated by blue ellipses) or via paracellular passive diffusion through disrupted tight junctions between capillary endothelial cells. Once released in the heart, MRN may diffuse passively into cardiomyocytes. Alternatively, MRN may be released from endosomes into the cytosol after receptor-mediated endocytosis of angiotensin II–modified MRN-loaded HSA nanoparticles via AT_1 receptors, although the mechanism of endosomal escape remains to be elucidated. MRN functions as a phosphodiesterase 3 (PDE3) inhibitor, preventing the hydrolysis of cyclic adenosine monophosphate (cAMP) into AMP. The accumulation of cAMP activates protein kinase A (PKA), which regulates cardiac contraction by modulating Ca^{2+} concentrations through sarcoplasmic reticulum Ca^{2+}-ATPase (SERCA2a) and ryanodine receptor 2 (RyR2). Calcium uptake from the cytosol into the sarcoplasmic reticulum (SR) occurs via SERCA2a, whereas calcium release from the SR into the cytosol is mediated by RyR2. As a result, MRN induces mild enhancement of cardiac contraction. MRN is indicated by a red circle, and PDE3 is indicated by a magenta ellipse.

(ii) A specific microRNA-1 inhibitor—anti-miR-1 oligonucleotides (AMO-1)—suppresses cytosolic microRNA-1 expression in ischemic myocardium. MicroRNA-1 is a muscle-specific miRNA preferentially expressed in adult cardiac and skeletal muscle tissues [43], and it is implicated in various heart diseases. Intravenous administration of AMO-1 encapsulated in a nanovector (AT1-PEG-DGL) conjugated with an AT1-targeting peptide led to rapid accumulation in the infarcted myocardium, significantly more than in the control group lacking the AT1-targeting peptide. In an in vivo mouse model of MI, this targeted delivery system demonstrated a pronounced anti-apoptotic effect and reduced the myocardial infarct size by 64.1% compared to the MI control group [44]. Although the exact mechanism by which AT1-PEG-DGL crossed the vascular endothelium remains unclear (Figure 7), it was found that AT1 receptor expression peaked at 24 h post-MI. Angiotensin II is known to activate cardiac fibroblasts via the AT1 receptor [39,45], and AT1 receptor overexpression is thought to be a compensatory response to myocardial injury. However, the mechanism of endosomal escape following AT1 receptor-mediated endocytosis remains to be elucidated (Figure 7).

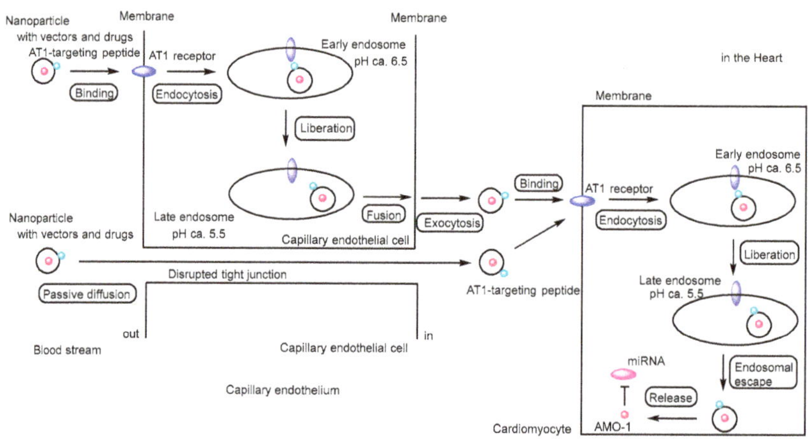

Figure 7. Schematic representation of the pathway by which anti-miR-1 oligonucleotide (AMO-1)–loaded nanoparticles exert their effects on cardiomyocytes. The nanoparticles protect the oligonucleotides from enzymatic degradation. The exact mechanism by which nanoparticles (AT1-PEG-DGL) cross the vascular endothelium remains unclear; however, receptor-mediated transcytosis via angiotensin II type 1 (AT1) receptors (indicated by blue ellipses) or paracellular passive diffusion through disrupted tight junctions between capillary endothelial cells are considered plausible pathways. Moreover, the mechanism of endosomal escape following AT_1 receptor–mediated endocytosis also remains to be elucidated. AMO-1 is indicated by a red circle, and miRNA is represented by a magenta ellipse.

(iii) Oxidative stress and inflammation are elevated in CHF [46]. Curcumin (Figure 5), a compound with potent anti-inflammatory and antioxidant properties, has shown therapeutic potential in slowing the progression of CHF. Additionally, a graphene quantum dot-based electrochemical biosensor has been reported to exhibit high sensitivity, specificity, and selectivity for the early diagnosis of MI by detecting cardiac biomarkers such as troponin and myoglobin [47]. Indeed, PEGylated graphene quantum dot nanoparticles loaded with curcumin (Cur-PEG-GQDs) have been shown to reduce myocardial infarct size, fibrosis, and left ventricular end-diastolic pressure following intraperitoneal administration in an in vivo rat model of MI induced by ligation of the left anterior descending artery. Although curcumin decreased antioxidant markers and increased oxidant markers in the heart in a dose-dependent manner, non-classical mechanisms may contribute to the cardioprotective effects of Cur-PEG-GQDs against MI-induced cardiac dysfunction [48]. The mechanisms by which these nanoparticles are internalized into cardiac tissue from the systemic circulation remain unclear, despite the known association of graphene quantum dots with MI. Curcumin release from the nanoparticles is presumed to occur in either the extracellular or intracellular regions of myocardial cells or associated cell types. If curcumin is released in the extracellular space, it is likely that a portion enters cells primarily via passive diffusion [49]. Curcumin is not considered a substrate for specific transporters. It exhibits antioxidant and anti-inflammatory activities in myocardial cells [50], and also exerts anti-inflammatory effects in immune cells, including macrophages, which secrete pro-inflammatory cytokines such as IL-6, IL-1β, and TNF-α [51]. Therefore, extracellular release of curcumin may be advantageous for exerting its therapeutic effects. Furthermore, to overcome barriers to clinical translation by minimizing off-target effects in the brain and reducing hepatic clearance and sequestration, drug-loaded nanoparticles should ideally reach the myocardial capillary endothelium via the coronary circulation earlier than they reach the brain or liver. However, direct injection into the pulmonary veins or coronary arteries is clinically dangerous and impractical. Although intravenously administered nanoparticles must inevitably undergo first-pass metabolism and dilution in the lungs, all types of intravenous injections follow the circulation pathway—right atrium, right ventricle, pulmonary circulation, left atrium, left ventricle, and aorta—before reaching the coronary circulation, and thus arrive there earlier than in the brain or liver. The pulmonary first-pass effect may be attenuated through appropriate nanoparticle design.

(iv) Moreover, glycerol monooleate (GMO)-based biodegradable nanoparticles loaded with curcumin demonstrated greater protective effects compared to curcumin alone and more effectively enhanced the activity of antioxidant enzymes under both in vitro and in vivo conditions [52]. However, the mechanism by which GMO-based biodegradable nanoparticles are internalized remains unclear.

Therefore, curcumin-encapsulated nanoparticles are expected to serve as promising formulations for the treatment of cardiovascular diseases through multiple effective mechanisms.

(v) Methotrexate (MTX) (Figure 5) is an anti-inflammatory drug that inhibits not only dihydrofolate reductase but also AICAR (5-aminoimidazole-4-carboxamide ribonucleotide) transformylase. Inhibition of AICAR transformylase promotes the accumulation of AICAR, leading to increased adenosine release. This adenosine exerts anti-inflammatory effects through G-protein-coupled adenosine receptor subtypes, including A1, A2A, and A2B [53]. Moreover, activation of A2A and A2B receptors is essential for reducing infarct size in the hearts of wild-type C57Bl/6N mice [54]. While MTX uptake is mediated by folate receptors, apolipoprotein E (ApoE) is recognized by low-density lipoprotein (LDL) receptors on the cell membrane, triggering LDL receptor-mediated endocytosis. A lipid core nanoparticle (LDE) was formulated from a lipid mixture containing 100 mg of cholesteryl oleate and 200 mg of egg phosphatidylcholine. MTX-loaded LDE nanoparticles (LDE-MTX) significantly enhanced MTX uptake via the LDL receptor-mediated endocytic pathway in an in vitro assay using myocardial tissue from rats with MI, compared to MTX alone. In an in vivo study using MI-induced male Wistar rats, intravenous administration of LDE-MTX via the tail vein resulted in a 40% improvement in left ventricular (LV) systolic function, along with reduced cardiac dilation and LV mass [55]. Nonetheless, it has been demonstrated that LDE enters the heart from the systemic circulation via LDL receptor-mediated endocytosis. The release of MTX from the nanoparticles is presumed to occur either in the extracellular or intracellular regions of myocardial cells or associated cell types. If MTX is released extracellularly, a portion may enter cells through passive diffusion or via folate receptor-mediated transport.

(vi) MicroRNAs (miRNAs) play a significant role in the progression of MI. Bone marrow-derived mesenchymal stem cells (BMSCs), which are multipotent and possess homing ability, can migrate to sites of injury. Exosomes derived from BMSCs are commonly utilized as carriers for drug delivery. Murine hypoxia-conditioned BMSC-derived exosomes enriched with miR-125b-5p have been shown to promote ischemic cardiac repair by reducing cardiomyocyte apoptosis in MI mouse models, compared with murine normoxia-conditioned BMSC-derived exosomes enriched in miR-125b-5p [56]. Conversely, knockdown of miR-125b-5p in these exosomes exacerbates ischemic injury by increasing cardiomyocyte apoptosis. MiR-125b-5p inhibits apoptosis by downregulating the expression of p53 and BAK1. Furthermore, intravenous administration of ischemic myocardium-targeting peptide (IMTP)-conjugated hypoxia-conditioned BMSC-derived exosomes demonstrated higher specificity for ischemic lesions in the injured heart compared with scramble peptide-conjugated or intact hypoxia-conditioned BMSC-derived exosomes, as shown by Cy5.5 fluorescence imaging. Nonetheless, IMTP-conjugated hypoxia-conditioned BMSC-derived exosomes labeled with Cy5.5 were also detected in the liver and kidneys, in addition to the heart. Exosomes are believed to be internalized by recipient cells through membrane fusion and/or endocytosis [57]. Additionally, standardization challenges in exosome production include (a) source variability of parental cells, such as heterogeneity among BMSCs and differences in cell culture conditions; (b) inconsistencies in isolation and purification, including co-isolation of contaminants and scalability issues; (c) difficulties in characterization and quantification, such as inadequate reference standards and measurement variability; and (d) cargo heterogeneity and limited functional reproducibility, including variable molecular content, uncontrolled drug-loading efficiency, and batch-to-batch functional differences [58,59].

(vii) Gold–selenium core–shell nanostructures (AS-I/S NCs), modified with near-infrared II (NIR-II) photoacoustic imaging capability, an ischemic myocardium-targeting peptide (IMTP), and the mitochondrial-targeted antioxidant peptide SS31, significantly improved myocardial function and promoted angiogenesis in an in vivo rat model. NIR-II photoacoustic imaging indicated selective distribution to cardiomyocytes. IMTP facilitated targeted cellular uptake by cardiomyocytes, while SS31 enhanced antioxidant capacity by inhibiting reactive oxygen species (ROS), thereby reducing oxidative damage in H9c2 cells subjected to oxygen-glucose deprivation/reperfusion (OGD/R) injury [60]. No specific toxic effects attributable solely to selenium have been identified in many selenium nanoparticles, which differ from the AS-I/S NCs mentioned above [61]. Representative IMTPs, also referred to as cardiac-homing peptides, include peptide motifs such as CSTSMLKAC, CKPGTSSYC, and CPDRSVNNC, which have been identified from ischemic heart tissue. However, CSTSMLKAC conjugated to Sumo-mCherry accumulated not only in the ischemic myocardium but also in the kidneys [62]. Sumo-mCherry is a fusion protein composed of Small ubiquitin-like modifier (Sumo) and the red fluorescent protein mCherry. The molecular weight of the Sumo-mCherry-CSTSMLKAC conjugate is approximately 40 kDa. The cyclic nonapeptide CSTSMLKAC mimics endogenous peptide sequences found in proteins such as titin, optic atrophy 1 (OPA-1), and dynamin-related protein 1 (DRP-1). Although the cellular uptake mechanism of IMTPs remains unclear [63], the receptors for other tumor-homing peptides have been identified. For instance, RGD peptides bind to integrin $\alpha v\beta 3$, while NGR peptides target the aminopeptidase N (APN/CD13) receptor. Nevertheless, IMTP-coated nanoparticles may facilitate targeted delivery to damaged cardiac tissue following heart injury.

(viii) 8P (PHWWEYRR), a peptide isolated from the stemless *Eleutherococcus senticosus*, is an antioxidant peptide that protects cells from oxidative stress-induced damage. PCM

(WLSEAGPVVTVRALRGTGSW) is a myocardium-specific targeting peptide that binds selectively to tenascin-X on myocardial cells. In an in vivo rodent model of doxorubicin-induced MI, PCM-modified liposomes encapsulating 8P were effectively delivered to myocardial cells via tail-vein injection and exhibited antioxidant activity. Although this was an in vivo study conducted in rodents, prolonged survival and reduced arrhythmia incidence could be anticipated in humans as long-term functional outcomes of 8P-mediated oxidative stress prevention. Polyethylene glycol (PEG) modification extended the circulation time of the liposomes by evading immune recognition and clearance. Consequently, PEGylated liposomes may repeatedly circulate through the bloodstream and potentially cross the capillary endothelium into the heart tissue, although the exact mechanism of transendothelial transport remains unclear [64]. Nonetheless, the tight junctions of the cardiac capillary endothelium can be compromised under pathological conditions such as CHF and MI, due to factors including inflammation, ischemia, oxidative stress, and cytokine release [65–68]. This disruption of the endothelial barrier may facilitate the delivery of drugs and nanoparticles into the myocardial interstitial space, in a manner analogous to the EPR effect observed in solid tumors [21,22].

Accordingly, nanoparticle-based drug therapy can be achieved through receptor-mediated transcytosis using appropriate targeting vectors, such as the angiotensin II peptide, AT1-targeting peptide, or PCM (WLSEAGPVVTVRALRGTGSW).

2.2. *Promising Strategies for Targeted Substance Delivery to the Heart*

Tight junctions in the cardiac capillary endothelium are frequently disrupted in conditions such as CHF and MI, leading to increased vascular permeability. This enhanced permeability facilitates the passive accumulation of drug-loaded nanoparticles in cardiac tissue, providing a therapeutic window for targeted delivery. However, this process relies primarily on passive distribution. Incorporating active targeting strategies in addition to passive delivery may further enhance the efficiency of cardiac drug delivery. Nonetheless, the limited expression of specific surface proteins on cardiac capillary endothelial cells poses a challenge for effective active targeting.

Nanoparticle blood removal pathways (NBRPs) encompass various mechanisms responsible for the clearance and sequestration of nanoparticles from the bloodstream. These pathways are broadly categorized into cell-dependent mechanisms—(i) the reticuloendothelial system (RES), (ii) the mononuclear phagocyte system (MPS), (iii) other leukocyte-mediated processes, and (iv) additional cellular mechanisms—and cell-independent mechanisms, including (i) glomerular filtration, (ii) hepatic sinusoidal uptake, (iii) splenic sinusoidal filtration, and (iv) other non-cellular processes. As a result of these clearance mechanisms, nanoparticles tend to accumulate in the following order of organs: liver > spleen > lymph nodes > lungs > bone marrow > skin > kidneys > tumor [69]. Thus, effective nanoparticle delivery to the heart remains challenging. Notably, the AT1 receptor is among the most highly overexpressed receptors in the myocardium 24 h after MI and in CHF [44]. In damaged cardiomyocytes, certain substances may enter cells via passive diffusion through disruptions in the cell membrane. However, for more effective heart-targeted delivery via intravenous administration, selectively expressed receptors on the surface of cardiac capillary endothelial cells should be exploited. Nonetheless, no ideal heart-specific receptors have been identified to date. (i) Insulin receptors are ubiquitously expressed and are particularly abundant in the heart and endothelial cells [70]. It is well established that insulin receptors on brain capillary endothelial cells are exploited for drug delivery across the BBB via receptor-mediated transcytosis using anti-insulin receptor antibodies [71]. The fundamental machinery of receptor-mediated endocytosis is common to many tissues and organs, including the heart, brain, and cancer tissues. Receptor-mediated endocytosis and transcytosis for drug delivery across the BBB are well established. A similar strategy could enable drug-loaded nanoparticles to enter the heart via receptor-mediated endocytosis or transcytosis. By analogy, nanoparticles functionalized with anti-insulin receptor antibodies may similarly traverse the cardiac endothelium through receptor-mediated transcytosis, representing a potential strategy for targeted delivery to the heart.

(ii) Similarly, transferrin receptors on capillary endothelial cells at the BBB are utilized to mediate receptor-dependent endocytosis and transcytosis for delivering substances into the brain. However, it remains unclear whether transferrin receptors are highly expressed on capillary endothelial cells in the heart. In a study involving lipopolysaccharide (LPS) administration, transferrin receptor mRNA levels were significantly upregulated in the lungs but significantly downregulated in the hearts of rats 4 h post-treatment [72]. Surprisingly, transferrin conjugated with oligonucleotides has been successfully delivered to skeletal muscle and heart tissue via transferrin receptor 1 (TfR1)-mediated transport [73]. Although the expression levels of transferrin receptors on cardiac capillary endothelial cells remain unclear, cardiac myocytes are known to express transferrin receptors abundantly. Nevertheless, transferrin–oligonucleotide conjugates were able to cross both the capillary endothelium and the plasma membrane of cardiac myocytes. Given that iron is endogenously abundant in the heart, nanoparticles functionalized with anti-transferrin receptor antibodies may also traverse the cardiac endothelium via receptor-mediated transcytosis, offering a potential strategy for targeted delivery to cardiac tissue.

(iii) As a promising non-invasive accumulation strategy, magnetic nanoparticles composed of magnetic materials such as iron can be directed to target sites using an external magnetic field [74]. In cases where toxicity arises due to excessive accumulation at the target site, the external magnetic field can be easily discontinued, similar to external treatments such as compresses or medicated patches. Since iron is naturally present in the heart, iron-based nanoparticles are expected to exhibit minimal toxicity. Indeed, Fe_3O_4 core-based nanoparticles, functionalized with PEG linked via hydrazone bonds to antibodies targeting CD63 antigens on exosomes and to antibodies specific for myosin light chain surface markers on injured cardiomyocytes, accumulated at the site of a locally applied magnetic field. An external magnet attracts the magnetic vesicle shuttle to the infarcted area. Under the acidic pH of injured cardiac tissue in rabbit and rat models of MI, the hydrazone bonds were cleaved, resulting in the release of exosomes [75]. However, it was unclear how much more effectively the external magnet attracted the magnetic vesicle shuttle compared with the absence of a magnet. Moreover, monodisperse Fe_3O_4 nanospheres administered via tail vein injection were effectively localized in the myocardial tissue through external magnetic targeting in an in vivo experiment. The application of a magnetic field increased uptake by approximately 15-fold compared with conditions without a magnetic field [76]. Although most nanoparticles enter cells via endocytosis, the precise mechanisms by which these magnetic nanoparticles traverse the capillary endothelium of the heart remain unclear. It is well known that IgG antibodies and albumin exhibit prolonged half-lives due to their ability to evade lysosomal degradation through binding to the neonatal Fc receptor (FcRn) within endosomes. Upon endosomal acidification, free IgGs or albumin bind to FcRn, and subsequently dissociate and are released into the systemic circulation at physiological pH [77]. Thus, it is suggested that spontaneous endocytosis in capillary endothelial cells occurs more frequently than previously assumed, leading to the uptake of bystander substances such as antibodies and albumin. Consequently, magnetic nanoparticles accumulating near the heart may enter cardiac tissue as bystanders via spontaneous endocytosis or transcytosis.
(iv) As an alternative strategy, graphene quantum dot nanoparticles carrying therapeutic

cargos, as mentioned above, have been shown to improve outcomes in MI [47,78]. These nanoparticles can serve as carriers for highly selective drug delivery to the heart.

3. Conclusions

Cardiovascular diseases, also referred to as heart diseases, are serious disorders affecting the heart and blood vessels. In addition to pharmacological therapies aimed at regulating abnormal heart rhythms, lowering high blood pressure, or modifying myocardial contractility (e.g., inotropes), surgical interventions such as heart transplantation, ventricular assist device implantation, coronary artery bypass grafting, and stent placement are commonly employed. However, open-heart procedures impose significant physical stress on patients. Therefore, the development of innovative, less invasive treatment strategies for cardiovascular diseases is urgently needed.

Currently, tissue-selective drug delivery using nanoparticles has been explored, enabling the selective and preferential transport of therapeutic cargos to target sites by leveraging endogenous biological mechanisms. Although this approach may appear materialistic, it is grounded in structuralism [3,4], whereby materials function within the body under the regulation of complex biological systems. Biological activity is inherently coordinated. In the heart, nutrients are absorbed via carrier-mediated transport or receptor-mediated endocytosis. Oxygen and nutrients are supplied to cardiomyocytes from the coronary arteries through the myocardial capillary endothelium. Similarly, substance-loaded nanocarriers may also be delivered to the heart via these physiological pathways.

Nonetheless, in current research, the development of nanoparticles for cardiac drug delivery has primarily focused on their materials and composition. As a result, the number of successful demonstrations of nanoparticle-mediated drug delivery to the heart remains limited. The design of molecular vectors and the biological mechanisms underlying nanoparticle transport across the blood–brain barrier have been well studied to elucidate the time evolution of drug dosage and the saturation dynamics of cellular receptors, such as transferrin receptors. In contrast, the design of molecular vectors and the biological mechanisms governing nanoparticle transport to cardiac tissues and cells have not been thoroughly investigated. Consequently, both the time-dependent behavior of drug dosage and the receptor saturation rate remain largely unknown. Given the relatively small size of the heart, strategies for selective and efficient distribution—similar to those developed for brain-targeted delivery—are needed. Although nanoparticle-based drug delivery to the heart has been achieved, reports remain scarce due to the challenges associated with achieving such targeted and effective distribution. During conditions such as CHF and MI, disruption of tight junctions increases endothelial permeability, allowing nanoparticles or drug carriers to passively extravasate into the myocardial interstitial space. However, this represents a passive form of distribution, and the extent of tight junction disruption is highly dependent on the specific pathological conditions. Therefore, combining active targeting with passive distribution may enable more effective drug delivery to the heart. A widely studied strategy involves the use of well-designed nanoparticles functionalized with targeting ligands that bind to receptors on the surface of capillary endothelial cells, facilitating transendothelial transport via receptor-mediated transcytosis (Table 2). Following transcytosis, some nanoparticles are further internalized into cardiomyocytes through receptor-mediated endocytosis and subsequently release their cargos within endosomes or the cytosol. However, in many cases of cardiac delivery, the efficiency of endosomal escape remains poorly characterized. On the other hand, some nanoparticles release their cargos, which then enter cardiomyocytes or immune cells via passive diffusion or carrier-mediated transport.

Table 2. The introduced applications of drug delivery into the heart using nanoparticls as a carrier.

#	Formulation	Vector	Cargo	Status	References
i	MRN-loaded nanoparticles based on human serum albumin (HSA), surface-modified with angiotensin II	HSA, angiotensin II	MRN	Basic research	[35]
ii	AMO-1 encapsulated in a nanovector (AT1-PEG-DGL) conjugated with an AT1-targeting peptide	AT1-targeting peptide	AMO-1	Basic research	[44]
iii	PEGylated graphene quantum dot nanoparticles loaded with curcumin (Cur-PEG-GQDs)	Graphene quantum dot	Curcumin	Basic research	[48]
iv	Glycerol monooleate (GMO)-based biodegradable nanoparticles loaded with curcumin	GMO	Curcumin	Basic research	[52]
v	Methotrexate (MTX)-loaded lipid core nanoparticle (LDE) nanoparticles (LDE-MTX)	LDE against LDL receptor	MTX	Basic research	[55]
vi	Bone marrow-derived mesenchymal stem cell (BMSC)-derived exosomes enriched with miR-125b-5p	BMSC-derived exosomes	MiR-125b-5p	Basic research	[56]
vii	Gold–selenium core–shell nanostructures (AS-I/S NCs), modified with near-infrared II (NIR-II) photoacoustic imaging capability, ischemic myocardium-targeting peptide (IMTP), and the mitochondrial-targeted antioxidant peptide SS31	IMTP	Antioxidant peptide SS31	Basic research	[60]
viii	myocardium-specific targeting peptide PCM-modified liposomes encapsulating antioxidant peptide 8P	PCM	8P	Basic research	[64]
x	PEGylated graphene quantum dot nanoparticles loaded with MTX and MRN	Graphene quantum dot	MTX, MRN	Under analysis in Tashima lab	-

Additionally, certain nanoparticles, such as graphene quantum dots, have been shown to cross the endothelium and accumulate in the infarcted heart, although the underlying mechanisms remain unclear. Magnetic nanoparticles composed of materials such as iron can be directed to target sites using an external magnetic field and subsequently enter the heart either via receptor-mediated transcytosis—when equipped with appropriate targeting vectors—or through passive diffusion facilitated by disrupted tight junctions. Ultimately, the development of well-designed nanoparticles guided by structuralist principles—where material behavior is regulated by biological machinery—holds great promise for achieving targeted drug delivery in the treatment of cardiovascular diseases.

Funding: This research received no external funding.

Institutional Review Board Statement: Not applicable.

Informed Consent Statement: Not applicable.

Data Availability Statement: No new data were created or analyzed in this study. Data sharing is not applicable to this article.

Acknowledgments: This review is just my opinion based on or inferred from available published articles and public knowledge. Thus, the intellectual property rights are not infringed upon.

Conflicts of Interest: The author declares no conflict of interest.

References

1. The World Health Organization. Cardiovascular Diseases. Available online: https://www.who.int/health-topics/cardiovascular-diseases#tab=tab_1 (accessed on 1 January 2025).
2. The World Health Organization. Cardiovascular Diseases (CVDs). 11 June 2021. Available online: https://www.who.int/news-room/fact-sheets/detail/cardiovascular-diseases-(cvds) (accessed on 1 January 2025).
3. Laughlin, C.D.; D'Aquili, E.G. *Biogenetic Structuralism*; Columbia University Press: New York, NY, USA, 1974.
4. Leavy, S.A. Biogenetic Structuralism. *Yale J. Biol. Med.* **1976**, *49*, 420–421. [PubMed Central]
5. Ho, S.Y.; Nihoyannopoulos, P. Anatomy, echocardiography, and normal right ventricular dimensions. *Heart* **2006**, *92* (Suppl. S1), i2–i13. [CrossRef] [PubMed]
6. Jalilinejad, N.; Rabiee, M.; Baheiraei, N.; Ghahremanzadeh, R.; Salarian, R.; Rabiee, N.; Akhavan, O.; Zarrintaj, P.; Hejna, A.; Saeb, M.R.; et al. Electrically conductive carbon-based (bio)-nanomaterials for cardiac tissue engineering. *Bioeng. Transl. Med.* **2022**, *8*, e10347. [CrossRef] [PubMed]
7. Brutsaert, D.L.; Fransen, P.; Andries, L.J.; De Keulenaer, G.W.; Sys, S.U. Cardiac endothelium and myocardial function. *Cardiovasc. Res.* **1998**, *38*, 281–290. [CrossRef]
8. Hausner, E.; Elmore, S.; Yang, X. Overview of the components of cardiac metabolism. *Drug Metab. Disposit.* **2019**, *47*, 673–688. [CrossRef]
9. Firth, J. Endothelial barriers: From hypothetical pores to membrane proteins. *J. Anat.* **2002**, *200*, 541–548. [CrossRef]
10. Nevola, R.; Villani, A.; Imbriani, S.; Alfano, M.; Criscuolo, L.; Beccia, D.; Ruocco, R.; Femine, A.D.; Gragnano, F.; Cozzolino, D.; et al. Sodium-Glucose Co-Transporters Family: Current Evidence, Clinical Applications and Perspectives. *Front. Biosci. (Landmark Ed)* **2023**, *28*, 103. [CrossRef]
11. Haddad, R.E.; Jurjus, A.R.; Ibrahim, M.Z.; Nahle, Z.A.; el-Kasti, M.M.; Bitar, K.M.; Kreydiyyeh, S.I.; Saadeh, F.A.; Bikhazi, A.B. Binding of ^{125}I-Insulin on Capillary Endothelial and Myofiber Cell Membranes in Normal and Streptozotocin-Induced Diabetic Perfused Rat Hearts. *Comp. Biochem. Physiol. Part A Physiol.* **1997**, *117*, 523–530. [CrossRef]
12. Xuan, Y.; Chen, C.; Wen, Z.; Wang, D.W. The Roles of Cardiac Fibroblasts and Endothelial Cells in Myocarditis. *Front. Cardiovasc. Med.* **2022**, *9*, 882027. [CrossRef]
13. Hennigs, J.K.; Matuszcak, C.; Trepel, M.; Körbelin, J. Vascular Endothelial Cells: Heterogeneity and Targeting Approaches. *Cells* **2021**, *10*, 2712. [CrossRef]
14. Mikkaichi, T.; Suzuki, T.; Onogawa, T.; Abe, T. Isolation and characterization of a digoxin transporter and its rat homologue expressed in the kidney. *Proc. Natl. Acad. Sci. USA* **2004**, *101*, 3569–3574. [CrossRef] [PubMed]
15. Igel, S.; Drescher, S.; Mürdter, T.; Hofmann, U.; Heinkele, G.; Tegude, H.; Glaeser, H.; Brenner, S.S.; Somogyi, A.A.; Omari, T.; et al. Increased absorption of digoxin from the human jejunum due to inhibition of intestinal transporter-mediated efflux. *Clin. Pharmacokinet.* **2007**, *46*, 777–785. [CrossRef] [PubMed]
16. Yao, H.M.; Chiou, W.L. The complexity of intestinal absorption and exsorption of digoxin in rats. *Int. J. Pharm.* **2006**, *322*, 79–86. [CrossRef] [PubMed]
17. Kimura, Y.; Morita, S.; Matsuo, M.; Ueda, K. Mechanism of multidrug recognition by MDR1/ABCB1. *Cancer Sci.* **2007**, *98*, 1303–1310. [CrossRef]
18. Thompson, S.E.; Bunting, K.V.; Townend, J.N. The modern-day role of digoxin in heart failure and atrial fibrillation—Benefits and limitations. *Br. J. Cardiol.* **2024**, *31* (Suppl. S3), S12–S18. [CrossRef]
19. Rahman, M.; Park, B.-Y. Na,K-ATPase β2 isoform (*atp1b2*) expressed in the retina of *Xenopus*. *J. Biomed. Res.* **2014**, *15*, 194–199. [CrossRef]
20. Afzal, O.; Altamimi, A.S.A.; Nadeem, M.S.; Alzarea, S.I.; Almalki, W.H.; Tariq, A.; Mubeen, B.; Murtaza, B.N.; Iftikhar, S.; Riaz, N.; et al. Nanoparticles in Drug Delivery: From History to Therapeutic Applications. *Nanomaterials* **2022**, *12*, 4494. [CrossRef]

21. Torrice, M. Does nanomedicine have a delivery problem? *ACS Cent. Sci.* **2016**, *2*, 434–437. [CrossRef]
22. Kobayashi, H.; Watanabe, R.; Choyke, P.L. Improving Conventional Enhanced Permeability and Retention (EPR) Effects; What Is the Appropriate Target? *Theranostics* **2014**, *4*, 81–89. [CrossRef]
23. Mhaske, A.; Shukla, S.; Ahirwar, K.; Singh, K.K.; Shukla, R. Receptor-Assisted Nanotherapeutics for Overcoming the Blood–Brain Barrier. *Mol. Neurobiol.* **2024**, *61*, 8702–8738. [CrossRef]
24. Terstappen, G.C.; Meyer, A.H.; Bell, R.D.; Zhang, W. Strategies for delivering therapeutics across the blood–brain barrier. *Nat. Rev. Drug Discov.* **2021**, *20*, 362–383. [CrossRef]
25. Abdullah, C.S.; Alam, S.; Aishwarya, R.; Miriyala, S.; Bhuiyan, M.A.N.; Panchatcharam, M.; Pattillo, C.B.; Orr, A.W.; Sadoshima, J.; Hill, J.A.; et al. Doxorubicin-induced cardiomyopathy associated with inhibition of autophagic degradation process and defects in mitochondrial respiration. *Sci. Rep.* **2019**, *9*, 2002. [CrossRef]
26. Swain, S.; Whaley, F.S.; Ewer, M.S. Congestive heart failure in patients treated with doxorubicin. *Cancer* **2003**, *97*, 2869–2879. [CrossRef]
27. Von Hoff, D.D.; Layard, M.W.; Basa, P.; Davis, H.L.; Von Hoff, A.L.; Rozencweig, M.; Muggia, F.M. Risk Factors for Doxorubicin-induced Congestive Heart Failure. *Ann. Intern. Med.* **1979**, *91*, 710–717. [CrossRef]
28. Christiansen, S.; Autschbach, R. Doxorubicin in experimental and clinical heart failure. *Eur. J. Cardiothorac. Surg.* **2006**, *30*, 611–616. [CrossRef] [PubMed]
29. Teran-Saavedra, N.G.; Sarabia-Sainz, J.A.; Velázquez-Contreras, E.F.; Ramos-Clamont Montfort, G.; Pedroza-Montero, M.; Vazquez-Moreno, L. Albumin-Albumin/Lactosylated Core-Shell Nanoparticles: Therapy to Treat Hepatocellular Carcinoma for Controlled Delivery of Doxorubicin. *Molecules* **2020**, *25*, 5432. [CrossRef] [PubMed]
30. Poelstra, K.; Prakash, J.; Beljaars, L. Drug targeting to the diseased liver. *J. Control. Release* **2012**, *161*, 188–197. [CrossRef]
31. Kabir, S.; Lingappa, N.; Mayrovitz, H. Potential Therapeutic Treatments for Doxorubicin-Induced Cardiomyopathy. *Cureus* **2022**, *14*, e21154. [CrossRef]
32. Beca, S.; Ahmad, F.; Shen, W.; Liu, J.; Makary, S.; Polidovitch, N.; Sun, J.; Hockman, S.; Chung, Y.W.; Movsesian, M.; et al. Phosphodiesterase type 3A regulates basal myocardial contractility through interacting with sarcoplasmic reticulum calcium ATPase type 2a signaling complexes in mouse heart. *Circ. Res.* **2013**, *112*, 289–297. [CrossRef] [PubMed]
33. Paradis, P.; Dali-Youcef, N.; Paradis, F.W.; Thibault, G.; Nemer, M. Overexpression of angiotensin II type I receptor in cardiomyocytes induces cardiac hypertrophy and remodeling. *Proc. Natl. Acad. Sci. USA* **2000**, *97*, 931–936. [CrossRef]
34. Xu, J.; Carretero, O.A.; Lin, C.X.; Cavasin, M.A.; Shesely, E.G.; Yang, J.J.; Reudelhuber, T.L.; Yang, X.P. Role of cardiac overexpression of angiotensin II in the regulation of cardiac function and remodeling post-myocardial infarction. *Am. J. Physiol. Heart Circ. Physiol.* **2007**, *293*, H1900–H1907. [CrossRef] [PubMed]
35. Lomis, N.; Sarfaraz, Z.K.; Alruwaih, A.; Westfall, S.; Shum-Tim, D.; Prakash, S. Albumin Nanoparticle Formulation for Heart-Targeted Drug Delivery: In Vivo Assessment of Congestive Heart Failure. *Pharmaceuticals* **2021**, *14*, 697. [CrossRef] [PubMed]
36. Liu, A.P.; Aguet, F.; Danuser, G.; Schmid, S.L. Local clustering of transferrin receptors promotes clathrin-coated pit initiation. *J. Cell Biol.* **2010**, *191*, 1381–1393. [CrossRef]
37. Cureton, D.K.; Harbison, C.E.; Cocucci, E.; Parrish, C.R.; Kirchhausen, T. Limited transferrin receptor clustering allows rapid diffusion of canine parvovirus into clathrin endocytic structures. *J. Virol.* **2012**, *86*, 5330–5340. [CrossRef]
38. Kibbey, R.G.; Rizo, J.; Gierasch, L.M.; Anderson, R.G. The LDL Receptor Clustering Motif Interacts with the Clathrin Terminal Domain in a Reverse Turn Conformation. *J. Cell Biol.* **1998**, *142*, 59–67. [CrossRef]
39. Bai, F.; Yang, G.; Eskew, J.; Wang, N.; Bose, H.; Zhao, Z. Antagonism of Angiotensin II AT1 Receptor and Silencing of CD44 Gene Expression Inhibit Cardiac Fibroblast Activation via Modulating TGF-β1/Smad Signaling Pathway. *Adv. Biosci. Biotechnol.* **2020**, *11*, 123–139. [CrossRef]
40. Schnitzer, J.E. gp60 is an albumin-binding glycoprotein expressed by continuous endothelium involved in albumin transcytosis. *Am. J. Physiol. Heart Circ. Physiol.* **1992**, *262 Pt 2*, H246–H254. [CrossRef]
41. Ishima, Y.; Maruyama, T. Human Serum Albumin as Carrier in Drug Delivery Systems. *Yakugaku Zasshi* **2016**, *136*, 39–47. [CrossRef]
42. Roncevic, D. Does angiotensin II cross the blood–brain barrier? *Hypertens. Res.* **2012**, *35*, 775. [CrossRef]
43. Zhao, Y.; Samal, E.; Srivastava, D. Serum response factor regulates a muscle-specific microRNA that targets Hand2 during cardiogenesis. *Nature* **2005**, *436*, 214–220. [CrossRef] [PubMed]
44. Xue, X.; Shi, X.; Dong, H.; You, S.; Cao, H.; Wang, K.; Wen, Y.; Shi, D.; He, B.; Li, Y. Delivery of microRNA-1 inhibitor by dendrimer-based nanovector: An early targeting therapy for myocardial infarction in mice. *Nanomedicine* **2018**, *14*, 619–631. [CrossRef]
45. Yang, B.; Li, D.; Phillips, M.I.; Mehta, P.; Mehta, J.L. Myocardial angiotensin II receptor expression and ischemia-reperfusion injury. *Vasc. Med.* **1998**, *3*, 121–130. [CrossRef]

46. Aimo, A.; Castiglione, V.; Borrelli, C.; Saccaro, L.F.; Franzini, M.; Masi, S.; Emdin, M.; Giannoni, A. Oxidative stress and inflammation in the evolution of heart failure: From pathophysiology to therapeutic strategies. *Eur. J. Prev. Cardiol.* **2020**, *27*, 494–510. [CrossRef] [PubMed]
47. Tabish, T.A.; Hayat, H.; Abbas, A.; Narayan, R.J. Graphene Quantum Dots-Based Electrochemical Biosensing Platform for Early Detection of Acute Myocardial Infarction. *Biosensors* **2022**, *12*, 77. [CrossRef]
48. Rostamzadeh, F.; Jafarinejad-Farsangi, S.; Ansari-Asl, Z.; Farrokhi, M.S.; Jafari, E. Treatment for Myocardial Infarction: In Vivo Evaluation of Curcumin-Loaded PEGylated-GQD Nanoparticles. *J. Cardiovasc. Pharmacol.* **2023**, *81*, 361–372. [CrossRef] [PubMed]
49. Perales-Salinas, V.; Purushotham, S.S.; Buskila, Y. Curcumin as a potential therapeutic agent for treating neurodegenerative diseases. *Neurochem. Int.* **2024**, *178*, 105790. [CrossRef] [PubMed]
50. Ahmed, S.; Khan, H.; Mirzaei, H. Mechanics insights of curcumin in myocardial ischemia: Where are we standing? *Eur. J. Med. Chem.* **2019**, *183*, 111658. [CrossRef]
51. Zhao, J.; Chen, Y.; Chen, Q.; Hong, T.; Zhong, Z.; He, J.; Ni, C. Curcumin Ameliorates Cardiac Fibrosis by Regulating Macrophage-Fibroblast Crosstalk via IL18-P-SMAD2/3 Signaling Pathway Inhibition. *Front. Pharmacol.* **2022**, *12*, 784041. [CrossRef]
52. Mishra, P.; Sahoo, D.K.; Mohanty, C.; Samanta, L. Curcumin-loaded nanoparticles effectively prevent T4-induced oxidative stress in rat heart. *Cell Biochem. Funct.* **2024**, *42*, e4070. [CrossRef]
53. Riksen, N.P.; Barrera, P.; van den Broek, P.H.; van Riel, P.L.; Smits, P.; Rongen, G.A. Methotrexate modulates the kinetics of adenosine in humans in vivo. *Ann. Rheum. Dis.* **2006**, *65*, 465–470. [CrossRef]
54. Methner, C.; Schmidt, K.; Cohen, M.V.; Downey, J.M.; Krieg, T. Both A2a and A2b adenosine receptors at reperfusion are necessary to reduce infarct size in mouse hearts. *Am. J. Physiol. Heart Circ. Physiol.* **2010**, *299*, H1262–H1264. [CrossRef]
55. Maranhão, R.C.; Guido, M.C.; de Lima, A.D.; Tavares, E.R.; Marques, A.F.; Tavares de Melo, M.D.; Nicolau, J.C.; Salemi, V.M.; Kalil-Filho, R. Methotrexate carried in lipid core nanoparticles reduces myocardial infarction size and improves cardiac function in rats. *Int. J. Nanomed.* **2017**, *12*, 3767–3784. [CrossRef]
56. Zhu, L.P.; Tian, T.; Wang, J.Y.; He, J.N.; Chen, T.; Pan, M.; Xu, L.; Zhang, H.X.; Qiu, X.T.; Li, C.C.; et al. Hypoxia-elicited mesenchymal stem cell-derived exosomes facilitates cardiac repair through miR-125b-mediated prevention of cell death in myocardial infarction. *Theranostics* **2018**, *8*, 6163–6177. [CrossRef]
57. Gonda, A.; Kabagwira, J.; Senthil, G.N.; Wall, N.R. Internalization of Exosomes through Receptor-Mediated Endocytosis. *Mol. Cancer Res.* **2019**, *17*, 337–347. [CrossRef] [PubMed]
58. Nelson, B.C.; Maragh, S.; Ghiran, I.C.; Jones, J.C.; DeRose, P.C.; Elsheikh, E.; Vreeland, W.N.; Wang, L. Measurement and Standardization Challenges for Exosome-Based Delivery Vectors. *Nanomedicine* **2020**, *15*, 2149–2170. [CrossRef] [PubMed]
59. Palakurthi, S.S.; Shah, B.; Kapre, S.; Charbe, N.; Immanuel, S.; Pasham, S.; Thalla, M.; Jain, A.; Palakurthi, S. A comprehensive review of challenges and advances in exosome-based drug delivery systems. *Nanoscale Adv.* **2024**, *6*, 5803–5826. [CrossRef]
60. Sun, Y.; Zhang, P.; Li, Y.; Hou, Y.; Yin, C.; Wang, Z.; Liao, Z.; Fu, X.; Li, M.; Fan, C.; et al. Light-Activated Gold–Selenium Core–Shell Nanocomposites with NIR-II Photoacoustic Imaging Performances for Heart-Targeted Repair. *ACS Nano* **2022**, *16*, 18667–18681. [CrossRef]
61. Ryabova, Y.V.; Sutunkova, M.P.; Minigalieva, I.A.; Shabardina, L.V.; Filippini, T.; Tsatsakis, A. Toxicological effects of selenium nanoparticles in laboratory animals: A review. *J. Appl. Toxicol.* **2024**, *44*, 4–16. [CrossRef]
62. Kanki, S.; Jaalouk, D.E.; Lee, S.; Yu, A.Y.; Gannon, J.; Lee, R.T. Identification of Targeting Peptides for Ischemic Myocardium by In Vivo Phage Display. *J. Mol. Cell. Cardiol.* **2011**, *50*, 841–848. [CrossRef] [PubMed]
63. Ferreira, M.P.; Ranjan, S.; Correia, A.M.; Mäkilä, E.M.; Kinnunen, S.M.; Zhang, H.; Shahbazi, M.A.; Almeida, P.V.; Salonen, J.J.; Ruskoaho, H.J.; et al. In vitro and in vivo assessment of heart-homing porous silicon nanoparticles. *Biomaterials* **2016**, *94*, 93–104. [CrossRef]
64. Zhao, M.; Liu, C.; Liu, Z.; Zuo, Y.; Chen, C.; Shi, S.; Shi, X.; Xie, Y.; Yang, H.; Chen, Y. Myocardium-targeted liposomal delivery of the antioxidant peptide 8P against doxorubicin-induced myocardial injury. *Int. J. Pharm.* **2024**, *663*, 124569. [CrossRef]
65. Grego, A.; Fernandes, C.; Fonseca, I.; Dias-Neto, M.; Costa, R.; Leite-Moreira, A.; Oliveira, S.M.; Trindade, F.; Nogueira-Ferreira, R. Endothelial dysfunction in cardiovascular diseases: Mechanisms and in vitro models. *Mol. Cell. Biochem.* **2025**, *480*, 4671–4695. [CrossRef]
66. Yun, J.H.; Park, S.W.; Kim, K.-J.; Bae, J.-S.; Lee, E.H.; Paek, S.H.; Kim, S.U.; Ye, S.; Kim, J.-H.; Cho, C.-H. Endothelial STAT3 Activation Increases Vascular Leakage Through Downregulating Tight Junction Proteins: Implications for Diabetic Retinopathy. *J. Cell. Physiol.* **2017**, *232*, 1123–1134. [CrossRef] [PubMed]
67. Alsaffar, H.; Martino, N.; Garrett, J.P.; Adam, A.P. Interleukin-6 promotes a sustained loss of endothelial barrier function via Janus kinase-mediated STAT3 phosphorylation and de novo protein synthesis. *Am. J. Physiol. Cell Physiol.* **2018**, *314*, C589–C602. [CrossRef]

68. Clark, P.R.; Kim, R.K.; Pober, J.S.; Kluger, M.S. Tumor Necrosis Factor Disrupts Claudin-5 Endothelial Tight Junction Barriers in Two Distinct NF-κB-Dependent Phases. *PLoS ONE* **2015**, *10*, e0120075. [CrossRef] [PubMed]
69. Wang, L.; Quine, S.; Frickenstein, A.N.; Lee, M.; Yang, W.; Sheth, V.M.; Bourlon, M.D.; He, Y.; Lyu, S.; Garcia-Contreras, L.; et al. Exploring and Analyzing the Systemic Delivery Barriers for Nanoparticles. *Adv. Funct. Mater.* **2023**, *34*, 2308446. [CrossRef]
70. Muniyappa, R.; Montagnani, M.; Koh, K.K.; Quon, M.J. Cardiovascular actions of insulin. *Endocr. Rev.* **2007**, *28*, 463–491. [CrossRef] [PubMed]
71. Tashima, T. Smart Strategies for Therapeutic Agent Delivery into Brain across the Blood–Brain Barrier Using Receptor-Mediated Transcytosis. *Chem. Pharm. Bull.* **2020**, *68*, 316–325. [CrossRef]
72. Upton, R.L.; Chen, Y.; Mumby, S. Variable tissue expression of transferrin receptors: Relevance to acute respiratory distress syndrome. *Eur. Respir. J.* **2003**, *22*, 335–341. [CrossRef]
73. Sugo, T.; Terada, M.; Oikawa, T.; Miyata, K.; Nishimura, S.; Kenjo, E.; Ogasawara-Shimizu, M.; Makita, Y.; Imaichi, S.; Murata, S.; et al. Development of antibody-siRNA conjugate targeted to cardiac and skeletal muscles. *J. Control. Release* **2016**, *237*, 1–13. [CrossRef]
74. Selim, M.M.; El-Safty, S.; Tounsi, A.; Shenashen, M. A review of magnetic nanoparticles used in nanomedicine. *APL Mater.* **2024**, *12*, 010601. [CrossRef]
75. Liu, S.; Chen, X.; Bao, L.; Liu, T.; Yuan, P.; Yang, X.; Qiu, X.; Gooding, J.J.; Bai, Y.; Xiao, J.; et al. Treatment of infarcted heart tissue via the capture and local delivery of circulating exosomes through antibody-conjugated magnetic nanoparticles. *Nat. Biomed. Eng.* **2020**, *4*, 1063–1075. [CrossRef] [PubMed]
76. Lim, B.-K.; Tighe, E.C.; Kong, S.D. The use of magnetic targeting for drug delivery into cardiac myocytes. *J. Magn. Magn. Mater.* **2019**, *473*, 21–25. [CrossRef]
77. Patel, D.D.; Bussel, J.B. Neonatal Fc receptor in human immunity: Function and role in therapeutic intervention. *J. Allergy Clin. Immunol.* **2020**, *146*, 467–478. [CrossRef]
78. Rostamzadeh, F.; Shadkam-Farrokhi, M.; Jafarinejad-Farsangi, S.; Najafipour, H.; Ansari-Asl, Z.; Yeganeh-Hajahmadi, M. PEGylated Graphene Quantum Dot Improved Cardiac Function in Rats with Myocardial Infarction: Morphological, Oxidative Stress, and Toxicological Evidences. *Oxid. Med. Cell. Longev.* **2021**, *2021*, 8569225. [CrossRef] [PubMed]

pharmaceutics

MDPI

Review

Polymeric Nanoparticles for Targeted Lung Cancer Treatment: Review and Perspectives

Devesh U. Kapoor [1], Sonam M. Gandhi [2], Sambhavi Swarn [2], Basant Lal [3], Bhupendra G. Prajapati [4,5,6], Supang Khondee [7], Supachoke Mangmool [8], Sudarshan Singh [8,9,*] and Chuda Chittasupho [8,*]

1 Department of Pharmaceutics, Dr. Dayaram Patel Pharmacy College, Bardoli 394601, Gujarat, India; dev7200@gmail.com
2 Department of Pharmaceutics, Shree Dhanvantary Pharmacy College, Dist-Kim, Olpad 394110, Gujarat, India; shena.saloni@gmail.com (S.M.G.); sambhaviamresh24@gmail.com (S.S.)
3 PQE Australia Pvt. Ltd., Melbourne, VIC 300, Australia; basant.tyagi63601@gmail.com
4 Department of Pharmaceutics, Parul Institute of Pharmacy, Faculty of Pharmacy, Parul University, Waghodia, Vadodara 391760, Gujarat, India; bhupendra.prajapati40731@paruluniversity.ac.in
5 Faculty of Pharmacy, Silpakorn University, Nakhon Pathom 73000, Thailand
6 Centre for Research Impact and Outcome, Chitkara College of Pharmacy, Chitkara University, Rajpura 140401, Punjab, India
7 School of Pharmaceutical Sciences, University of Phayao Maek, Muaeng, Phayao 56000, Thailand; supang.kh@up.ac.th
8 Faculty of Pharmacy, Chiang Mai University, Chiang Mai 50200, Thailand; supachoke.man@cmu.ac.th
9 Office of Research Administration, Chiang Mai University, Chiang Mai 50200, Thailand
* Correspondence: sudarshan.s@cmu.ac.th (S.S.); chuda.c@cmu.ac.th (C.C.)

Abstract: Lung cancer remains a foremost cause of cancer-related impermanence globally, demanding innovative and effective therapeutic strategies. Polymeric nanoparticles (NPs) have turned up as a promising transport system for drugs due to their biodegradability, biocompatibility, and capability to provide controlled and targeted release of therapeutic agents. This review offers a thorough examination of different polymeric NP platforms, such as chitosan, gelatin, alginate, poly (lactic acid), and polycaprolactone, highlighting their mechanisms, formulations, and applications in the treatment of lung cancer. These NPs facilitate the delivery of chemotherapeutic agents, gene therapies, and immune modulators, with enhanced bioavailability and reduced systemic toxicity. Additionally, advanced formulations such as ligand-conjugated, stimuli-responsive, and multifunctional NPs demonstrate improved tumor-specific accumulation and cellular uptake. The review also discusses quantum dots, magnetic and lipid-based NPs, and green-synthesized metallic polymeric hybrids, emphasizing their potential in theranostics and combination therapies. Preclinical studies show promising results, yet clinical translation faces challenges; for example, large-scale production, long-term toxicity, and regulatory hurdles. Overall, polymeric NPs represent a powerful platform for advancing personalized lung cancer therapy, with future prospects rooted in multifunctional, targeted, and patient-specific nanomedicine.

Keywords: polymeric nanoparticles; lung cancer; anticancer; molecular pathways; targeted therapy

1. Introduction

Lung cancer continues to be the most fatal cancer globally, accounting for approximately 1.8 million deaths in 2020, alongside 2.2 million newly diagnosed cases [1,2].

Regions such as Europe, Eastern Asia, and Southeast Asia report the highest prevalence, with the use of tobacco identified as the key risk factor, contributing to nearly 85% of all cases [3]. However, non-smoking-related lung cancer, often linked to genetic mutations (e.g., epidermal growth factor receptor (EGFR), anaplastic lymphoma kinase [4] receptors) and environmental exposures (e.g., radon, air pollution), accounts for 10–25% of cases globally [5–9].

Age-standardized incidence rates are higher in men, though increasing trends among women, particularly in high-income countries, are concerning [10]. Socioeconomic disparities also influence outcomes, with lower survival rates due to delayed detection and inadequate treatment availability, which low- and middle-income nations face [1,11,12]. Advances in early detection (e.g., low-dose computerized tomography (CT) screening) and targeted therapies have improved survival in high-resource settings, yet global disparities persist (National Lung Screening Trial Research Team 2011). Figure 1 illustrates the trends, disparities, and emerging risks of lung cancer globally.

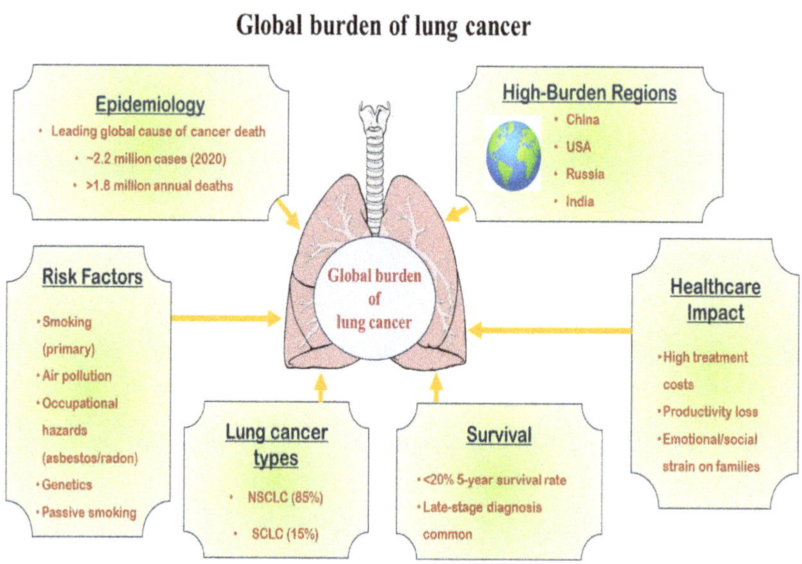

Figure 1. Diagrammatic representation of the global burden of lung cancer.

Lung cancer arises from a complicated interaction of behavioral, environmental, and genetic factors. Tobacco smoking remains the primary cause, contributing to 80–90% of cases, with carcinogens like polycyclic aromatic hydrocarbons and nitrosamines inducing DNA mutations [1]. However, 15–20% of lung cancers occur in people who do not have a smoking habit, with risk factors including radon gas, secondhand smoke, air pollution (e.g., $PM_{2.5}$), and occupational exposures (asbestos, arsenic, silica) [13] (Figure 2).

Genetic susceptibility plays a crucial role, with mutations in EGFR, KRAS (Kirsten rat sarcoma virus), ALK, and Tumor suppressor (TP) 53 driving oncogenesis, particularly in non-smokers [14]. Additionally, chronic inflammation from conditions like chronic obstructive pulmonary disease (COPD) and pulmonary fibrosis elevates lung cancer risk [15]. Emerging evidence suggests dietary factors (low fruit/vegetable intake) and hormonal influences (estrogen signaling) may contribute, though further research is needed [15]. In recent years, electronic cigarette (e-cigarette) use has raised concerns due to potential carcinogenic effects, though long-term studies are still lacking [15]. Understanding these multifactorial causes is essential for preventive strategies and personalized treatment approaches.

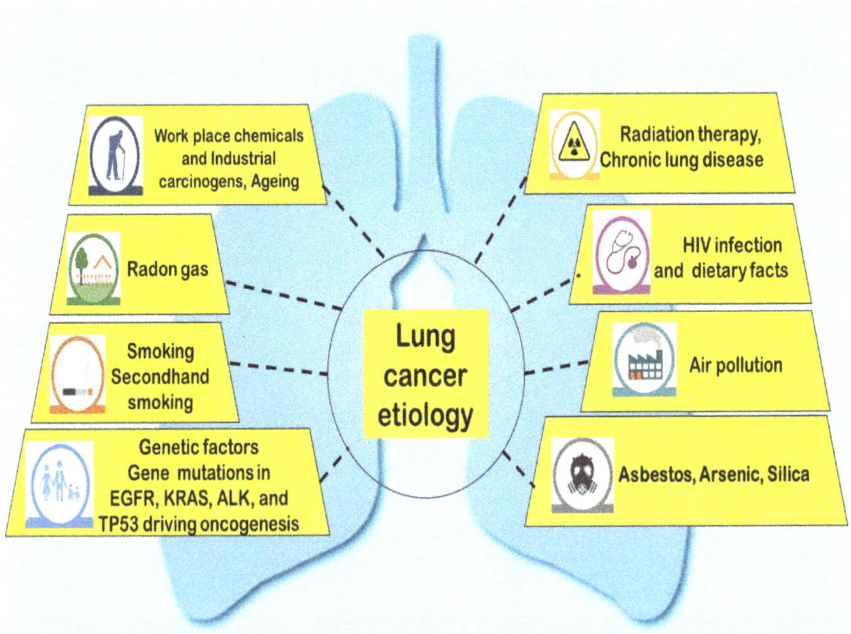

Figure 2. Illustration of the etiology of lung cancer.

Lung cancer pathogenesis involves the dysregulation of key molecular pathways that drive tumor initiation, progression, and metastasis. The EGFR/RAS/RAF/MEK/ERK and PI3K/AKT/mTOR pathways are frequently activated, promoting cell proliferation and survival [16]. Mutations in EGFR, KRAS, and BRAF are common in non-small cell lung cancer (NSCLC), while ALK, ROS1, and RET rearrangements define distinct molecular subtypes [15]. In small-cell lung cancer (SCLC), TP53 and RB1 inactivation are nearly universal, leading to unchecked cell cycle progression [17]. Epigenetic alterations, such as DNA methylation and histone modifications, further contribute to oncogenic signaling [18]. The immune checkpoint pathways (PD-1/PD-L1, CTLA-4) play a critical role in tumor immune evasion, making them key targets for immunotherapy [18]. Emerging research highlights the role of tumor microenvironment [19] components, including cancer-associated fibroblasts and immune cells, in modulating therapeutic resistance [20]. Additionally, WNT/β-catenin and Hippo-YAP/TAZ signaling pathways are implicated in lung cancer stemness and metastasis [20].

Polymeric nanoparticle systems can be strategically designed not only to deliver therapeutic agents but also to actively remodel the tumor microenvironment [19] to overcome resistance mechanisms. Beyond stimuli-responsive release, PNPs can be engineered to co-deliver immunomodulators, such as checkpoint inhibitors or cytokines that reprogram immunosuppressive TMEs into immunostimulatory ones, enhancing antitumor immunity. Surface functionalization with ligands targeting cancer-associated fibroblasts or tumor-associated macrophages can disrupt stromal barriers, reduce extracellular matrix (ECM) density, and improve drug penetration [21]. Incorporating enzymes like hyaluronidase or collagenase within PNPs enables localized ECM degradation, lowering interstitial fluid pressure and enhancing vascular perfusion. Polymeric NPs can also deliver agents that inhibit pro-survival signaling pathways (e.g., STAT3, HIF-1α) or modulate hypoxia by carrying oxygen-generating compounds, thereby sensitizing tumors to chemotherapy or radiotherapy [22]. Additionally, NPs can be designed to release reactive oxygen species modulators to disrupt redox balance, making cancer cells more vulnerable to therapy. By integrating these TME-modulating capabilities with precise drug delivery, polymeric NPs offer a multifunctional approach to simultaneously overcome physical, biochemical, and immunological resistance barriers, ultimately improving therapeutic efficacy in lung cancer management [23].

2. Targeted Therapies in Lung Cancer

Current progress in nanotechnology has led to the development of advanced drug delivery methods, enhancing the effectiveness of cancer therapy while reducing harmful adverse effects [24–27]. Among these, liposomal formulations remain clinically dominant due to their biocompatibility and ability to encapsulate both hydrophilic and hydrophobic drugs [28]. Modern liposomes incorporate surface modifications, such as PEGylation, to extend circulation times and antibody conjugation for active tumor targeting.

Polymeric NPs, particularly those using biodegradable poly(lactic-co-glycolic acid) (PLGA), demonstrate excellent controlled-release properties through tunable polymer degradation rates [29,30]. Recent innovations include stimuli-responsive designs that release payloads specifically in acidic tumor microenvironments [31,32]. Dendrimers offer unparalleled structural precision for high drug-loading capacity and functionalization, though toxicity concerns require further investigation [33].

Inorganic NPs, such as gold nanostructures, provide dual diagnostic and therapeutic capabilities through photothermal ablation and imaging contrast enhancement [34]. Meanwhile, natural exosome-based systems show promise for their innate biocompatibility and homing capabilities to specific tissues [35]. Current research focuses on hybrid systems combining multiple nanocarrier advantages while addressing manufacturing scalability and regulatory requirements for clinical translation.

However, translating polymeric NPs from bench to bedside in lung cancer therapy faces multifaceted challenges. Scalability remains a primary hurdle, as achieving reproducible size, drug loading, and surface characteristics during large-scale manufacturing is technically demanding and cost-intensive [36]. Regulatory considerations add complexity, with evolving guidelines for nanomedicines requiring exhaustive characterization, toxicological profiling, and demonstration of long-term safety. Safety evaluation is critical, given the potential for unforeseen immunogenicity, off-target accumulation, or toxicity from degradation products, particularly with repeated dosing [37]. Comprehensive in vitro and in vivo studies, alongside predictive models, are essential for risk assessment. Eventually, cost-effectiveness poses a barrier; despite promising therapeutic outcomes, high production and quality control costs can limit commercial viability and patient accessibility. Addressing these challenges will require standardized manufacturing protocols, robust regulatory frameworks, advanced safety analytics, and cost-reduction strategies to ensure clinical translation and broader adoption of PNP-based lung cancer therapeutics [38].

Comparison Between Conventional and Novel Therapies Involved in the Management of Lung Cancer

Traditional approaches to managing lung carcinoma, such as surgical resection, systemic chemotherapy, and radiotherapy, have historically served as the primary treatment options. Nevertheless, these methods frequently exhibit poor precision, resulting in severe adverse effects and reduced effectiveness, particularly in late-stage disease [39]. Emerging treatment strategies, including molecularly targeted agents, offer promising alternatives. Therapy, immunotherapy, and nanotechnology-based drug delivery offer improved selectivity, reduced systemic toxicity, and enhanced therapeutic outcomes. Targeted therapies inhibit specific molecular pathways, while immunotherapy boosts the body's immune response against cancer. Nanocarriers enhance drug accumulation at tumor sites [40]. Although novel drug delivery suppresses the disadvantages of conventional therapies, the mucociliary clearance is indeed a critical physiological barrier limiting the pulmonary residence time of inhalable therapeutics. In the formulation, the polymeric NPs were engineered with a muco-inert surface by incorporating PEGylated polymers, which reduce adhesive interactions with mucus glycoproteins, thereby enabling deeper penetration into the lung epithelium. Furthermore, the NPs possess a narrow size distribution (~200 nm) to avoid rapid entrapment in mucus mesh pores while facilitating alveolar deposition [41]. The hydrophilic polymer coating also minimizes opsonization and phagocytic uptake by alveolar macrophages, supporting prolonged drug retention. In addition, the controlled-release matrix of the polymer enables sustained drug liberation at the target site, countering clearance-driven loss. Collectively, these features synergistically mitigate mucociliary removal, enhance deep lung bioavailability, and ensure prolonged therapeutic levels at the tumor site, thereby maximizing the clinical potential of our inhalable polymeric NP system for localized lung cancer therapy [42]. Therefore, overall, novel approaches provide personalized and more effective management, especially for NSCLC, though they may still face challenges like resistance and high cost. A comparative data on conventional and novel therapies for lung cancer management is presented in Table 1.

Table 1. Comparing conventional and novel therapies for lung cancer management.

Therapy Aspect	Conventional Therapies	Novel Therapies	References
Primary treatment types	- Surgery (lobectomy, pneumonectomy) - Chemotherapy (platinum-based) - Radiation therapy (external beam)	- Targeted therapy (EGFR inhibitor/PI3K inhibitor) - Immunotherapy (PD-1/PD-L1 inhibitors) - Stereotactic body radiotherapy (SBRT)	[40,43,44]
Mechanism of action	- Cytotoxic drugs kill rapidly dividing cells - Radiation induces DNA damage	- Targeted drugs block specific oncogenic pathways - Immunotherapy enhances T-cell antitumor activity	[13]
Response rates	- 20–30% for chemotherapy - 30–50% for radiation in early stages	- 60–80% for targeted therapy in mutation-positive cases - 20–40% durable responses with immunotherapy	[14]
Adverse reactions	- Bone marrow suppression - Nausea/vomiting - Radiation pneumonitis	- Immune-related adverse events (colitis, pneumonitis) - Skin rash with EGFR inhibitors	[45–48]
Personalization level	- Largely based on cancer stage and histology	- Requires molecular profiling (NGS) - Biomarker-driven (PD-L1, TMB)	[49–52]
5-Year survival benefit	- 15–20% for stage III chemo-radiation - 40–50% for early-stage surgery	- 25–30% for stage IV with targeted therapy - 15–25% long-term survivors with immunotherapy	[52–55]

3. Advanced Nanoparticle Platforms for Lung Cancer Theranostics and Targeted Therapy

3.1. Magnetic Nanoparticles

Magnetic NPs have emerged as efficient tools in lung cancer diagnosis and treatment because of their unique physicochemical properties. Superparamagnetic iron oxide NPs are extensively studied, enabling magnetic resonance imaging (MRI) enhancement for early tumor detection with high sensitivity [56]. Functionalized MNPs conjugated with targeting ligands (e.g., EGFR antibodies) improve tumor-specific drug delivery, reducing systemic toxicity [57].

In hyperthermia therapy, alternating magnetic fields heat Magnetic NPs to 42–46 °C, inducing selective cancer cell apoptosis while sparing healthy tissue [58]. Combined with chemotherapy, e.g., doxorubicin (DOX)-loaded MNPs, this approach enhances tumor penetration and treatment efficacy [57]. MNPs also facilitate gene therapy by delivering siRNA to silence oncogenes (e.g., KRAS) [59]. Recent advances include multifunctional MNPs integrating imaging, therapy, and real-time monitoring [60]. For example, superparamagnetic iron oxide NPs coated with PD-L1 inhibitors enable immunotherapy alongside MRI tracking [61]. Challenges remain in optimizing biocompatibility, biodistribution, and large-scale production [62].

3.2. Solid Lipid Nanoparticles

Solid lipid nanoparticles (SLNs) have gained attention as efficient drug transport systems for lung cancer due to their stability, biocompatibility, and ability to encapsulate both water-loving and hydrophobic drugs [63]. Composed of physiological lipids, SLNs increased tumor targeting while reducing the systemic toxicity [64]. Their small size (50–300 nm) facilitates passive accumulation in tumors via the enhanced permeability and retention (EPR) effect, improving drug bioavailability [64]. SLNs are particularly valuable for delivering chemotherapeutic agents (e.g., paclitaxel, docetaxel) with reduced side effects compared to conventional formulations [65]. Surface modification with ligands (e.g., folate, transferrin) enables active targeting of cancer cells overexpressing specific receptors [66]. Additionally, SLNs can co-deliver siRNA or miRNA to silence oncogenic pathways (e.g., EGFR, KRAS), enhancing therapeutic efficacy [67]. Recent studies highlight SLNs in combination therapy, such as co-loading chemotherapeutics with immunomodulators (e.g., PD-1 inhibitors) to synergize treatment [64]. SLNs also show promise in inhalable formulations, enabling localized lung delivery and bypassing first-pass metabolism [68].

3.3. Polymeric Reduced Green Metallic Nanoparticles

Polymeric reduced green metallic (PRGM) NPs represent a convergence of green chemistry and nanotechnology, offering an eco-friendly alternative, sustainable to conventional nanoparticle synthesis methods. These NPs are synthesized using biocompatible polymers and reducing agents derived from natural plant extracts, microbes, or other biotic resources, eliminating the need for hazardous chemicals [69]. PRGM NPs have emerged as a novel and sustainable approach in lung cancer therapy, integrating the principles of nanotechnology, green chemistry, and polymer science. These NPs are synthesized using either natural or synthetic polymeric materials as reducing agents, offering an eco-friendly alternative to conventional chemical methods [69]. The green synthesis minimizes toxicity while improving biocompatibility, essential traits for cancer therapeutics. Incorporating biopolymers such as chitosan, PLGA, or polyethylene glycol (PEG) into the nanoparticle matrix enhances stability, targeted delivery, and controlled drug release. Further, the variability in process of fabrication of NPs can be minimized through authenticated sourcing, purity, and standardized synthesis process in case of synthetic polymers and extraction protocols for natural gums. Furthermore, the process standardization includes fixed parameters (pH, temperature, isolation-to-precursor ratio) and quality control checkpoints to monitor precursor quality, extraction yield, and NPs characteristics. This approach enhances reproducibility, comparability, and therapeutic predictability while reducing batch-to-batch variability. PRGM NPs can be engineered to carry chemotherapeutic agents or imaging molecules, enabling simultaneous treatment and diagnosis (theranostics). Their surface can be functionalized for tumor-specific targeting, allowing for improved accumulation in cancerous lung tissues via the EPR effect or active targeting ligands.

In lung cancer, where drug resistance and systemic toxicity pose significant challenges, polymeric reduced green metallic NPs provide enhanced cellular uptake, reduced off-target effects, and the potential to bypass multidrug resistance pathways [70]. Additionally, their antioxidant and anti-inflammatory properties derived from the plant extracts used in synthesis can contribute to synergistic therapeutic effects. Overall, polymeric reduced green metallic NPs represent a promising and sustainable nanoplatform for advancing lung cancer treatment, offering high efficacy with minimized side effects and ecological impact [71]. Table 2 illustrates the comprehensive tabular summary of PRGM NPs in lung cancer management.

Table 2. Comprehensive tabular summary of polymeric reduced green metallic nanoparticles (PRGM NPs) in lung cancer management.

Aspect	Key Findings	References
Targeted imaging	Folic acid-conjugated chitosan nanoparticles (fCNA) significantly enhanced protoporphyrin IX (PpIX) accumulation in colorectal cancer cells (HT29, Caco-2) via folate receptor-mediated endocytosis, improving tumor detection.	[72]
Chemotherapy	Mitochondria-targeted green AuNPs increased reactive oxygen species (ROS) 4.2-fold in H460 cells and reduced tumor volume by 72% in xenografts (5 mg/kg dose).	[73]
siRNA delivery	Chitosan-coated AuNPs delivered EGFR siRNA to A549 xenografts, reducing tumor growth by 65% (0.5 mg/kg, q3d × 4 weeks).	[74]
Combo therapy	Fol-LSMO NPs enable targeted hyperthermia with DOX release in breast cancer. Triggers apoptosis and autophagy via caspase/LC3-II pathways. Folate conjugation enhances uptake, reduces off-target effects.	[75]

3.4. Quantum Dots

Quantum dots (QDs) are semiconductor nanocrystals, typically 2–10 nm in diameter, that exhibit unique size-dependent optical and electronic properties due to quantum confinement effects [76]. They possess high photostability, tunable emission wavelengths, and exceptional brightness, making them highly suitable for biomedical imaging, targeted drug delivery, and multifunctional theranostic applications [77]. Their surfaces can be functionalized with biomolecules, drugs, or polymers, enabling simultaneous diagnostic and therapeutic functions within a single nanoplatform.

Many QDs especially those with cadmium selenide (CdSe), cadmium telluride (CdTe), or lead-based cores pose significant heavy metal toxicity risks. Cadmium ions, released through core degradation under physiological conditions, can induce oxidative stress, mitochondrial damage, and DNA strand breaks, leading to mutagenesis and carcinogenesis. These effects are exacerbated by the small size and large surface area of QDs, which facilitate cellular uptake and interaction with biomolecules [77]. Chronic exposure or incomplete clearance may result in accumulation in organs such as the liver, spleen, kidneys, and brain, raising concerns over nephrotoxicity, hepatotoxicity, and neurotoxicity. Long-term persistence could also trigger chronic inflammation and immune dysregulation, potentially contributing to fibrosis or secondary malignancies. While surface coatings (e.g., PEG, silica, or polymer shells) and heavy-metal-free alternatives (e.g., carbon or indium phosphide QDs) can reduce immediate toxicity, their long-term safety profiles remain underexplored [78]. Comprehensive in vivo studies covering biodistribution, degradation pathways, clearance rates, and delayed toxic effects are essential before clinical translation. Without such evaluation, the full risk profile of QD-based lung cancer theranostics remains uncertain, particularly for repeated or high-dose applications [78].

QDs exhibit substantial promise in revolutionizing lung cancer theranostics through multifunctional capabilities. Research demonstrates that QDs functionalized with targeting ligands, such as EGFR antibodies or RGD peptides, significantly enhance the selective delivery of chemotherapeutic agents (e.g., DOX, cisplatin) to lung tumor cells in vitro and animal models, improving efficacy while minimizing off-target effects compared to free drugs [79]. Their exceptional optical properties are leveraged for superior imaging; specifically, near-infrared-II (NIR-II) window QDs enable deep-tissue, high-resolution tumor visualization, precise lymphatic mapping, and real-time monitoring of drug distribution and therapeutic response, surpassing conventional imaging agents in sensitivity and photostability [80]. The inherent theranostic potential of QDs is further exploited in integrated platforms where QDs co-loaded with drugs and photosensitizers facilitate combined chemotherapy and photodynamic therapy, generating reactive oxygen species (ROS) upon light activation for synergistic tumor cell killing, as confirmed in recent studies [81]. Addressing biocompatibility, significant research focuses on mitigating potential toxicity: surface engineering with dense PEG layers or biocompatible polymers reduces opsonization and prolongs circulation [82]. Furthermore, QDs are being explored as carriers for siRNA targeting oncogenes (e.g., survivin, KRAS) and for immune checkpoint inhibitors (e.g., anti-PD-L1), demonstrating enhanced gene silencing efficacy and potentiation of antitumor immunity in lung cancer models [83]. Despite compelling preclinical evidence for enhanced targeting, imaging, combination therapy, and improved safety profiles of novel compositions, the translation of QD-based systems necessitates rigorous large-scale toxicity assessments and scalable manufacturing solutions [84].

4. Application of Polymeric Nanoparticles in Lung Cancer Management

Polymeric nanoparticles are colloidal systems composed of synthetic or natural polymers, typically measuring between 10 and 1000 nm in size. These NPs are engineered to encapsulate, protect, and transport therapeutic agents to specific target areas, enabling controlled drug release and improved bioavailability. Due to their biocompatibility and tunable surface characteristics, they are extensively utilized in targeted drug delivery applications [85]. In lung cancer therapy, polymeric NPs significantly improve treatment effectiveness while lowering systemic toxicity. They facilitate the targeted delivery of chemotherapeutic drugs, immune checkpoint inhibitors, or gene therapies directly to tumor sites, enhancing drug concentration in cancerous tissues while sparing healthy lung cells [86]. Additionally, polymeric NPs can overcome biological barriers and release drugs in response to tumor-specific triggers, such as acidic pH or elevated enzyme levels. This targeted approach enhances therapeutic efficiency while reducing adverse effects often seen with traditional lung cancer treatments [87]. Different polymers exhibit distinct degradation mechanisms and by-products, influencing long-term safety in pulmonary delivery. Chitosan degrades enzymatically via lysozyme to glucosamine and N-acetylglucosamine, both naturally occurring metabolites with low toxicity [88]. Alginate undergoes depolymerization by alginate lyase to guluronic and mannuronic acid residues, which are generally well tolerated [89]. PLGA hydrolyzes into lactic and glycolic acid, entering the Krebs cycle, though rapid degradation in poorly ventilated lung regions may cause transient local acidity [90]. A comparative overview of key polymers, targeting approaches, and delivery routes for polymeric NPs in lung cancer therapy has been presented in Table 3.

Table 3. Comparative overview of key polymers, targeting approaches, and delivery routes for polymeric NPs in lung cancer therapy.

Category	Examples	Advantages	Limitations	References
Polymers	PLGA	Biodegradable and biocompatible; FDA-approved; tunable degradation; suitable for controlled and sustained release; compatible with multiple drugs	Acidic degradation products may cause local irritation; relatively slow drug release for hydrophilic drugs	[90]
	PEG	Improves hydrophilicity; prolongs circulation time by reducing opsonization; reduces immunogenicity; enhances stability	Can cause "accelerated blood clearance" on repeated dosing; non-biodegradable (requires clearance); potential hypersensitivity reactions	[91]
	Chitosan	Natural, biocompatible and biodegradable; mucoadhesive; enhances permeation; pH-sensitive release; modifiable for active targeting	Poor solubility at neutral pH; batch-to-batch variability; limited mechanical strength	[92]
	Alginate	Biocompatible; gel-forming; modifiable; good for protein/peptide encapsulation; supports multiple routes	Low mechanical stability without crosslinking; limited cell uptake without modifications	[93]
	Gelatin	Biodegradable; non-toxic; versatile for surface modification; good for protein drugs	Enzymatic degradation may be too rapid; requires crosslinking for stability	[94]
	PCL	Slow degradation—good for long-term release; high drug loading for hydrophobic drugs	Not ideal for rapid drug release; long-term persistence may cause accumulation	[95]
Targeting approach	Passive targeting (EPR effect)	Simple design; no need for complex ligand chemistry; good for solid tumors with leaky vasculature	Tumor heterogeneity limits EPR efficiency; significant off-target uptake (liver/spleen)	[96]
	Active targeting (ligands, antibodies, peptides)	Higher tumor specificity; receptor-mediated uptake; can bypass some resistance mechanisms	Requires detailed knowledge of tumor receptor profile; ligand conjugation adds complexity and cost; receptor heterogeneity may reduce efficacy	[97]
Delivery Routes	Intravenous (IV)	Enables systemic delivery; suitable for metastatic disease; can exploit both passive and active targeting	Risk of systemic toxicity; opsonization and clearance by MPS	[98]
	Pulmonary (inhalation)	Localized delivery to lung tumors; bypasses first-pass metabolism; reduces systemic toxicity	Formulation stability in aerosol form; variability in lung deposition; mucociliary clearance	[99]
	Oral	Patient compliance; potential for chronic therapy	Poor bioavailability for many drugs; degradation in GI tract; first-pass metabolism	[91]
	Intratumoral	High local concentration; minimal systemic exposure	Invasive; limited use for inaccessible tumors	[97]

Polymeric NPs can be engineered for multimodal theranostic applications in lung cancer by integrating various imaging agents. Magnetic resonance imaging labels, such as superparamagnetic iron oxide or gadolinium chelates, provide high soft-tissue contrast without ionizing radiation, though sensitivity is lower than nuclear techniques. Computed tomography (CT) contrast can be achieved by incorporating high-Z elements (gold, bismuth, iodine), enabling excellent spatial resolution for lung anatomy but at the cost of radiation exposure [100]. Positron emission tomography and single photon emission computed tomography use radio-labeled NPs for highly sensitive, quantitative molecular imaging, with PET offering higher resolution but shorter isotope half-lives. Optical imaging, particularly near-infrared modalities, enables real-time, high-resolution visualization for surgical guidance, although tissue penetration is limited in the thoracic region. Ultrasound and photoacoustic imaging employ NP-encapsulated acoustic or optical absorbers for real-time, radiation-free imaging, but pulmonary air spaces limit deep-tissue access. Hybrid PNP systems combining multiple imaging modalities can leverage complementary strengths for improved diagnosis, image-guided therapy, and treatment monitoring in lung cancer theranostics [101]. A summary on different imaging modalities integrated with polymeric nanoparticles for lung cancer theranostics is presented in Table 4.

Table 4. Imaging modalities integrated with polymeric nanoparticles for lung cancer theranostics.

Imaging Modality	Key Advantages and Limitations in Lung Cancer Theranostics	References
MRI	High soft-tissue contrast, non-ionizing; sensitive tracking of NP biodistribution; lower molecular sensitivity; gadolinium safety concerns.	[102]
CT	Excellent spatial resolution, widely available; uses ionizing radiation, low molecular sensitivity, possible metal-agent toxicity.	[103]
PET	Highest molecular sensitivity, quantitative biodistribution, detects metastases; short isotope half-lives, radiation exposure, complex radiochemistry.	[104]
SPECT	Good sensitivity, longer-lived isotopes, whole-body tracking; lower spatial resolution than PET, regulatory hurdles.	[94]
Optical (NIR, NIR-II)	High sensitivity, real-time surgical guidance; limited tissue penetration, autofluorescence, mainly preclinical use.	[105]
Ultrasound/Photoacoustic	Real-time, no ionizing radiation, depth improvement with PA; lung air limits access, microbubble stability issues.	[106]

4.1. Chitosan-Based Nanoparticles

Traditional chemotherapy methods face numerous challenges, such as rapid drug clearance, poor drug targeting, multidrug resistance, and limited therapeutic efficiency. Chitosan, a polymer naturally derived from chitin via deacetylation, comprises β-(1-4)-linked d-glucosamine and N-acetyl-*d*-glucosamine units [92]. Its favorable properties are biocompatibility, biodegradability, and mucoadhesiveness, which make it a suitable candidate for developing nanoparticulate drug delivery systems. Due to the presence of free amino groups, it is modifiable and allows functionalization according to specific therapeutic needs, specifically for lung cancer treatment. Recent advancements in chitosan NPs (CNPs) demonstrate their promising application in targeted drug delivery [107].

Pawar and Jaganathan illustrated the high loading capacity of glycol-modified CNPs for hepatitis B vaccines, supporting their ability in mucosal vaccine delivery [108]. Encapsulation of DOX in CNPs notably improved drug targeting and drug permeability, as demonstrated by Zare et al., showing a 12.7-fold increase in intestinal absorption. Paclitaxel-loaded CNPs demonstrated stronger anticancer effects than free paclitaxel in MDA-MB-231 breast cancer cells, with sustained release and improved cytotoxicity. Studies also show pH-responsive release and enhanced uptake of 5-FU-loaded CNPs in colorectal cancer models [109].

A study by Shali et al. [110] examined the co-delivery of insulin-like growth factor 1 receptor (IGF-1R)-specific small interfering RNA (siRNA) and DOX using chitosan-based NPs to improve anticancer efficacy in the A549 lung cancer cell line. The CNPs had an average size of approximately 176 nm along with a zeta potential of about 11 mV and a polydispersity index of 0.3, indicating their suitability for drug delivery. The results illustrated that the combination of IGF-1R siRNA and DOX encapsulated in CNPs possessed a synergistic effect, notably maximizing cytotoxicity and inducing apoptosis in A549 cells. The co-delivery system significantly downregulated cell migration and reduced the expression of vascular endothelial growth factor (VEGF), matrix metalloproteinase 9 (MMP9), and signal transducer and activator of transcription 3 (STAT3). All of these are associated with tumor progression and metastasis. These findings suggested that the chitosan-based NP system effectively increases the therapeutic efficacy of DOX by simultaneously silencing IGF-1R, a receptor implicated in cancer cell survival and proliferation. This dual-target approach holds beneficial results for improving lung cancer treatment outcomes by overcoming limitations associated with conventional chemotherapy.

Mahmood et al. [111] also demonstrated the synthesis and anticancer potential of selenium NPs (Se NPs) combined with CNPs (Se-CNPs) for lung cancer therapy. These NPs were synthesized by using grape seed extract, applying a green synthesis approach that is environmentally friendly and sustainable. Characterization of the NPs revealed spherical shapes with average sizes of 55.285 nm for SeNPs and 30.9 nm for Se-CNPs. Both types possessed dose-dependent cytotoxicity against the A549 lung cancer cell line, having IC50 values of 24.09 μg/mL for CNPs and 18.56 μg/mL for SeNPs. Also, normal human kidney (HK-2) cells were not significantly affected, which indicated the selective toxicity toward cancer cells. Mechanistic studies showed that treatment with SeNPs and Se-CNPs led to amplified generation of ROS, resulting in loss of membrane potential and mitochondrial dysfunction. These effects resulted in apoptosis of A549 cells via ROS-mediated pathways. The findings suggest that Se NPs and Se-CNPs can modulate ROS signaling, leading to cancer cell death. These results emphasize the potential of biogenic Se NPs and Se-CNPs as promising nanotherapeutic agents for lung cancer treatment, leveraging an eco-friendly alternative to conventional therapies.

Mahmoud and colleagues [112] investigated the development and properties of berberine-loaded chitosan nanoparticles (BBR-CNPs) and evaluated their efficacy in protecting male albino mice from urethane-induced lung cancer. The synthesized BBR-COSNPs exhibited a spherical morphology with an average size of 45.56 nm and a zeta potential of 39.82 mV, indicating excellent stability and suitability for drug delivery. In vivo studies demonstrated that BBR-COSNPs effectively inhibited lung tumor progression and promoted apoptosis by regulating serum BAX levels and caspase 9 gene expression in lung tissue. Additionally, these NPs reduced tumor angiogenesis by lowering serum VEGFR2 and suppressing HIF-1 gene expression in the lungs. The study suggests that BBR-COSNPs could be a promising oral anticancer therapy for lung cancer, offering an environmentally friendly alternative to traditional treatments. These findings underscore the potential of BBR-COSNPs as an effective nanotherapeutic approach for lung cancer management.

Zhu et al. [113] examined the targeting efficacy of T7 peptide-modified (CBT) NPs was examined through a fluorescence uptake study involving six experimental groups (non-modified CB NPs and targeted CBT NPs at 30 and 60 min, and pretreated CB/CBT with free T7 at 60 min). Flow cytometry amplified fluorescence uptake, thereby revealing that CBT at 60 min possessed the highest intensity. To confirm the specificity of T7-mediated targeting, cells were pretreated with excess free T7 peptide, that notably reduced CBT uptake. Immunofluorescence imaging of lung cancer cells (A549 and H1299) validated the findings. The fluorescence distribution pattern supported increased cellular uptake due to the T7 targeting peptide.

Patel et al. [114] conducted a study aimed at developing a targeted pulmonary drug delivery system for lung cancer treatment using silibinin-loaded CNPs functionalized with folic acid NPs. The researchers aimed to increase the therapeutic efficacy and site-specific delivery of silibinin by utilizing a quality by design approach, a natural anticancer agent, by creating inhalable NPs that actively target folate receptors, which are highly expressed in many lung cancer cells. Folic acid was electrostatically conjugated to CNPs to achieve targeted delivery. The optimized formulation demonstrated promising physicochemical characteristics, including a spherical morphology, an average particle size of 275 ± 1.20 nm, a polydispersity index of 0.234 ± 0.07, and a positive surface charge (ζ-potential) of 32.50 ± 0.21 mV, which is favorable for cellular uptake. The drug-loading and entrapment efficiency were satisfactory, with $75.52 \pm 0.87\%$ of silibinin encapsulated within the NPs. In vitro release studies showed a cumulative drug release of $63.25 \pm 1.21\%$ over 48 h, indicating sustained drug delivery. The aerodynamic behavior of the formulation was assessed to ensure its suitability for pulmonary administration with a mass median aerodynamic diameter of 2.75 ± 1.02 µm and a geometric size distribution of 3.15 ± 0.88 µm in the optimal range for deep lung deposition. The cytotoxicity assay conducted on a lung cancer cell-bearing rat model indicated an IC_{50} value of 24.5 µg/mL after incubating for 24 h, suggesting notable anticancer potential. The biodistribution studies confirmed that silibinin-loaded CNPs functionalized with folic acid NPs led to a higher accumulation of silibinin in lung tissues compared to non-targeted formulations. These results collectively show that the developed NPs system not only offers a non-invasive delivery method but also confirms improved localization of the therapeutic agent at the tumor site due to active targeting, making it a promising candidate for lung cancer therapy.

4.2. Sodium Alginate-Based Nanoparticles

Polymeric biomaterials like alginate (ALG) are widely used in medical applications due to their compatibility with physiological conditions and lesser adverse effects. ALG is highly valued for its biocompatible and hemostatic properties, as well as its capacity for chemical modification. Its hydroxyl and carboxyl groups allow hydrogen bonding, making it suitable for gel formation, muco-adhesion, and improving transdermal drug delivery [115]. ALG NPs are effective drug carriers, allowing biodegradability, high drug-loading potential, and low toxicity. These NPs can seal a variety of therapeutic agents, including oligosaccharides, proteins, and anticancer compounds. They are adaptable for multiple administration routes—oral, nasal, intravenous, and ocular [115]. ALG's properties can be improved through blending with other polymers, chemical or physical crosslinking, or surface modifications using targeting agents such as antibodies, peptides, or aptamers. These modifications enhance specificity and minimize required drug doses [115].

Huang et al. [116] formulated an innovative nanocomposite comprising silver NPs (Ag NPs) supported on alginate-modified magnetic NPs (Fe_3O_4/Alg). The material was synthesized to target human lung carcinoma cells. The use of sodium alginate, which is a natural anionic polysaccharide, provided stability and biocompatibility during the reduction of Ag(I) ions to Ag NPs. The resulting nanocomposite combines the therapeutic potential of the magnetic targeting capability of Fe_3O_4 with silver. Given the high mortality rate of lung cancer and limitations in current treatments, this study shows a promising magnetically controllable, biocompatible, and cytotoxic agent, particularly designed to oppose lung cancer cells without harming normal cells. The synthesis of the Fe_3O_4/Alg-Ag nanocomposite began with the preparation of magnetite (Fe_3O_4) particles, served as the magnetic core. These particles were coated with sodium alginate, which functioned as both a stabilizer and reducing agent for silver ions. Then, after $AgNO_3$ was introduced, AgNPs were reduced in situ, resulting in their stable deposition on the alginate-modified Fe_3O_4 surface. The resulting composite was characterized by various techniques; Fourier Transform Infrared (FTIR) confirmed chemical bonding and functional groups Field Emission Scanning Electron Microscopy (FE-SEM) and Transmission electron microscopy [117] offered morphological insights, Energy Dispersive X-ray (EDX) analysis validated elemental composition, Vibrating Sample Magnetometer (VSM) evaluated magnetic properties, and inductively coupled plasma–optical emission spectrometer (ICP-OES) provided quantitative elemental analysis. To evaluate biological effects, cytotoxicity studies were conducted using the 3-(4,5-dimethylthiazolyl-2)-2,5-diphenyltetrazolium bromide (MTT) assay. Non-small cell lung carcinoma cell lines NCI-H1975, NCI-H1563, and NCI-H1299 were treated with varying concentrations of the nanocomposite. The MTT assay assessed mitochondrial activity by measuring absorbance at 570 nm. Human umbilical vein endothelial cells (HUVECs) were used as a control to detect toxicity toward normal cells. The nanocomposite's antioxidant activity was analyzed by using the 2,2-(2,2-diphenyl-1-picrylhydrazyl (DPPH) assay, in which half of the free radicals were neutralized at a concentration of 194 μg/mL, demonstrating moderate antioxidant capacity, potentially linked to its anticancer effects.

Recently, Işıklan et al. [118] formulated an innovative alginate-based bio-nanocomposite for targeted delivery of the chemotherapeutic drug etoposide and improved photothermal therapy in lung cancer treatment. The research highlights the central role of alginate in creating a stable, responsive delivery platform. Chemical modifications in alginate were performed by grafting it with poly (2-hydroxypropyl methacrylamide), forming a copolymer (SA-g-poly (2-hydroxypropyl methacrylamide) that improved the structural and functional features of the nanocomposite. The (SA-g-poly (2-hydroxypropyl methacrylamide) matrix was further combined with magnetite-functionalized graphene oxide to enhance responsiveness to external stimuli. However, it is the alginate backbone that provides the primary scaffold, ensuring the hydrogel maintains its integrity and allows for environmentally sensitive drug release. The nanocomposite was synthesized by using an emulsion method and underwent comprehensive physicochemical characterization using FT-IR, ultraviolet–visible (UV–Vis) spectroscopy, X-ray diffractometer (XRD), dynamic light scattering (DLS), TEM, FE-SEM, and atomic force microscope (AFM) analyses. These techniques confirmed the successful formation of a nanostructured network with favorable morphological and surface properties. Upon exposure to near-infrared laser light (808 nm, 1 W/cm^2, 10 min), the composite possessed a notable temperature rise of over 29 °C. This thermal response facilitated a controlled release of etoposide, further regulated by pH, magnetic fields, and magnetite-functionalized graphene oxide concentration. The alginate-based platform played a vital role in enabling stimulus-responsive drug release, making it especially effective for site-specific lung cancer therapy. Cell studies using H1299 lung cancer lines revealed that the nanocomposite, when combined with near-infrared irradiation, notably increased cell death, demonstrating the synergistic potential of combined chemotherapy and photothermal treatment.

4.3. Gelatin-Based Nanoparticles

Gelatin-based NPs (GNPs) offer significant benefits in the management of lung cancer due to their biodegradability, biocompatibility, and ease of surface modification. Derived from collagen, gelatin is a naturally occurring polymer that is well tolerated and non-toxic in biological systems, making it an ideal material for drug delivery applications. GNPs can encapsulate a vast range of therapeutic agents, including targeted therapies, chemotherapeutics, and immunomodulators, allowing for controlled and sustained release at tumor sites [119]. In lung cancer, targeted delivery is especially critical to reduce damage to healthy lung tissue and minimize systemic side effects. One major benefit of GNPs is their ability to increase drug solubility and stability, thus improving the bioavailability of poorly soluble anticancer drugs. Their flexible surface properties allow for functionalization with antibodies and ligands, enabling active targeting of cancer cells or tumor-associated microenvironments [120]. This targeted approach enhances therapeutic efficacy and minimizes off-target toxicity. GNPs can be engineered to respond to specific stimuli such as pH or enzymes present in tumor tissues, allowing site-specific drug release. Their small size promotes penetration into tumors and increases cellular uptake. Overall, GNPs display a promising and versatile platform for safer, more effective lung cancer therapy [121].

Ali et al. [122] performed a novel approach to improve cancer immunotherapy by using GNPs as carriers for nivolumab, which is a PD-1 checkpoint inhibitor, to promote the cytotoxic activity of effector Jurkat T-cells against A549 lung carcinoma cells. The researchers developed GNPs to increase the delivery of nivolumab for improved cancer immunotherapy. The optimized GNPs (191.9 nm, PDI 0.027, 54.67% entrapment) showed increased cytotoxicity in a Jurkat T-cell–A549 lung carcinoma co-culture. Nivolumab-loaded GNPs achieved an 87.88% inhibition rate versus 60.53% for the free drug and a lower IC_{50} (0.41 μM vs. 1.22 μM), indicating increased potency. This nano-formulation allows a promising, more effective, and safer alternative for targeted cancer immunotherapy.

Gu et al. [94] formulated paclitaxel-loaded PEGylated GNPs (PEG-PTX-GNPs) to improve treatment efficacy in NSCLC. First, PTX-GNPs were synthesized using glutaraldehyde crosslinking and then PEGylated with PEG 400. These NPs were evaluated for surface morphology, size, drug loading, in vitro release, zeta potential, cytotoxicity, cellular uptake, in vivo antitumor activity, histopathology, pulmonary deposition, and immunohistochemistry. The PEG-PTX-GNPs were spherical (90–115 nm) with a drug loading of 20.18–32.11% and stable encapsulation efficiency. In vitro, they showed the highest antiproliferative effect and cellular uptake. In vivo, they achieved 100% survival, maximum tumor inhibition, highest pulmonary deposition (6.5–12.55 μg/g), and superior histological outcomes, confirming their therapeutic potential in NSCLC.

Vaghasiya et al. [123] formulated an enzyme-responsive, advanced, and receptor-targeted drug delivery system using GNPs for effective lung cancer therapy. These NPs were designed to respond particularly to the TME by targeting overexpressed mannose receptors and matrix metalloproteinase-2 (MMP-2) enzymes commonly found in NSCLC tissues. The GNPs were surface-functionalized with concanavalin A, a plant lectin that binds to mannose-rich glycoproteins on cancer cells or tumor cells membranes, increasing receptor-mediated endocytosis. Cisplatin was encapsulated within the gelatin matrix to form CG-NP, and when coated with concanavalin-A, the final construct was termed CCG-NP. This NP system possessed a dual-targeting mechanism as concanavalin-A allowed cell-specific uptake via mannose receptor recognition, and the gelatin matrix was engineered for MMP-2-sensitive degradation, leveraging enzyme-triggered and localized release of cisplatin. In vitro studies using A549 lung cancer cells showed that CCG-NPs possessed superior cellular internalization after 12 h compared to non-coated particles. The presence of MMP-2 boosted drug release, enhancing therapeutic outcomes. Also, CCG-NPs notably induced ROS production, induced apoptosis, and caused cell cycle arrest in S and G2/M phases. Overall, this stimuli-responsive, inhalable, NP system provides a highly selective and efficient platform for lung cancer treatment.

Kononenko et al. [124] examined a novel lung cancer treatment strategy using GNPs as carriers for APS7, which is a synthetic analog of a marine toxin that functions as a nicotinic acetylcholine receptor (nAChR) antagonist. Their study aimed to improve chemotherapy efficacy while reducing off-target effects, mainly in nicotine-exposed lung cancer cells, which often show high resistance to therapy due to nAChR activation. Nicotine, by stimulating these receptors, supports cancer cell proliferation and inhibits apoptosis. By administering APS7 through GNPs, the researchers targeted these resistance mechanisms in A549 lung adenocarcinoma cells, displaying a potential dual-action therapeutic nanoplatform. APS7 was encapsulated within GNPs (APS7-GNPs) and their biological effects were tested on A549 lung cancer cells, including in nicotine-stimulated conditions. The study assessed calcium influx, proliferation, and chemosensitivity to cisplatin. Both APS7 and APS7-GNPs blocked nicotine-induced calcium influx and improved cisplatin efficacy. APS7-GNPs notably outperformed free APS7 by more effectively reducing proliferation and demonstrating higher selectivity, sparing normal lung epithelial BEAS-2B cells. The primary method to measure reproductive survival and colony formation was a clonogenic assay, assessing the influence of nicotine, APS7, and cisplatin on cancer cell viability over 13 days. The panel in Figure 3A evaluated the impact of nicotine on untreated and pretreated A549 cells utilizing a clonogenic assay. Cells exposed to 0.5 μM nicotine displayed notably higher surviving fractions (SFs) as compared to those grown in nicotine-free medium, indicating nicotine's pro-survival effect. This suggests that nicotine improves the reproductive viability of lung cancer cells, which may contribute to chemoresistance. Figure 3B shows the effects of pre-treatment with GNPs, APS7, and APS7-GNPs on reproductive survival were examined. Pre-treatment with empty GNPs had no significant effect on SF in either medium. Free APS7 reduced the SF of nicotine-stimulated cells but not in standard medium. Significantly, APS7-GNPs notably minimized SF in both media, suggesting increased efficacy and a more generalized antiproliferative effect. Figure 3C displays a panel that assessed cisplatin's impact, alone and in combination with APS7 or APS7-GNPs. Cisplatin alone reduced SF in normal conditions (SF = 0.56), but its effect was negligible under nicotine exposure (SF = 0.94). When combined with either APS7 or APS7-GNPs, this nicotine-induced resistance was attenuated. APS7-GNPs were more effective, restoring cisplatin sensitivity even in the presence of nicotine. Figure 3D presents imaging of colonies, further validating APS7-GNPs' superior performance.

In 2020, Chen et al. [125] performed a study to address the challenge of cisplatin resistance in lung cancer therapy by developing a dual-drug delivery system using gelatin-based NPs. Cisplatin is a standard chemotherapeutic agent in treating lung cancer, but its efficacy is hindered by the development of resistance. To counter the limitation, the researchers designed a nanocarrier system incorporating both cisplatin and epigallocatechin gallate, a naturally occurring polyphenol known for its anticancer properties. Gelatin was selected as the polymeric matrix due to its inherent biocompatibility, biodegradability, and capacity for drug encapsulation, making it an ideal vehicle for targeted drug delivery. The self-assembled gelatin/epigallocatechin gallate (GE) NPs were synthesized first, followed by the incorporation of cisplatin to create GE NPs. These particles were spherical with an average size of 75 nm and a positive zeta potential of +19.83 ± 0.25 mV, as determined by dynamic light scattering and confirmed through TEM. The NPs showed high drug-loading efficiency, approximately 63.7% for cisplatin and 89% for epigallocatechin gallate. In vitro experiments using A549 human lung adenocarcinoma cells showed that GE-Pt NPs at low concentrations, i.e., epigallocatechin gallate 5 μg/mL and cisplatin 2 μg/mL, possessed notably improved cytotoxicity as compared to cisplatin alone. Further analysis via inductively coupled plasma mass spectrometry displayed higher intracellular accumulation of cisplatin when delivered through the NPs, attributed to efficient endocytic uptake, and this improved delivery led to increased therapeutic efficacy. The combination of epigallocatechin gallate and cisplatin within a gelatin-based nanocarrier possessed a synergistic effect, suggesting a promising approach to overcome cisplatin resistance. The study emphasizes the advantages of gelatin as a nanocarrier, not only for its safe biological profile but also for its ability to improve drug stability and promote controlled release. Overall, the research supports the potential of GE-Pt NPs as an effective strategy for enhancing lung cancer treatment outcomes.

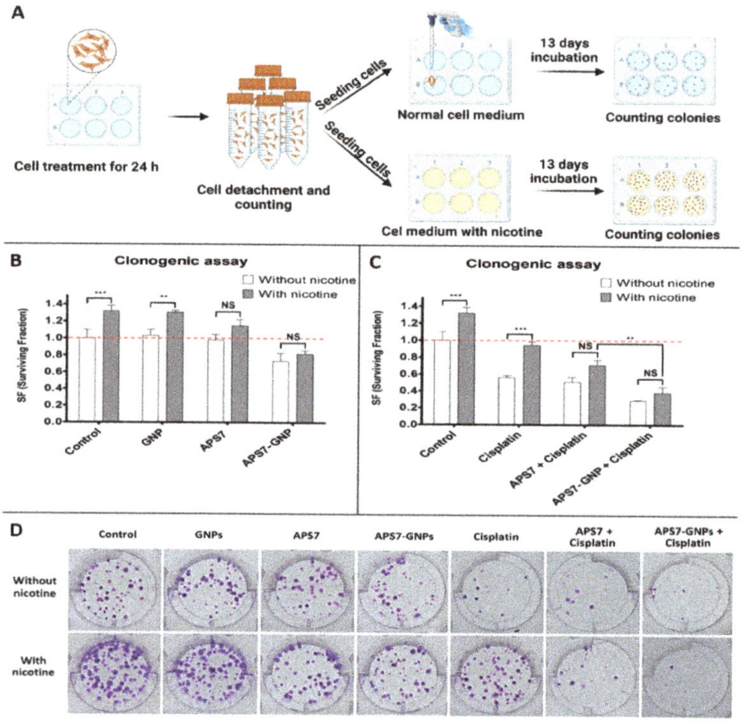

Figure 3. Effects of APS7-GNPs and cisplatin on colony formation of A549 lung cancer cells in normal and nicotine-containing conditions. (**A**) Schematic illustration of the clonogenic assay procedure. Cells were treated with gold nanoparticles (GNPs), APS7, APS7-GNPs, or cisplatin for 24 h, with or without nicotine, followed by seeding and incubation for 13 days before colony counting. (**B**) Quantitative analysis showing that APS7-GNPs significantly reduced colony formation compared to GNPs or APS7 alone, especially under nicotine exposure. (**C**) Combination of APS7-GNPs with cisplatin further reduced surviving fraction (SF), even in nicotine-containing medium. (**D**) Representative images of colony formation under each treatment condition. ** equals $p < 0.01$; *** equals $p < 0.001$; NS equals $p > 0.05$—not significant; ANOVA with Bonferroni's post test). Data from [124], published by Elsevier, 2024.

4.4. Poly Lactic Acid-Based Nanoparticles

Poly lactic acid (PLA)-based NPs are biocompatible and biodegradable carriers derived from renewable sources, mainly used in drug delivery applications, including cancer therapy. In the management of lung cancer, PLA NPs offer a major platform for the targeted delivery of gene therapies, chemotherapeutic drugs, and immune-modulating agents. Their biodegradability ensures safe breakdown into lactic acid, a naturally occurring metabolite, minimizing long-term toxicity [126]. The primary advantage of PLA-based NPs lies in their ability to provide controlled and sustained drug release, which increases therapeutic efficiency while minimizing dosing frequency. Their small size facilitates deep tumor penetration and cellular uptake, while surface modification allows active targeting of lung cancer cells or tumor-specific receptors. PLA NPs can also protect sensitive drugs from premature degradation, improving their bioavailability [126]. Overall, PLA-based NPs systems increase treatment precision, reduce systemic side effects, and support the development of safer and more effective therapies for lung cancer patients.

Wang et al. [127] introduced a novel targeted drug delivery system using transferrin-conjugated polymeric NPs (CD-RES TNPs) to increase resveratrol (RES) delivery for the treatment of NSCLC. RES is a natural polyphenol with anticancer properties, but it has poor solubility and bioavailability. By complexing RES with cyclodextrin (CD) and encapsulating it in poly lactic glycolic acid (PLGA) NPs, functionalized with transferrin ligands, the researchers aimed to improve RES targeting to NSCLC cells via transferrin receptor (TfR)-mediated endocytosis, thereby enhancing therapeutic outcomes and reducing off-target toxicity. The researchers formulated cyclodextrin-RES-loaded NPs (CD-RES NPs) and CD-RES TNPs using PLGA and tested them in H1299 NSCLC cells. NPs demonstrated high encapsulation efficiency ~90%. Cytotoxicity assays indicated that CD-RES TNPs notably reduced IC_{50} values and colony growth compared to plain RES or non-targeted NPs. Apoptosis induction and anti-migration effects were improved in CD-RES TNPs. Hemolysis assays confirmed biosafety. Three-dimensional spheroid culture studies mimicking tumor architecture validated the superior efficacy of CD-RES TNPs in tumor penetration and growth inhibition over 15 days, under both single and multiple dosing regimens. Figure 4A shows single-dose spheroid images: After one dose of 7 μM treatment on day 0, spheroids were imaged over 15 days. Control spheroids showed continuous growth, while CD-RES NP and CD-RES TNP treatments significantly inhibited spheroid expansion. CD-RES TNPs notably reduced spheroid size more than plain RES, indicating enhanced tumor inhibition from receptor-mediated delivery. Figure 4B shows single-dose spheroid volume data: Quantitative data showed control spheroids grew 2.59 ± 0.30-fold ($\approx$258.7%), while RES-treated spheroids grew 1.48 ± 0.30-fold ($\approx$148.4%). CD-RES NP and CD-RES TNP-treated spheroids shrank notably, with volumes decreasing to $15.6 \pm 14.0\%$ and $67.1 \pm 3.1\%$ of the original size, respectively. CD-RES TNPs outperformed plain RES significantly ($p < 0.05$), though no statistical difference was found between CD-RES NP and CD-RES TNP groups. Figure 4C displays multiple-dose spheroid images: In a repeated-dose setup, spheroids treated with all formulations (RES, CD-RES NP, CD-RES TNP) possessed visibly reduced growth compared to controls. Visual observation confirmed a consistent pattern of volume suppression by NP treatments over 15 days. Figure 4D displays multiple-dose spheroid volume data: Control spheroids expanded 2.21 ± 0.27-fold (~221.5%), while RES, CD-RES NP, and CD-RES TNP groups reduced volumes to $58.3 \pm 3.9\%$, $52.3 \pm 4.0\%$, and $61.9 \pm 3.6\%$, respectively ($p < 0.0001$ vs. control). This confirms the superior therapeutic efficiency of NP-based formulations in sustained tumor growth inhibition.

Yao et al. [128] conducted a study to improve lung cancer treatment by developing a novel nano-delivery system using PLGA NPs to co-deliver resveratrol and phosphatase and tensin homolog deleted on chromosome ten (PTEN) siRNA. NSCLC contributes to a majority of lung cancer cases are mainly metastatic, and survival rates remain low because of limited therapeutic efficacy and cancer cell resistance. Resveratrol, a naturally occurring polyphenol, possesses significant anticancer properties, including prevention of carcinogenesis, inhibition of tumor proliferation, and induction of apoptosis. However, its clinical application is restricted because of poor bioavailability and short biological half-life. To overcome this, the researchers encapsulated resveratrol along with PTEN-targeting siRNA into PLGA NPs modified with polyethylene imine

(PEI). PLGA was selected due to its favorable characteristics—biocompatibility, biodegradability, excellent drug-loading capacity, and extended systemic circulation, making it an ideal vehicle for controlled and passive tumor-targeted drug delivery. The NPs were synthesized using a solvent-free evaporation technique without stabilizers and were then characterized using electron microscopy. Resveratrol and siRNA were fluorescently labeled with Oregon Green and Cy5, respectively, to help in tracking and cellular uptake analysis. The final NPs complexes, termed PLGA-PEI-RES-PTEN, measured approximately 80 nm in size. Notably, NPs' size increased upon PEI surface modification and siRNA loading. Functional studies, including Western blotting and CCK8 assays, indicated that co-delivery of resveratrol and PTEN siRNA enhanced cytotoxicity against A549 and cisplatin-resistant A549/T12 lung cancer cell lines. The knockdown of PTEN notably amplified resveratrol's toxicity toward cancer cells, suggesting a synergistic anticancer effect. Additionally, the nanocomplex was effectively internalized by cancer cells, facilitating PTEN gene silencing and thereby disrupting tumor cell survival and proliferation. The study demonstrated that this PLGA-based nanoplatform not only improved the therapeutic efficacy of resveratrol but also provided a dual-action approach by combining chemotherapeutic and gene silencing effects.

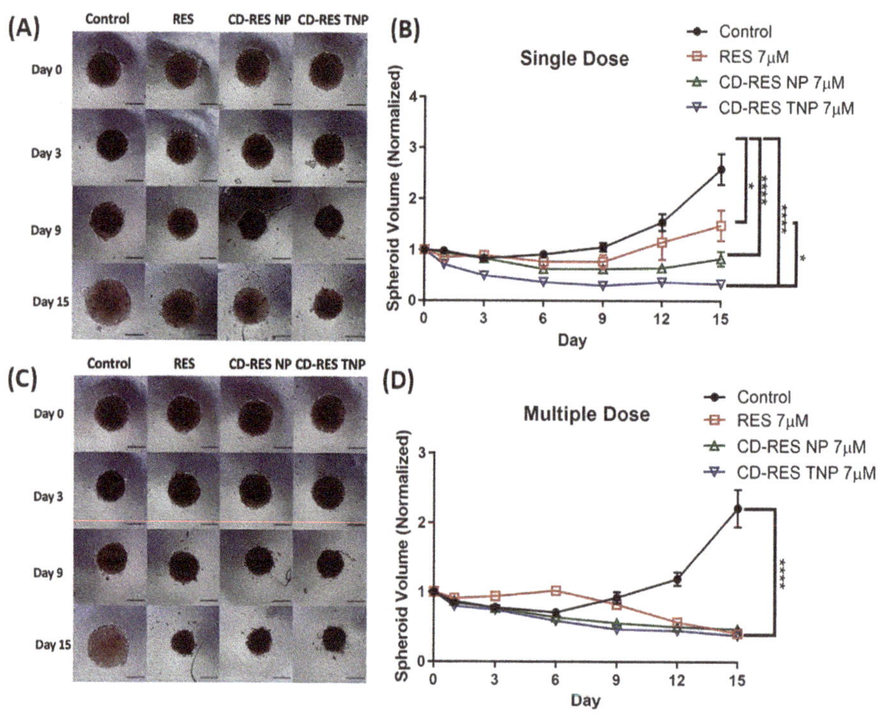

Figure 4. Three-dimensional spheroid assay evaluating the impact of treatments on H1299 tumor spheroid growth. (**A**) Representative images from a single dose. (**B**) Quantitative analysis of spheroid volume, normalized to day 0, for the single-dose group over 15 days. (**C**) Multiple-dose experiments captured at 10× magnification on days 0, 3, 9, and 15 (scale bar = 500 μm). (**D**) Volume quantification for multiple-dose spheroids during the same period. Significance between groups was analyzed by one-way ANOVA and Tukey's multiple comparisons test. * $p < 0.05$ and **** $p < 0.0001$. Reproduced with permission from Xuechun Wang et al., Journal of Drug Delivery Science and Technology; published by Elsevier, 2025 [127].

Chittasupho and colleagues demonstrated that conjugating the CXCR4-targeting peptide LFC131 onto PLGA NPs significantly enhanced the delivery of doxorubicin to CXCR4-expressing A549 lung cancer cells. These LFC131-DOX NPs exhibited faster and greater cellular uptake, CXCR4-specific internalization confirmed via competitive inhibition, and sustained drug release, highlighting their promise as a targeted and controlled delivery system [83]. In another study, the authors demonstrated that cLABL-conjugated PLGA NPs specifically target ICAM-1-expressing A549 lung epithelial cells, enhancing doxorubicin uptake and cytotoxicity via ICAM-1-mediated internalization. The system showed sustained drug release and confirmed potential as a targeted and controlled-release drug delivery platform [84].

4.5. Poly Caprolactone-Based Nanoparticles

Polycaprolactone (PCL)-based NPs are a class of biodegradable and biocompatible polymeric carriers known for their slow degradation rate and excellent drug encapsulation capabilities. In the context of lung cancer management, PCL NPs have emerged as a valuable platform for the sustained and targeted delivery of anticancer agents. Their hydrophobic nature allows effective loading of poorly water-soluble chemotherapeutic drugs, enhancing their solubility and therapeutic impact [129]. The key advantage of PCL NPs is their long-term drug release profile, which maintains therapeutic drug levels over extended periods, reducing the need for frequent dosing. This sustained delivery helps minimize systemic toxicity and enhances patient compliance. Additionally, PCL NPs can be surface-functionalized for active targeting of lung tumor cells, improving drug accumulation at the tumor site while sparing healthy tissue. Their stability and compatibility with a range of drugs make PCL NPs a promising tool for improving treatment outcomes in lung cancer therapy [130].

Cabeza et al. [131] developed DOX-loaded PCL NPs as a strategy to enhance antitumor efficacy while reducing systemic toxicity in lung and breast cancer treatments. Recognizing the severe side effects of free DOX, especially cardiotoxicity, the researchers designed PCL NPs with improved physicochemical and biocompatibility profiles. Their goal was to assess these NPs' ability to deliver DOX more effectively to tumor sites while sparing healthy tissues, using both in vitro and in vivo models, including human and mouse breast and lung cancer cell lines, as well as immunocompetent mice bearing tumor xenografts. In their experimental setup, the team synthesized DOX-loaded PCL NPs via a modified nanoprecipitation method and characterized them for cellular uptake, cytotoxicity, and in vivo therapeutic effect. The NPs displayed excellent biocompatibility, enhanced DOX uptake by cancer cells, and induced up to a 98% reduction in the IC_{50} in E0771 breast cancer cells. In vivo studies confirmed a significant reduction in tumor volume (~36%) in both breast and lung tumor models. Importantly, mice treated with DOX-PCL NPs maintained normal body weight and showed fewer toxic side effects, particularly reduced cardiotoxicity, compared to free DOX-treated groups. Figure 5A demonstrates tumor volume in C57BL/6 mice. No significant difference was seen between saline and blank PCL NP groups. However, mice treated with DOX-PCL NPs showed a ~36% greater tumor volume reduction than those treated with free DOX, in both lung and breast cancer models ($p < 0.05$), indicating superior antitumor efficacy of the nano-formulation. Figure 5B highlights survival data. In lung tumor-bearing mice, DOX-PCL NP treatment significantly prolonged survival compared to free DOX. However, no survival difference was observed between the DOX and DOX-PCL NP groups in breast cancer models, as shown by overlapping Kaplan–Meier curves. Figure 5C shows organ-specific toxicity. Histological examination of liver and lung tissues revealed no notable injuries across all treatment groups, confirming the safety of

PCL NPs at the administered dose. Figure 5D focuses on cardiotoxicity markers. Free DOX elevated CRP levels by 49.7% ($p < 0.05$), indicating inflammation, while DOX-PCL NPs caused no such increase. Other markers like MMP-9 and sVACM-1 remained stable across all groups. Figure 5E depicts body weight trends during a 14-day study. No significant weight changes were observed in any group, further supporting the minimal systemic toxicity of DOX-loaded PCL NPs.%

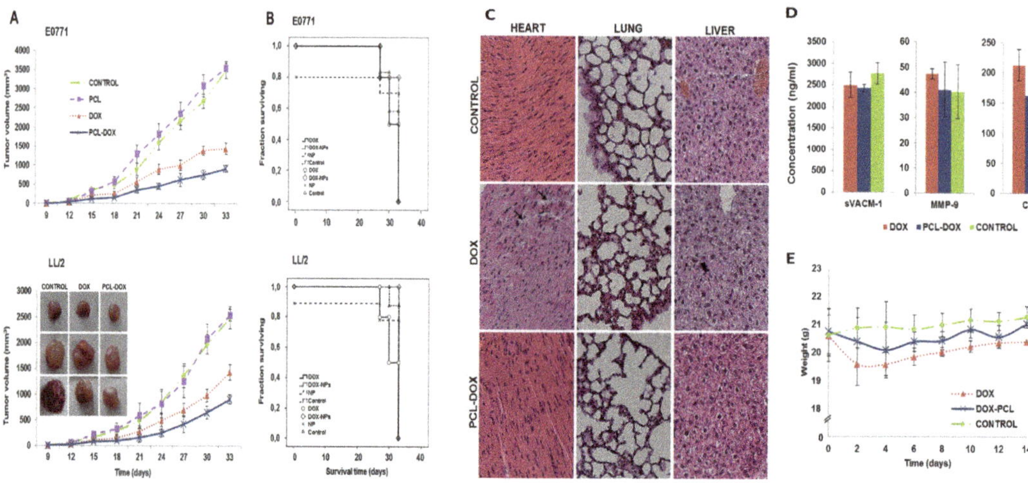

Figure 5. (**A**) Tumor volume progression in C57BL/6 mice with subcutaneous tumors derived from E0771 breast cancer and LL/2 lung cancer cells is shown after intravenous treatment with DOX-loaded PCL NPs, free DOX, and blank PCL NPs. Tumor growth inhibition is graphically presented as mean ± S.D. (n = 10), with representative images of LL/2 tumors. (**B**) Kaplan–Meier survival curves depict the survival outcomes of mice following the same treatments. Statistical differences between groups were analyzed using the log-rank test (n = 10). (**C**) Hematoxylin and eosin staining of heart, liver, and lung tissues reveals cardiotoxic effects in mice treated with free DOX. (**D**) Blood biomarkers of cardiovascular injury show elevated CRP levels in the free DOX group. (**E**) Body weight trends over 14 days post-treatment show weight loss in mice given free DOX. Reproduced with permission from Laura Cabeza et al., European Journal of Pharmaceutical Sciences; published by Elsevier, 2017 [131].

Akbari et al. [132] analyzed a novel dual-drug nanocarrier system to improve the therapeutic efficacy of lung cancer treatment by co-encapsulating cisplatin and methotrexate (MTX) within a PLA-PEG copolymer-based NPs platform. The study aimed to harness the complementary mechanisms of these two clinically established anticancer agents. Cisplatin induces DNA damage leading to apoptosis, while MTX disrupts folate metabolism essential for DNA synthesis and repair. The combination of these drugs within a single delivery system offers a strategic advantage in targeting multiple pathways involved in cancer progression. The selected PLA-PEG polymeric matrix was chosen due to its biodegradable, biocompatible, and environment-responsive properties, making it an ideal candidate for controlled and sustained drug release in cancer therapy. The NPs, characterized as spherical via scanning electron microscopy, effectively encapsulated both drugs. In vitro evaluation against A549 human lung carcinoma cells demonstrated that the methotrexate-*co*-encapsulating cisplatin-loaded NPs notably reduced cell viability, as confirmed by MTT assay, and induced significant levels of apoptosis, as observed through DAPI nuclear staining. Further, qRT-PCR analysis supported these findings by revealing enhanced expression of apoptosis-related genes. This dual-drug-loaded nanocarrier exhibited superior antitumor activity compared to single-drug formulations, highlighting its potential in overcoming chemoresistance and enhancing cytotoxic effects. The synergistic action of MTX and co-encapsulating cisplatin, delivered through a PLA-PEG platform, not only maximizes therapeutic impact but also reduces systemic toxicity by allowing for targeted and localized drug release. The study underlines the critical role of PLA-PEG copolymers in nanomedicine, as they provide a stable, customizable, and efficient vehicle for delivering combination chemotherapy agents. These findings suggest that PLA-PEG-based NPs co-loaded with methotrexate and cisplatin represent a promising and effective approach for the treatment of NSCLC, warranting further preclinical and clinical exploration.

4.6. Poly(amidoamine)-Based Nanoparticles

Poly(amidoamine) (PAA or PAMAM) could be stepwise synthesized to yield dendrimers that have a hyper-branched, tree-like structure with active functional groups and well-defined size. Dendrimers are biocompatible, water soluble, and redox sensitive. Regrettably, uncoated dendrimers were reported to have certain cellular toxicity; therefore, surface modification, such as PEGylation, is essential to improve their safety [133].

Guo et al. [134] synthesized PAA-iRGD/siEGFR polyplexes from PAA dendrimers that were functionalized with iRGD tumor penetration peptide and complexed with siEGFR, siRNA that silences EGFR for lung cancer therapy. PAA-iRGD showed an average size of 223.7 $\pm$ 3.1 nm and zeta potential of 35.2 $\pm$ 3.2 mV. PAA-iRGD/siEGFR showed higher gene silencing ability compared to PAA and PEI, significantly inhibited H1299 cells' proliferation and migration. PAA-iRGD/siEGFR inhibited lung tumor growth in BALB/c mice compared to PAA/siEGFR and PEI/siEGFR. These findings indicated that PAA-iRGD is a promising gene carrier in cancer treatment.

Bai et al. [135] prepared docetaxel (DTX)-alendronate (ALN)-PAMAM (DTX@ALN-PAMAM) to treat bone metastases of lung cancer. The drug loading, size, and zeta potential of DTX@ALN-PAMAM were 5.2 $\pm$ 0.4%, 99 $\pm$ 13 nm, and 4.35 $\pm$ 0.28 nm, respectively. DTX@ALN-PAMAM significantly enhanced DTX anticancer activity and inhibited osteoclast formation in in vitro bone metastases model. DTX@ALN-PAMAM showed the highest cancer inhibition rate compared to ALN and free DTX in vivo mouse model. The findings suggested that DTX@ALN-PAMAM exhibited synergistic effects that could be used on bone metastases of lung cancer.

4.7. Polymethacrylate-Based Nanoparticles

NPs from poly (methyl methacrylate) (PMMA) are easy to synthesize and functionalize, biocompatible. Core–shell PMMA NPs were developed to resolve the agglomeration issue of PMMA. Yin et al. [136] prepared polymethacrylic acid (PMAA) NPs loaded with Fe(III) and cypate dye coated with mesenchymal stem cell membrane (MSCs) or red blood cell membranes (RBCs) for photothermal-enhanced radiotherapy of NSCLC. The average size and zeta potential of Cyp-PMAA-Fe@MSCs were 248.4 and −22.3 mV. Cyp-PMAA-Fe@MSCs exhibited a strong photothermal hyperthermia effect when exposed to 808 nm laser irradiation. The fluorescence signal of Cyp-PMAA-Fe@MSCs in tumor-bearing mice was 21% stronger than that in the RBCs group, indicating more tumor accumulation. Tumors on mice treated with photothermal therapy and radiotherapy shrank 32% more than those treated with only radiotherapy. The findings suggested that Cyp-PMAA-Fe@MSCs could be applied for tumor diagnosis and precise treatment of NSCLC. Table 5 illustrates the recent investigation of polymeric NPs for the treatment of lung cancer.

Table 5. Studies for polymeric NPs for lung cancer management.

Type of Polymeric NPs	Drug Encapsulated/Cell Line Used	In Vitro/In Vivo Studies	Outcome	References
Chitosan	Carbamazepine	In vivo	Enhanced brain targeting and therapeutic efficacy at 219 nm and 80% drug entrapment	[92]
Glycol-modified chitosan	Hepatitis B vaccines	In vivo	Enhanced mucosal delivery	[108]
Chitosan	DOX	In vivo	Enhanced intestinal absorption	[137]
Chitosan	Paclitaxel—MDA-MB-231 breast cancer cells	Both	Enhanced anticancer activity compared to free paclitaxel in MDA-MB-231 breast cancer cells	[137]
Chitosan	DOX, siRNA, IGF-1R—A549 lung cancer cell line	In vitro	Improve anticancer efficacy in A549 lung cancer cell line	[110]
Chitosan	Berberine—urethane-induced lung cancer in male albino mice	In vivo	Prevented tumor angiogenesis by reducing levels of serum VEGFR2 and lung HIF-1 gene expression, oral administration and eco-friendly alternative	[112]
Gelatin	Nivolumab—A549 lung carcinoma cell	In vitro	87.88% inhibition rate, increased potency	[122]
PEGylated gelatin	Paclitaxel NSCLC	Both	Highest antiproliferative effect, 100% survival, maximum tumor inhibition	[94]
Gelatin	Concanavalin A—A549 lung cancer cells	In vitro	Increased apoptosis by boosted drug release	[123]
Gelatin	APS7 and cisplatin—A549 cells	In vitro	Improved cisplatin efficacy by blocking nicotine-induced calcium influx	[124]
Polycaprolactone	DOX—E0771 breast cancer cells and lung models	Both	Reduced DOX cardiotoxicity and improved anticancer activity	[131]
Poly(amidoamine)	Docetaxel (DTX)-alendronate (ALN)—A549 cells and mouse model	Both	Enhanced DTX anticancer activity Suppressed bone resorption, pain response, and growth of bone metastases	[135]

5. Clinical Trials Associated with Polymeric NPs-Based Drug Delivery for Lung Cancer

Clinical trials investigating polymeric NPs for lung cancer treatment are gaining momentum due to their potential to enhance targeted drug delivery and reduce systemic toxicity. Several ongoing and completed trials have evaluated polymeric NP-based formulations such as albumin-bound paclitaxel (Abraxane) and PEGylated drug carriers for improved therapeutic outcomes [138]. These trials focus on evaluating safety, pharmacokinetics, and tumor response in patients with NSCLC. Early results indicate improved drug accumulation at tumor sites and reduced adverse effects compared to conventional therapies. However, larger-scale, randomized studies are needed to establish their efficacy and enable widespread clinical adoption. Table 6 illustrates the summary of clinical trials on polymeric NPs for lung cancer management.

Table 6. Summary of clinical trials on polymeric NPs for lung cancer management.

Clinical Trial ID/Registration No.	Polymeric NP Type/Formulation	Drug/Therapeutic Agent	Study Phase	Study Design	Target Population	Primary Outcome Measures	Status/Sponsor	Reference
NCT00729612	Albumin based NP	Paclitaxel	Phase II	Interventional	NSCLC, Advanced Stage	PFS, OS	Completed/Greg Otterson, Ohio State University Comprehensive Cancer Center	[139]
NCT02240238	Cisplatin NP NC 6004	Gemcitabine	Phase I and II	Interventional	NSCLC, solid tumors	Safety, MTD	Completed/Nano Carrier Co., Ltd.	[140]
NCT00073723	Protein formulation	ABI-007, Protein formulation of paclitaxel	Phase I and II	Interventional	NSCLC	Safety, MTD	Completed/Celgene Corporation	[141]
NCT04314895	NanoPac (sterile nanoparticulate) intratumoral injection	Paclitaxel	Phase II	Interventional	NSCLC, SCLC, Neoplasm of lung	Safety, MTD	Completed/NanOlogy, LLC.	[140]
NCT02740985	Capsule nanoparticle suspension	Durvalumab AZD4635	Phase Ib	Interventional	NSCLC Advanced solid malignancies	PFS, ORR, Safety	Completed/AstraZeneca	[142]

6. Challenges, Limitations, and Future Perspective

Despite the promising advancements in polymeric NPs for targeted lung cancer therapy, several critical challenges and limitations hinder their full clinical translation. These challenges are associated with physicochemical properties, biological barriers, safety concerns, manufacturing scalability, and regulatory approval. The major obstacle for polymeric NPs mediated lung cancer therapy is the complex biological environment that impedes efficient drug delivery [143]. After systemic administration, NPs must navigate through the bloodstream, avoid clearance by the mononuclear phagocyte system, and overcome endothelial barriers to reach the tumor site. Even when utilizing passive targeting via the EPR effect, many NPs are sequestered by the liver and spleen, reducing lung-specific accumulation. Active targeting using ligands such as antibodies or peptides improves specificity, but receptor heterogeneity in tumors can limit their efficacy [144].

The EPR effect, while foundational to passive tumor targeting, is highly heterogeneous and influenced by multiple factors. Tumor type significantly affects vascular permeability, with highly angiogenic and leaky vasculature in some solid tumors (e.g., hepatocellular carcinoma) favoring stronger EPR, whereas desmoplastic or poorly vascularized tumors (e.g., pancreatic cancer) exhibit limited NP accumulation. Tumor size also plays a role—smaller, rapidly growing tumors often have immature vasculature that enhances permeability, while larger tumors may develop necrotic or hypoxic cores with reduced perfusion, limiting drug delivery [145]. Additionally, tumor location impacts EPR efficacy; for example, tumors in highly vascularized organs (like the liver) may exhibit different NP uptake patterns than those in low-perfusion sites or protected areas (such as the brain, due to the blood–brain barrier). These variabilities present challenges in clinical translation, as preclinical models often overestimate EPR-driven delivery efficiency, compared to human tumors, which are more heterogeneous. Understanding and accounting for these factors is essential for designing nanomedicines with consistent and predictable therapeutic outcomes [146] (Table 7).

Table 7. Factors influencing EPR effect variability and their impact on clinical translation.

Factor	Influence on EPR Effect	Impact on Clinical Translation	References
Tumor type	Highly angiogenic tumors (e.g., liver, some breast cancers) exhibit greater vascular permeability; desmoplastic tumors (e.g., pancreatic, some lung tumors) have restricted NP extravasation.	Nanomedicine accumulation varies widely; EPR-based delivery more effective in leaky tumors, less so in fibrotic or poorly vascularized tumors.	[147]
Tumor size	Small, rapidly growing tumors have immature, leaky vasculature; large tumors may have hypoxic or necrotic cores with poor perfusion.	Drug penetration is reduced in large tumors; dosing strategies may need to be tailored to tumor growth stage.	[148]
Tumor location	Location determines local blood flow and vessel permeability; tumors in highly vascularized organs differ from those in low-perfusion or protected areas (e.g., brain).	EPR effect is less pronounced in tumors with restricted access (e.g., behind the BBB), limiting passive targeting efficacy.	[149]
Tumor microenvironment	Dense extracellular matrix, high interstitial fluid pressure, and stromal barriers reduce NP penetration.	May necessitate combination strategies (e.g., stromal modulation, vascular normalization) to enhance delivery.	[147]
Patient-specific physiology	Variations in vascular density, permeability, and immune response affect NP clearance and distribution.	Personalized nanomedicine design and patient stratification may improve therapeutic predictability.	[148]

While pulmonary (inhalation) delivery is ideal for localized lung cancer treatment, achieving uniform drug deposition in deep lung tissues remains difficult. Factors such as mucociliary clearance, enzymatic degradation, and variable breathing patterns impact drug retention and efficacy. Moreover, the formulation must maintain aerodynamic properties compatible with inhalation devices without compromising NP stability or drug activity [144].

Maintaining the physical and chemical stability of polymeric NPs during formulation, storage, and administration is another challenge. Premature drug leakage, aggregation, and degradation can reduce therapeutic efficiency. Ensuring controlled, stimulus-responsive drug release triggered by pH, enzymes, or redox conditions in the TME requires precise engineering, which is often difficult to reproduce consistently across batches [150].

Although polymeric materials like PLGA, chitosan, and PEG are generally biocompatible, the degradation products of some polymers may cause inflammation or toxicity. Furthermore, limited data exist on the long-term fate and clearance of NPs, especially when repeatedly administered. Accumulation in non-target organs and the risk of immunogenicity remain unresolved concerns. Surface modifications intended to prolong circulation or enhance targeting can inadvertently increase toxicity or alter pharmacokinetics unpredictably [151].

Translating lab-scale synthesis to large-scale production poses major technical and economic challenges. Reproducibility in NPs size, drug-loading efficiency, surface charge, and batch consistency is essential for clinical success. Sophisticated techniques such as microfluidics or nanoprecipitation, though precise, are often expensive or not easily scalable, limiting their commercial viability [152]. Moreover, from a health economics standpoint, cost-effectiveness is influenced by quality-adjusted life years gained. Preliminary pharmacoeconomic modeling in oncology suggests that targeted nanomedicine, despite higher upfront costs, can achieve favorable cost per quality-adjusted life years ratios when accounting for extended survival and improved quality of life. Additionally, the potential for multifunctional theranostic NPs combining therapy and real-time imaging could streamline diagnostic workflows, further reducing cumulative costs [153]. Nonetheless, manufacturing scalability, supply chain logistics, and regulatory compliance remain cost drivers. Without economies of scale and optimized production, these therapies risk being unaffordable in low- and middle-income countries. Health systems may require value-based pricing models and risk-sharing agreements to ensure accessibility [154]. In conclusion, while novel NP-based lung cancer therapies currently carry higher sticker prices than conventional modalities, their potential to deliver durable clinical benefits with fewer downstream costs positions them as economically viable in the long term provided manufacturing and policy innovations support widespread adoption [155].

The regulatory pathway for NPs-based drug delivery systems is still evolving. These systems often blur the lines between drugs and devices, complicating approval processes. Agencies such as the FDA and EMA require comprehensive characterization of physicochemical properties, stability, pharmacokinetics, biodistribution, and long-term safety. Demonstrating consistent manufacturing under Good Manufacturing Practice conditions is essential [156]. Additionally, lack of standardized evaluation protocols for NPs complicates the approval process. Regulatory submissions must address potential immunogenicity, accumulation in non-target tissues, and variability in patient responses due to tumor heterogeneity. Bridging preclinical promise to clinical reality will require harmonized guidelines, robust scalability strategies, and cost-optimized manufacturing pipelines [97]. Additionally, patient variability, tumor heterogeneity, and limited predictive animal models further complicate clinical trials [99].

The patient variability and tumor heterogeneity in lung cancer stem from diverse genetic mutations (e.g., EGFR, KRAS, ALK) and phenotypic traits, as well as differences in tumor microenvironments. These variations can significantly impact therapeutic response, making one-size-fits-all NP designs less effective. To address this, PNPs with active targeting ligands can be engineered for patient-specific profiles [151]. Molecular profiling through next-generation sequencing can guide the selection of ligands—such as antibodies, peptides, or aptamers—that bind receptors overexpressed in an individual's tumor subtype. Multifunctional PNPs can also incorporate multiple ligands to simultaneously target heterogeneous cell populations within the same tumor, reducing the risk of resistant subclones. Additionally, stimuli-responsive PNPs can be tuned to exploit unique tumor microenvironment features (pH, enzymes, redox gradients) that differ between patients [157]. Modular PNP platforms further allow rapid swapping of targeting moieties and payloads, enabling personalized therapy adaptation during treatment. This precision design approach improves drug accumulation in genetically diverse tumors, enhances therapeutic efficacy, and mitigates off-target toxicity, ultimately bridging the gap between nanomedicine innovation and real-world patient variability [158].

The integration of polymeric NPs into lung cancer therapy holds immense promise for transforming current treatment paradigms. Future research must prioritize the formulation of multifunctional, stimuli-responsive NPs effective in co-delivering chemotherapeutics, gene therapies, and immunomodulators in a single platform [145]. Innovations in ligand engineering, particularly tumor-specific peptides and antibodies, will enhance active targeting, overcome tumor heterogeneity, and improve drug accumulation at tumor sites. Additionally, the design of inhalable NP formulations offers a non-invasive, localized approach to maximize therapeutic outcomes while minimizing systemic toxicity, a critical advancement for lung cancer care [159].

Emerging trends in bioinspired and green-synthesized NPs, including hybrid polymer-metallic systems, also warrant extensive exploration, given their potential for synergistic imaging and therapeutic applications [160]. Long-term safety evaluations, pharmacokinetics, and scalable manufacturing techniques must be refined to bridge the gap between laboratory innovation and clinical translation. Moreover, integrating artificial intelligence and machine learning for predictive modeling of NP behavior and personalized therapy optimization could redefine precision oncology in lung cancer. Ultimately, multidisciplinary collaboration among materials scientists, oncologists, and regulatory agencies will be essential to realize the full clinical potential of polymeric nanomedicine for effective, patient-tailored lung cancer management [161].

7. Conclusions

Lung cancer remains a leading therapeutic challenge due to its aggressive nature, late diagnosis, and limitations of conventional treatments. Polymeric NPs have turned up as promising tools for targeted drug delivery, offering ameliorated efficacy, tumor specificity, and reduced systemic toxicity. This review explores polymeric NPs, including chitosan, gelatin, alginate, poly lactic acid, and polycaprolactone, which enhance the delivery of chemotherapeutics, gene therapies, and immunomodulators. These nanocarriers protect drugs from degradation and improve bioavailability. Advanced designs, including ligand-conjugated, stimuli-responsive, and multifunctional NPs, have further improved therapeutic outcomes. Incorporating green-synthesized metallic NPs and quantum dots into polymeric platforms has introduced novel theranostics opportunities, combining imaging and treatment. However, barriers including NPs stability, biological interactions, long-term toxicity, and production scalability remain key challenges. Future directions include optimizing material properties, enhancing biocompatibility, and exploring inhalable formulations. Artificial intelligence-based therapy design may also accelerate personalized treatment development. In conclusion, polymeric NPs represent a transformative approach in lung cancer therapy. With continued interdisciplinary research and clinical validation, these systems could significantly improve treatment outcomes, survival rates, and quality of life for lung cancer patients globally.

Author Contributions: Conceptualization, S.S. (Sudarshan Singh) and C.C.; methodology, S.K., D.U.K., S.M.G., S.S. (Sambhavi Swarn) and B.L.; software, B.G.P.; validation, S.K., S.S. (Sudarshan Singh) and C.C.; formal analysis, B.G.P.; investigation, S.M.G., S.S. (Sambhavi Swarn) and B.L.; resources, B.G.P.; data curation, D.U.K.; writing—original draft preparation, S.K., S.M., C.C., S.M.G., S.S. (Sambhavi Swarn) and B.L.; writing—review and editing, S.S. (Sudarshan Singh), S.K., S.M. and C.C.; visualization, C.C.; supervision, B.G.P. and S.S. (Sudarshan Singh); project administration, C.C. All authors have read and agreed to the published version of the manuscript.

Funding: This research received no external funding.

Institutional Review Board Statement: Not applicable.

Informed Consent Statement: Not applicable.

Data Availability Statement: No new data were created.

Acknowledgments: This work was partially supported by CMU Proactive Researcher Scheme (2023), Chiang Mai University, for Sudarshan Singh.

Conflicts of Interest: The author Basant Lal was employed by PQE Australia Pvt. Ltd. The remaining authors declare that the research was conducted in the absence of any commercial or financial relationships that could be construed as a potential conflict of interest.

References

1. Allemani, C.; Matsuda, T.; Di Carlo, V.; Harewood, R.; Matz, M.; Nikšić, M.; Bonaventure, A.; Valkov, M.; Johnson, C.J.; Estève, J.; et al. Global Surveillance of Trends in Cancer Survival 2000–14 (Concord-3): Analysis of Individual Records for 37,513,025 Patients Diagnosed with One of 18 Cancers from 322 Population-Based Registries in 71 Countries. *Lancet* **2018**, *391*, 1023–1075. [CrossRef] [PubMed]
2. Sung, H.; Ferlay, J.; Siegel, R.L.; Laversanne, M.; Soerjomataram, I.; Jemal, A.; Bray, F. Global Cancer Statistics 2020: Globocan Estimates of Incidence and Mortality Worldwide for 36 Cancers in 185 Countries. *CA Cancer J. Clin.* **2021**, *71*, 209–249. [CrossRef]
3. Malhotra, J.; Malvezzi, M.; Negri, E.; La Vecchia, C.; Boffetta, P. Risk Factors for Lung Cancer Worldwide. *Eur. Respir. J.* **2016**, *48*, 889–902. [CrossRef]
4. Brinkløv, S.; Kalko, E.K.V.; Surlykke, A. Intense Echolocation Calls from Two 'Whispering' Bats, *Artibeus jamaicensis* and *Macrophyllum macrophyllum* (*Phyllostomidae*). *J. Exp. Biol.* **2009**, *212*, 11–20. [CrossRef]
5. Bade, B.C.; Cruz, C.S.D. Lung Cancer 2020: Epidemiology, Etiology, and Prevention. *Clin. Chest Med.* **2020**, *41*, 1–24. [CrossRef]
6. Li, Y.; Wang, N.; Huang, Y.; He, S.; Bao, M.; Wen, C.; Wu, L. Circmybl1 Suppressed Acquired Resistance to Osimertinib in Non-Small-Cell Lung Cancer. *Cancer Genet.* **2024**, *284–285*, 34–42. [CrossRef]
7. Wang, S.; Liu, G.; Yu, L.; Zhang, C.; Marcucci, F.; Jiang, Y. Fluorofenidone Enhances Cisplatin Efficacy in Non-Small Cell Lung Cancer: A Novel Approach to Inhibiting Cancer Progression. *Transl. Lung Cancer Res.* **2024**, *13*, 3175–3188. [CrossRef]
8. Gao, F.; Yan, G.; Sun, L.; Xia, H.; Guo, Z.; Lin, H.; Du, G. The Real Mechanisms of Emodin Treating Lung Cancer Based on System Pharmacology. *Eurasian J. Med. Oncol.* **2023**, *7*, 10–23. [CrossRef]
9. Wang, J.; Dong, X.; Liu, Y.; Lin, K.; Chen, J. Copy Number Gain of Met Gene with Low Level in a Metastatic Lung Adenocarcinoma Patient Represents Response to Salvage Treatment with Savolitinib and Osimertinib: A Case Report. *Front. Oncol.* **2025**, *15*, 1507677. [CrossRef] [PubMed]
10. Ferlay, J.; Colombet, M.; Soerjomataram, I.; Parkin, D.M.; Piñeros, M.; Znaor, A.; Bray, F. Cancer Statistics for the Year 2020: An Overview. *Int. J. Cancer* **2021**, *149*, 778–789. [CrossRef] [PubMed]
11. Zhang, Y.; Liu, C.; Chen, X.; Zhang, Y.; Li, Y.; Hu, X. Effects of Web-Based Acceptance and Commitment Therapy on Health-Related Outcomes among Patients with Lung Cancer: A Feasibility Randomized Controlled Trial. *Psychooncology* **2024**, *33*, e70045. [CrossRef]
12. El-Megharbel, S.M.; Albogami, B.; Hassoubah, S.A.; Beyari, E.A.; Albaqami, N.M.; Alsolami, K.; Hamza, R.Z. Spectral Analysis of Novel Minocycline/Zn Complex with Promising Anticancer Activities against Large Lung Cancer Cells (H460), Antibacterial and Antioxidant Activities against Acrylamide-Induced Pulmonary Toxicity in Male Rats. *Int. J. Pharmacol.* **2024**, *20*, 1247–1270. [CrossRef]

13. Samet, J.M.; Avila-Tang, E.; Boffetta, P.; Hannan, L.M.; Olivo-Marston, S.; Thun, M.J.; Rudin, C.M. Lung Cancer in Never Smokers: Clinical Epidemiology and Environmental Risk Factors. *Clin. Cancer Res.* **2009**, *15*, 5626–5645. [CrossRef]
14. Herbst, R.S.; Morgensztern, D.; Boshoff, C. The Biology and Management of Non-Small Cell Lung Cancer. *Nature* **2018**, *553*, 446–454. [CrossRef]
15. McGarry, H.A. Mechanistic Links between Copd and Lung Cancer. *Nat. Rev. Cancer* **2013**, *13*, 233–245. [CrossRef] [PubMed]
16. Skoulidis, F.; Heymach, J.V. Co-Occurring Genomic Alterations In non-Small-Cell Lung Cancer Biology and Therapy. *Nat. Rev. Cancer* **2019**, *19*, 495–509. [CrossRef] [PubMed]
17. George, J.; Lim, J.S.; Jang, S.J.; Cun, Y.; Ozretić, L.; Kong, G.; Leenders, F.; Lu, X.; Fernández-Cuesta, L.; Bosco, G.; et al. Comprehensive Genomic Profiles of Small Cell Lung Cancer. *Nature* **2015**, *524*, 47–53. [CrossRef]
18. Baylin, S.B.; Jones, P.A. Epigenetic Determinants of Cancer. *Cold Spring Harb. Perspect. Biol.* **2016**, *8*, a019505. [CrossRef] [PubMed]
19. New York State Department of Environmental Conservation. *Guidelines for Conducting Bird and Bat Studies at Commercial Wind Energy Projects;* Division of Fish Wildlife and Marine Resources, Ed.; New York State Department of Environmental Conservation: Albany, NY, USA, 2009.
20. Hinshaw, D.C.; Shevde, L.A. The Tumor Microenvironment Innately Modulates Cancer Progression. *Cancer Res.* **2019**, *79*, 4557–4566. [CrossRef]
21. Arandhara, A.; Bhuyan, P.; Das, B.K. Exploring Lung Cancer Microenvironment: Pathways and Nanoparticle-Based Therapies. *Discov. Oncol.* **2025**, *16*, 159. [CrossRef]
22. Zhang, X.; Zhang, X.; Yong, T.; Gan, L.; Yang, X. Boosting Antitumor Efficacy of Nanoparticles by Modulating Tumor Mechanical Microenvironment. *eBioMedicine* **2024**, *105*, 105200. [CrossRef]
23. Sabit, H.; Pawlik, T.M.; Radwan, F.; Abdel-Hakeem, M.; Abdel-Ghany, S.; Wadan, A.-H.S.; Elzawahri, M.; El-Hashash, A.; Arneth, B. Precision Nanomedicine: Navigating the Tumor Microenvironment for Enhanced Cancer Immunotherapy and Targeted Drug Delivery. *Mol. Cancer* **2025**, *24*, 160. [CrossRef]
24. Mitchell, M.J.; Billingsley, M.M.; Haley, R.M.; Wechsler, M.E.; Peppas, N.A.; Langer, R. Engineering Precision Nanoparticles for Drug Delivery. *Nat. Rev. Drug Discov.* **2021**, *20*, 101–124. [CrossRef]
25. Li, L.; Li, J.; Zhong, M.; Wu, Z.; Wan, S.; Li, X.; Zhang, Y.; Lv, K. Nanozyme-Enhanced Tyramine Signal Amplification Probe for Preamplification-Free Myocarditis-Related Mirnas Detection. *Chem. Eng. J.* **2025**, *503*, 158093. [CrossRef]
26. Feng, C.; Wang, Y.; Xu, J.; Zheng, Y.; Zhou, W.; Wang, Y.; Luo, C. Precisely Tailoring Molecular Structure of Doxorubicin Prodrugs to Enable Stable Nanoassembly, Rapid Activation, Potent Antitumor Effect. *Pharmaceutics* **2024**, *16*, 1582. [CrossRef]
27. Gao, Y.; Wang, Y.; Jiang, J.; Wei, P.; Sun, H. Triggered "on/Off" Luminescent Polypeptide Bowl-Shaped Nanoparticles for Selective Lighting of Tumor Cells. *Small* **2025**, *21*, e2411432. [CrossRef]
28. Rani, V.; Venkatesan, J.; Prabhu, A. Liposomes—A Promising Strategy for Drug Delivery in Anticancer Applications. *J. Drug Deliv. Sci. Technol.* **2022**, *76*, 103739. [CrossRef]
29. Kumari, P.; Ghosh, B.; Biswas, S. Nanocarriers for Cancer-Targeted Drug Delivery. *J. Drug Target.* **2016**, *24*, 179–191. [CrossRef]
30. Dong, Q.; Jiang, Z. Platinum–Iron Nanoparticles for Oxygen-Enhanced Sonodynamic Tumor Cell Suppression. *Inorganics* **2024**, *12*, 331. [CrossRef]
31. Mi, P. Stimuli-Responsive Nanocarriers for Drug Delivery, Tumor Imaging, Therapy and Theranostics. *Theranostics* **2020**, *10*, 4557–4588. [CrossRef] [PubMed]
32. Wang, Y.; Xu, Y.; Song, J.; Liu, X.; Liu, S.; Yang, N.; Wang, L.; Liu, Y.; Zhao, Y.; Zhou, W.; et al. Tumor Cell-Targeting and Tumor Microenvironment-Responsive Nanoplatforms for the Multimodal Imaging-Guided Photodynamic/Photothermal/Chemodynamic Treatment of Cervical Cancer. *Int. J. Nanomed.* **2024**, *19*, 5837–5858. [CrossRef]
33. Madaan, K.; Kumar, S.; Poonia, N.; Lather, V.; Pandita, D. Dendrimers in Drug Delivery and Targeting: Drug-Dendrimer Interactions and Toxicity Issues. *J. Pharm. Bioallied Sci.* **2014**, *6*, 139–150. [CrossRef]
34. Jafari, S.; Derakhshankhah, H.; Alaei, L.; Fattahi, A.; Varnamkhasti, B.S.; Saboury, A.A. Mesoporous Silica Nanoparticles for Therapeutic/Diagnostic Applications. *Biomed. Pharmacother.* **2019**, *109*, 1100–1111. [CrossRef] [PubMed]
35. Aqil, F.; Gupta, R.C. Exosomes in Cancer Therapy. *Cancers* **2022**, *14*, 500. [CrossRef] [PubMed]
36. Anselmo, A.C.; Prabhakarpandian, B.; Pant, K.; Mitragotri, S. Clinical and Commercial Translation of Advanced Polymeric Nanoparticle Systems: Opportunities and Material Challenges. *Transl. Mater. Res.* **2017**, *4*, 014001. [CrossRef]
37. Gaspar, R.; Duncan, R. Polymeric Carriers: Preclinical Safety and the Regulatory Implications for Design and Development of Polymer Therapeutics. *Adv. Drug Deliv. Rev.* **2009**, *61*, 1220–1231. [CrossRef]
38. Feng, J.; Zhang, P.; Wang, D.; Li, Y.; Tan, J. New Strategies for Lung Cancer Diagnosis and Treatment: Applications and Advances in Nanotechnology. *Biomark. Res.* **2024**, *12*, 136. [CrossRef]
39. García-Fernández, C.; Fornaguera, C.; Borrós, S. Nanomedicine in Non-Small Cell Lung Cancer: From Conventional Treatments to Immunotherapy. *Cancers* **2020**, *12*, 1609. [CrossRef] [PubMed]

40. Owen, D.H.; Wei, L.; Bertino, E.M.; Edd, T.; Villalona-Calero, M.A.; He, K.; Shields, P.G.; Carbone, D.P.; Otterson, G.A. Incidence, Risk Factors and Effect on Survival of Immune-Related Adverse Events in Patients with Non-Small-Cell Lung Cancer. *Clin. Lung Cancer* **2018**, *19*, e893–e900. [CrossRef]
41. Chen, D.; Liu, J.; Wu, J.; Suk, J.S. Enhancing Nanoparticle Penetration through Airway Mucus to Improve Drug Delivery Efficacy in the Lung. *Expert Opin. Drug Deliv.* **2021**, *18*, 595–606. [CrossRef]
42. Sarma, K.; Akther, M.H.; Ahmad, I.; Afzal, O.; Altamimi, A.S.A.; Alossaimi, M.A.; Jaremko, M.; Emwas, A.-H.; Gautam, P. Adjuvant Novel Nanocarrier-Based Targeted Therapy for Lung Cancer. *Molecules* **2024**, *29*, 1076. [CrossRef]
43. Antonia, S.J.; Villegas, A.; Daniel, D.; Vicente, D.; Murakami, S.; Hui, R.; Yokoi, T.; Chiappori, A.; Lee, K.H.; de Wit, M.; et al. Durvalumab after Chemoradiotherapy in Stage Iii Non-Small-Cell Lung Cancer. *N. Engl. J. Med.* **2017**, *377*, 1919–1929. [CrossRef]
44. Hellmann, M.D.; Ciuleanu, T.E.; Pluzanski, A.; Lee, J.S.; Otterson, G.A.; Audigier-Valette, C.; Minenza, E.; Linardou, H.; Burgers, S.; Salman, P.; et al. Nivolumab Plus Ipilimumab in Lung Cancer with a High Tumor Mutational Burden. *N. Engl. J. Med.* **2018**, *378*, 2093–2104. [CrossRef]
45. Soria, J.C.; Ohe, Y.; Vansteenkiste, J.; Reungwetwattana, T.; Chewaskulyong, B.; Lee, K.H.; Dechaphunkul, A.; Imamura, F.; Nogami, N.; Kurata, T.; et al. Osimertinib in Untreated Egfr-Mutated Advanced Non-Small-Cell Lung Cancer. *N. Engl. J. Med.* **2018**, *378*, 113–125. [CrossRef] [PubMed]
46. Rittmeyer, A.; Barlesi, F.; Waterkamp, D.; Park, K.; Ciardiello, F.; von Pawel, J.; Gadgeel, S.M.; Hida, T.; Kowalski, D.M.; Dols, M.C.; et al. Atezolizumab Versus Docetaxel in Patients with Previously Treated Non-Small-Cell Lung Cancer (Oak): A Phase 3, Open-Label, Multicentre Randomised Controlled Trial. *Lancet* **2017**, *389*, 255–265. [CrossRef] [PubMed]
47. Gourd, Alectinib Shows Cns Efficacy in Alk-Positive Nsclc. *Lancet Oncol.* **2018**, *19*, e520. [CrossRef]
48. Ando, K.; Manabe, R.; Kishino, Y.; Kusumoto, S.; Yamaoka, T.; Tanaka, A.; Ohmori, T.; Sagara, H. Comparative Efficacy of Alk Inhibitors for Treatment-Naïve Alk-Positive Advanced Non-Small Cell Lung Cancer with Central Nervous System Metastasis: A Network Meta-Analysis. *Int. J. Mol. Sci.* **2023**, *24*, 2242. [CrossRef]
49. Lindeman, N.I.; Cagle, P.T.; Aisner, D.L.; Arcila, M.E.; Beasley, M.B.; Bernicker, E.H.; Colasacco, C.; Dacic, S.; Hirsch, F.R.; Kerr, K.; et al. Updated Molecular Testing Guideline for the Selection of Lung Cancer Patients for Treatment with Targeted Tyrosine Kinase Inhibitors: Guideline from the College of American Pathologists, the International Association for the Study of Lung Cancer, and the Association for Molecular Pathology. *Arch. Pathol. Lab. Med.* **2018**, *142*, 321–346. [CrossRef] [PubMed]
50. Rolfo, C.; Mack, P.; Scagliotti, G.V.; Aggarwal, C.; Arcila, M.E.; Barlesi, F.; Bivona, T.; Diehn, M.; Dive, C.; Dziadziuszko, R.; et al. Liquid Biopsy for Advanced Nsclc: A Consensus Statement from the International Association for the Study of Lung Cancer. *J. Thorac. Oncol.* **2021**, *16*, 1647–1662. [CrossRef]
51. Schilsky, R.L. Personalized Medicine in Oncology: The Future Is Now. *Nat. Rev. Drug Discov.* **2010**, *9*, 363–366. [CrossRef]
52. Pascual, J.; Attard, G.; Bidard, F.C.; Curigliano, G.; De Mattos-Arruda, L.; Diehn, M.; Italiano, A.; Lindberg, J.; Merker, J.D.; Montagut, C.; et al. Esmo Recommendations on the Use of Circulating Tumour DNA Assays for Patients with Cancer: A Report from the Esmo Precision Medicine Working Group. *Ann. Oncol.* **2022**, *33*, 750–768. [CrossRef]
53. Tsuboi, M.; Herbst, R.S.; John, T.; Kato, T.; Majem, M.; Grohé, C.; Wang, J.; Goldman, J.W.; Lu, S.; Su, W.C.; et al. Overall Survival with Osimertinib in Resected Egfr-Mutated Nsclc. *N. Engl. J. Med.* **2023**, *389*, 137–147. [CrossRef] [PubMed]
54. Reck, M.; Remon, J.; Hellmann, M.D. First-Line Immunotherapy for Non-Small-Cell Lung Cancer. *J. Clin. Oncol.* **2022**, *40*, 586–597. [CrossRef]
55. Ramalingam, S.S.; Vansteenkiste, J.; Planchard, D.; Cho, B.C.; Gray, J.E.; Ohe, Y.; Zhou, C.; Reungwetwattana, T.; Cheng, Y.; Chewaskulyong, B.; et al. Overall Survival with Osimertinib in Untreated, Egfr-Mutated Advanced Nsclc. *N. Engl. J. Med.* **2020**, *382*, 41–50. [CrossRef]
56. Kievit, F.M.; Zhang, M. Cancer Nanotheranostics: Improving Imaging and Therapy by Targeted Delivery across Biological Barriers. *Adv. Mater.* **2011**, *23*, H217–H247. [CrossRef]
57. Wahajuddin; Arora, S. Superparamagnetic Iron Oxide Nanoparticles: Magnetic Nanoplatforms as Drug Carriers. *Int. J. Nanomed.* **2012**, *7*, 3445–3471. [CrossRef]
58. Manescu, V.; Antoniac, I.; Paltanea, G.; Nemoianu, I.V.; Mohan, A.G.; Antoniac, A.; Rau, J.V.; Laptoiu, S.A.; Mihai, P.; Gavrila, H.; et al. Magnetic Hyperthermia in Glioblastoma Multiforme Treatment. *Int. J. Mol. Sci.* **2024**, *25*, 10065. [CrossRef] [PubMed]
59. Mykhaylyk, O.; Sanchez-Antequera, Y.; Vlaskou, D.; Cerda, M.B.; Bokharaei, M.; Hammerschmid, E.; Anton, M.; Plank, C. Magnetic Nanoparticle and Magnetic Field Assisted Sirna Delivery in Vitro. *Methods Mol. Biol.* **2015**, *1218*, 53–106.
60. Wei, X.; Song, M.; Li, W.; Huang, J.; Yang, G.; Wang, Y. Multifunctional Nanoplatforms Co-Delivering Combinatorial Dual-Drug for Eliminating Cancer Multidrug Resistance. *Theranostics* **2021**, *11*, 6334–6354. [CrossRef]
61. Li, X.; Younis, M.H.; Wei, W.; Cai, W. Pd-L1 - Targeted Magnetic Fluorescent Hybrid Nanoparticles: Illuminating the Path of Image-Guided Cancer Immunotherapy. *Eur. J. Nucl. Med. Mol. Imaging* **2023**, *50*, 2240–2243. [CrossRef] [PubMed]

62. Arami, H.; Khandhar, A.; Liggitt, D.; Krishnan, K.M.; Delivery, I.V. Pharmacokinetics, Biodistribution and Toxicity of Iron Oxide Nanoparticles. *Chem. Soc. Rev.* **2015**, *44*, 8576–8607. [CrossRef]
63. Müller, R.H.; Mäder, K.; Gohla, S. Solid Lipid Nanoparticles (Sln) for Controlled Drug Delivery—A Review of the State of the Art. *Eur. J. Pharm. Biopharm.* **2000**, *50*, 161–177. [CrossRef]
64. Sivadasan, D.; Ramakrishnan, K.; Mahendran, J.; Ranganathan, H.; Karuppaiah, A.; Rahman, H. Solid Lipid Nanoparticles: Applications and Prospects in Cancer Treatment. *Int. J. Mol. Sci.* **2023**, *24*, 6199. [CrossRef]
65. Song, Y.; Cai, H.; Yin, T.; Huo, M.; Ma, P.; Zhou, J.; Lai, W. Paclitaxel-Loaded Redox-Sensitive Nanoparticles Based on Hyaluronic Acid-Vitamin E Succinate Conjugates for Improved Lung Cancer Treatment. *Int. J. Nanomed.* **2018**, *13*, 1585–1600. [CrossRef]
66. Chai, Z.; Ran, D.; Lu, L.; Zhan, C.; Ruan, H.; Hu, X.; Xie, C.; Jiang, K.; Li, J.; Zhou, J.; et al. Ligand-Modified Cell Membrane Enables the Targeted Delivery of Drug Nanocrystals to Glioma. *ACS Nano* **2019**, *13*, 5591–5601. [CrossRef]
67. Subhan, M.A.; Filipczak, N.; Torchilin, V.P. Advances with Lipid-Based Nanosystems for Sirna Delivery to Breast Cancers. *Pharmaceuticals* **2023**, *16*, 970. [CrossRef]
68. Al Khatib, A.O.; El-Tanani, M.; Al-Obaidi, H. Inhaled Medicines for Targeting Non-Small Cell Lung Cancer. *Pharmaceutics* **2023**, *15*, 2777. [CrossRef] [PubMed]
69. Pechyen, C.; Tangnorawich, B.; Toommee, S.; Marks, R.; Parcharoen, Y. Green Synthesis of Metal Nanoparticles, Characterization, and Biosensing Applications. *Sens. Int.* **2024**, *5*, 100287. [CrossRef]
70. Tinajero-Díaz, E.; Salado-Leza, D.; Gonzalez, C.; Velázquez, M.M.; López, Z.; Bravo-Madrigal, J.; Knauth, P.; Flores-Hernández, F.Y.; Herrera-Rodríguez, S.E.; Navarro, R.E.; et al. Green Metallic Nanoparticles for Cancer Therapy: Evaluation Models and Cancer Applications. *Pharmaceutics* **2021**, *13*, 1719. [CrossRef]
71. Abdel-Aziz, M.M.; Elella, M.H.A.; Mohamed, R.R. Green Synthesis of Quaternized Chitosan/Silver Nanocomposites for Targeting Mycobacterium Tuberculosis and Lung Carcinoma Cells (a-549). *Int. J. Biol. Macromol.* **2020**, *142*, 244–253. [CrossRef] [PubMed]
72. Yang, S.J.; Lin, F.H.; Tsai, K.C.; Wei, M.F.; Tsai, H.M.; Wong, J.M.; Shieh, M.J. Folic Acid-Conjugated Chitosan Nanoparticles Enhanced Protoporphyrin Ix Accumulation in Colorectal Cancer Cells. *Bioconjug. Chem.* **2010**, *21*, 679–689. [CrossRef]
73. Huang, M.; Myers, C.R.; Wang, Y.; You, M. Mitochondria as a Novel Target for Cancer Chemoprevention: Emergence of Mitochondrial-Targeting Agents. *Cancer Prev. Res.* **2021**, *14*, 285–306. [CrossRef]
74. Baghani, L.; Heris, N.N.; Khonsari, F.; Dinarvand, S.; Dinarvand, M.; Atyabi, F. Trimethyl-Chitosan Coated Gold Nanoparticles Enhance Delivery, Cellular Uptake and Gene Silencing Effect of Egfr-Sirna in Breast Cancer Cells. *Front. Mol. Biosci.* **2022**, *9*, 871541. [CrossRef] [PubMed]
75. Kulkarni-Dwivedi, N.; Patel, P.R.; Shravage, B.V.; Umrani, R.D.; Paknikar, K.M.; Jadhav, S.H. Hyperthermia and Doxorubicin Release by Fol-Lsmo Nanoparticles Induce Apoptosis and Autophagy in Breast Cancer Cells. *Nanomedicine* **2022**, *17*, 1929–1949. [CrossRef]
76. Quesada-González, D.; Merkoçi, A. Quantum Dots for Biosensing: Classification and Applications. *Biosens. Bioelectron.* **2025**, *273*, 117180. [CrossRef] [PubMed]
77. Pareek, A.; Kumar, D.; Pareek, A.; Gupta, M.M. Advancing Cancer Therapy with Quantum Dots and Other Nanostructures: A Review of Drug Delivery Innovations, Applications, and Challenges. *Cancers* **2025**, *17*, 878. [CrossRef]
78. Ren, L.; Wang, L.; Rehberg, M.; Stoeger, T.; Zhang, J.; Chen, S. Applications and Immunological Effects of Quantum Dots on Respiratory System. *Front. Immunol.* **2021**, *12*, 795232. [CrossRef]
79. Emami, F.; Duwa, R.; Banstola, A.; Woo, S.M.; Kwon, T.K.; Yook, S. Dual Receptor Specific Nanoparticles Targeting Egfr and Pd-L1 for Enhanced Delivery of Docetaxel in Cancer Therapy. *Biomed. Pharmacother.* **2023**, *165*, 115023. [CrossRef]
80. Yi, Y.; Zhao, H. Revolutionizing Tissue Clearing and 3-Dimensional Imaging: Transparent Embedding Solvent System for Uniform High-Resolution Imaging. *BME Front.* **2025**, *6*, 95. [CrossRef] [PubMed]
81. Merlin, J.P.J.; Crous, A. Abrahamse, Combining Photodynamic Therapy and Targeted Drug Delivery Systems: Enhancing Mitochondrial Toxicity for Improved Cancer Outcomes. *Int. J. Mol. Sci.* **2024**, *25*, 10796. [CrossRef]
82. Mohkam, M.; Sadraeian, M.; Lauto, A.; Gholami, A.; Nabavizadeh, S.H.; Esmaeilzadeh, H.; Alyasin, S. Exploring the Potential and Safety of Quantum Dots in Allergy Diagnostics. *Microsyst. Nanoeng.* **2023**, *9*, 145. [CrossRef]
83. Getz, T.; Qin, J.; Medintz, I.L.; Delehanty, J.B.; Susumu, K.; Dawson, P.E.; Dawson, G. Quantum Dot-Mediated Delivery of Sirna to Inhibit Sphingomyelinase Activities in Brain-Derived Cells. *J. Neurochem.* **2016**, *139*, 872–885. [CrossRef]
84. Woutersen, M.; Muller, A.; Pronk, M.E.J.; Cnubben, N.H.P.; Hakkert, B.C. Regulating Human Safety: How Dose Selection in Toxicity Studies Impacts Human Health Hazard Assessment and Subsequent Risk Management Options. *Regul. Toxicol. Pharmacol.* **2020**, *114*, 104660. [CrossRef]
85. Jin, M.; Hou, Y.; Quan, X.; Chen, L.; Gao, Z.; Huang, W. Smart Polymeric Nanoparticles with Ph-Responsive and Peg-Detachable Properties (Ii): Co-Delivery of Paclitaxel and Vegf Sirna for Synergistic Breast Cancer Therapy in Mice. *Int. J. Nanomed.* **2021**, *16*, 5479–5494. [CrossRef]

86. Tiwari, R.; Patil, A.; Verma, R.; Deva, V.; Rudrangi, S.R.S.; Bhise, M.R.; Vinukonda, A. Biofunctionalized Polymeric Nanoparticles for the Enhanced Delivery of Erlotinib in Cancer Therapy. *J. Biomater. Sci. Polym. Ed.* **2025**, *36*, 817–842. [CrossRef] [PubMed]
87. Beach, M.A.; Nayanathara, U.; Gao, Y.; Zhang, C.; Xiong, Y.; Wang, Y.; Such, G.K. Polymeric Nanoparticles for Drug Delivery. *Chem. Rev.* **2024**, *124*, 5505–5616. [CrossRef]
88. Thambiliyagodage, C.; Jayanetti, M.; Mendis, A.; Ekanayake, G.; Liyanaarachchi, H.; Vigneswaran, S. Recent Advances in Chitosan-Based Applications—A Review. *Materials* **2023**, *16*, 2073. [CrossRef] [PubMed]
89. Sharfi, L.; Nowroozi, M.R.; Smirnova, G.; Fedotova, A.; Babarykin, D.; Mirshafiey, A. The Safety Properties of Sodium Alginate and Its Derivatives. *Br. J. Healthc. Med. Res.* **2024**, *11*, 263–274.
90. Mahar, R.; Chakraborty, A.; Nainwal, N.; Bahuguna, R.; Sajwan, M.; Jakhmola, V. Application of Plga as a Biodegradable and Biocompatible Polymer for Pulmonary Delivery of Drugs. *AAPS PharmSciTech* **2023**, *24*, 39. [CrossRef]
91. Moharir, K.; Kale, V.; Ittadwar, A.; Paul, M.K. Polymeric Nanoparticle-Based Drug–Gene Delivery for Lung Cancer. In *Handbook of Lung Targeted Drug Delivery Systems*; CRC Press: Boca Raton, FL, USA, 2021; pp. 193–207.
92. Kuen, C.Y.; Masarudin, M.J. Chitosan Nanoparticle-Based System: A New Insight into the Promising Controlled Release System for Lung Cancer Treatment. *Molecules* **2022**, *27*, 473. [CrossRef]
93. Yasamineh, S.; Gholizadeh, O.; Kalajahi, H.G.; Yasamineh, P.; Firouzi-Amandi, A.; Dadashpour, M. Future Prospects of Natural Polymer-Based Drug Delivery Systems in Combating Lung Diseases. In *Natural Polymeric Materials Based Drug Delivery Systems in Lung Diseases*; Springer: Berlin/Heidelberg, Germany, 2023; pp. 465–482.
94. Gu, M.; Luan, J.; Song, K.; Qiu, C.; Zhang, X.; Zhang, M. Development of Paclitaxel Loaded Pegylated Gelatin Targeted Nanoparticles for Improved Treatment Efficacy in Non-Small Cell Lung Cancer (Nsclc): An in Vitro and in Vivo Evaluation Study. *Acta Biochim. Pol.* **2021**, *68*, 583–591. [CrossRef]
95. Dhuri, A.; Sriram, A.; Aalhate, M.; Mahajan, S.; Parida, K.K.; Singh, H.; Gupta, U.; Maji, I.; Guru, S.K.; Singh, P.K. Chitosan Functionalized Pcl Nanoparticles Bearing Tyrosine Kinase Inhibitor Osimertinib Mesylate for Effective Lung Cancer Therapy. *Pharm. Dev. Technol.* **2023**, *28*, 460–478. [CrossRef] [PubMed]
96. Vagena, I.-A.; Malapani, C.; Gatou, M.-A.; Lagopati, N.; Pavlatou, E.A. Enhancement of Epr Effect for Passive Tumor Targeting: Current Status and Future Perspectives. *Appl. Sci.* **2025**, *15*, 3189. [CrossRef]
97. Sharma, A.; Shambhwani, D.; Pandey, S.; Singh, J.; Lalhlenmawia, H.; Kumarasamy, M.; Singh, S.K.; Chellappan, D.K.; Gupta, G.; Prasher, P. Advances in Lung Cancer Treatment Using Nanomedicines. *ACS Omega* **2022**, *8*, 10–41. [CrossRef] [PubMed]
98. Hani, U.; Begum, M.Y.; Wahab, S.; Siddiqua, A. Riyaz Ali M Osmani, Mohamed Rahamathulla, A Comprehensive Review of Current Perspectives on Novel Drug Delivery Systems and Approaches for Lung Cancer Management. *J. Pharm. Innov.* **2022**, *17*, 1530–1553. [CrossRef]
99. Cojocaru, E.; Petriș, O.R.; Cojocaru, C. Nanoparticle-Based Drug Delivery Systems in Inhaled Therapy: Improving Respiratory Medicine. *Pharmaceuticals* **2024**, *17*, 1059. [CrossRef] [PubMed]
100. Dahlsgaard-Wallenius, S.E.; Hildebrandt, M.G.; Johansen, A.; Vilstrup, M.H.; Petersen, H.; Gerke, O.; Høilund-Carlsen, P.F.; Morsing, A.; Andersen, T.L. Hybrid Pet/Mri in Non-Small Cell Lung Cancer (Nsclc) and Lung Nodules—A Literature Review. *Eur. J. Nucl. Med. Mol. Imaging* **2021**, *48*, 584–591. [CrossRef]
101. Batouty, N.M.; Saleh, G.A.; Sharafeldeen, A.; Kandil, H.; Mahmoud, A.; Shalaby, A.; Yaghi, M.; Khelifi, A.; Ghazal, M.; El-Baz, A. State of the Art: Lung Cancer Staging Using Updated Imaging Modalities. *Bioengineering* **2022**, *9*, 493. [CrossRef]
102. Anani, T.; Rahmati, S.; Sultana, N.; David, A.E. Mri-Traceable Theranostic Nanoparticles for Targeted Cancer Treatment. *Theranostics* **2021**, *11*, 579–601. [CrossRef]
103. Najdian, A.; Beiki, D.; Abbasi, M.; Gholamrezanezhad, A.; Ahmadzadehfar, H.; Amani, A.M.; Ardestani, M.S.; Assadi, M. Exploring Innovative Strides in Radiolabeled Nanoparticle Progress for Multimodality Cancer Imaging and Theranostic Applications. *Cancer Imaging* **2024**, *24*, 127. [CrossRef]
104. Freitas, L.F.; Ferreira, A.H.; Thipe, V.C.; Varca, G.H.C.; Lima, C.S.A.; Batista, J.G.S.; Riello, F.N.; Nogueira, K.; Cruz, C.P.C.; Mendes, G.O.A.; et al. The State of the Art of Theranostic Nanomaterials for Lung, Breast, and Prostate Cancers. *Nanomaterials* **2021**, *11*, 2579. [CrossRef]
105. Zhao, M.; Leggett, E.; Bourke, S.; Poursanidou, S.; Carter-Searjeant, S.; Po, S.; Carmo, M.P.D.; Dailey, L.A.; Manning, P.; Ryan, S.G.; et al. Theranostic near-Infrared-Active Conjugated Polymer Nanoparticles. *ACS Nano* **2021**, *15*, 8790–8802. [CrossRef] [PubMed]
106. Wen, Y.; Guo, D.; Zhang, J.; Liu, X.; Liu, T.; Li, L.; Jiang, S.; Wu, D.; Jiang, H. Clinical Photoacoustic/Ultrasound Dual-Modal Imaging: Current Status and Future Trends. *Front. Physiol.* **2022**, *13*, 1036621. [CrossRef]
107. Shaikh, K.I.M.A.J.; Afzal, O.; Altamimi, A.S.A.; Almalki, W.H.; Alzarea, S.I.; Al-Abbasi, F.A.; Pandey, M.; Dureja, H.; Singh, S.K.; Dua, K.; et al. Chitosan-Based Nano Drug Delivery System for Lung Cancer. *J. Drug Deliv. Sci. Technol.* **2023**, *81*, 104196. [CrossRef]

108. Pawar, D.; Jaganathan, K.S. Mucoadhesive Glycol Chitosan Nanoparticles for Intranasal Delivery of Hepatitis B Vaccine: Enhancement of Mucosal and Systemic Immune Response. *Drug Deliv.* **2016**, *23*, 185–194. [CrossRef]
109. Zare, M.; Samani, S.M.; Sobhani, Z. Enhanced Intestinal Permeation of Doxorubicin Using Chitosan Nanoparticles. *Adv. Pharm. Bull.* **2018**, *8*, 411–417. [CrossRef]
110. Shali, H.; Shabani, M.; Pourgholi, F.; Hajivalili, M.; Aghebati-Maleki, L.; Jadidi-Niaragh, F.; Baradaran, B.; Akbari, A.A.M.-S.; Younesi, V.; Yousefi, M. Co-Delivery of Insulin-Like Growth Factor 1 Receptor Specific Sirna and Doxorubicin Using Chitosan-Based Nanoparticles Enhanced Anticancer Efficacy in A549 Lung Cancer Cell Line. *Artif. Cells Nanomed. Biotechnol.* **2018**, *46*, 293–302. [CrossRef]
111. Mahmood, R.I.; Al-Taie, A.; Al-Rahim, A.M.; Mohammed-Salih, H.S.; Ibrahim, H.A.; Albukhaty, S.; Jawad, S.F.; Jabir, M.S.; Salem, M.M.; Bekhit, M.M. Biogenic Synthesized Selenium Nanoparticles Combined Chitosan Nanoparticles Controlled Lung Cancer Growth Via Ros Generation and Mitochondrial Damage Pathway. *Nanotechnol. Rev.* **2025**, *14*, 20250142. [CrossRef]
112. Mahmoud, M.A.; El-Bana, M.A.; Morsy, S.M.; Badawy, E.A.; Farrag, A.-E.; Badawy, A.M.; Abdel-Wahhab, M.A.; El-Dosoky, M.A. Synthesis and Characterization of Berberine-Loaded Chitosan Nanoparticles for the Protection of Urethane-Induced Lung Cancer. *Int. J. Pharm.* **2022**, *618*, 121652. [CrossRef]
113. Zhu, X.; Yu, Z.; Feng, L.; Deng, L.; Fang, Z.; Liu, Z.; Li, Y.; Wu, X.; Qin, L.; Guo, R.; et al. Chitosan-Based Nanoparticle Co-Delivery of Docetaxel and Curcumin Ameliorates Anti-Tumor Chemoimmunotherapy in Lung Cancer. *Carbohydr. Polym.* **2021**, *268*, 118237. [CrossRef] [PubMed]
114. Patel, P.; Raval, M.; Airao, V.; Ali, N.; Shazly, G.A.; Khan, R.; Prajapati, B. Formulation of Folate Receptor-Targeted Silibinin-Loaded Inhalable Chitosan Nanoparticles by the Qbd Approach for Lung Cancer Targeted Delivery. *ACS Omega* **2024**, *9*, 10353–10370. [CrossRef]
115. Rostami, E. Recent Achievements in Sodium Alginate-Based Nanoparticles for Targeted Drug Delivery. *Polym. Bull.* **2022**, *79*, 6885–6904. [CrossRef]
116. Huang, J.; Guo, J.; Zhu, J.; Zou, X. Supported Silver Nanoparticles over Alginate-Modified Magnetic Nanoparticles: Synthesis, Characterization and Treat the Human Lung Carcinoma. *J. Saudi Chem. Soc.* **2022**, *26*, 101393. [CrossRef]
117. Veysi, A.; Yaghoobi-Ershadi, M.R.; Rassi, Y.; Hosseini-Vasoukolaei, N.; Jeddi-Tehrani, M.; Rezaee-Node, A.; Gholampour, F.; Saeidi, Z.; Fatemi, M.; Arandian, M.H.; et al. Rearing and Biology of Phlebotomus Sergenti, the Main Vector of Anthroponotic Cutaneous Leishmaniasis in Iran. *J. Arthropod-Borne Dis.* **2017**, *11*, 504–514. [PubMed]
118. Işıklan, N.; Geyik, G.; Güncüm, E. Alginate-Based Bio-Nanocomposite Reinforced with Poly(2-Hydroxypropyl Methacrylamide) and Magnetite Graphene Oxide for Delivery of Etoposide and Photothermal Therapy. *Mater. Today Chem.* **2024**, *41*, 102323. [CrossRef]
119. Parashar, A.K.; Saraogi, G.K.; Tiwari, B.K.; Tyagi, L.K.; Sethi, V.A.; Shrivastava, V. Polymeric Nanoparticles-Based Strategies for Cancer Immunotherapy. In *Nanotechnology Based Strategies for Cancer Immunotherapy: Concepts, Design, and Clinical Applications*; Sharma, R., Pandey, V., Mishra, N., Eds.; Springer Nature: Singapore, 2024; pp. 355–378.
120. Milano, F.; Masi, A.; Madaghiele, M.; Sannino, A.; Salvatore, L.; Gallo, N. Current Trends in Gelatin-Based Drug Delivery Systems. *Pharmaceutics* **2023**, *15*, 1499. [CrossRef]
121. Wu, D.; Cao, X.-H.; Jia, P.-Z.; Zeng, Y.-J.; Feng, Y.-X.; Tang, L.-M.; Zhou, W.-X.; Chen, K.-Q. Excellent Thermoelectric Performance in Weak-Coupling Molecular Junctions with Electrode Doping and Electrochemical Gating. *Sci. China Phys. Mech. Astron.* **2020**, *63*, 276811. [CrossRef]
122. Ali, D.S.; Gad, H.A.; Hathout, R.M. Enhancing Effector Jurkat Cell Activity and Increasing Cytotoxicity against A549 Cells Using Nivolumab as an Anti-Pd-1 Agent Loaded on Gelatin Nanoparticles. *Gels* **2024**, *10*, 352. [CrossRef]
123. Vaghasiya, K.; Ray, E.; Singh, R.; Jadhav, K.; Sharma, A.; Khan, R.; Katare, O.P.; Verma, R.K. Efficient, Enzyme Responsive and Tumor Receptor Targeting Gelatin Nanoparticles Decorated with Concanavalin-a for Site-Specific and Controlled Drug Delivery for Cancer Therapy. *Mater. Sci. Eng. C* **2021**, *123*, 112027. [CrossRef]
124. Kononenko, V.; Joukhan, A.; Bele, T.; Križaj, I.; Kralj, S.; Turk, T.; Drobne, D. Gelatin Nanoparticles Loaded with 3-Alkylpyridinium Salt Aps7, an Analog of Marine Toxin, Are a Promising Support in Human Lung Cancer Therapy. *Biomed. Pharmacother.* **2024**, *177*, 117007. [CrossRef]
125. Chen, Y.-J.; Wang, Z.-W.; Lu, T.-L.; Gomez, C.B.; Fang, H.-W.; Wei, Y.; Tseng, C.-L. The Synergistic Anticancer Effect of Dual Drug-(Cisplatin/Epigallocatechin Gallate) Loaded Gelatin Nanoparticles for Lung Cancer Treatment. *J. Nanomater.* **2020**, *2020*, 9181549. [CrossRef]
126. Mehrotra, N.; Anees, M.; Tiwari, S.; Kharbanda, S.; Singh, H. Polylactic Acid Based Polymeric Nanoparticle Mediated Co-Delivery of Navitoclax and Decitabine for Cancer Therapy. *Nanomedicine* **2023**, *47*, 102627. [CrossRef] [PubMed]

127. Wang, X.; Choudhary, S.M.; Chauhan, G.; Muth, A.; Gupta, V. Transferrin-Conjugated Polymeric Nanoparticles for Receptor-Mediated Delivery of Resveratrol-Cyclodextrin Complex in Non-Small Cell Lung Cancer (Nsclc) Cells. *J. Drug Deliv. Sci. Technol.* **2025**, *105*, 106588. [CrossRef]
128. Yao, F.; Lin, L.; Shi, W.; Li, C.; Liang, Z.; Huang, C. Inhibitory Effect of Poly(Lactic-Co-Glycolic Acid) Nanoparticles Loaded with Resveratrol and Phosphatase and Tensin Homolog Deleted on Chromosome Ten (Pten) Sirna on Lung Cancer Cells. *Sci. Adv. Mater.* **2022**, *14*, 810–817. [CrossRef]
129. Niza, E.; Ocaña, A.; Castro-Osma, J.A.; Bravo, I.; Alonso-Moreno, C. Polyester Polymeric Nanoparticles as Platforms in the Development of Novel Nanomedicines for Cancer Treatment. *Cancers* **2021**, *13*, 3387. [CrossRef]
130. Lombardo, R.; Ruponen, M.; Rautio, J.; Lampinen, R.; Kanninen, K.M.; Koivisto, A.M.; Penttilä, E.; Löppönen, H.; Demartis, S.; Giunchedi, P.; et al. A Technological Comparison of Freeze-Dried Poly-ε-Caprolactone (Pcl) and Poly (Lactic-Co-Glycolic Acid) (Plga) Nanoparticles Loaded with Clozapine for Nose-to-Brain Delivery. *J. Drug Deliv. Sci. Technol.* **2024**, *93*, 105419. [CrossRef]
131. Cabeza, L.; Ortiz, R.; Prados, J.; Delgado, Á.V.; Martín-Villena, M.J.; Clares, B.; Perazzoli, G.; Entrena, J.M.; Melguizo, C.; Arias, J.L. Improved Antitumor Activity and Reduced Toxicity of Doxorubicin Encapsulated in Poly(ε-Caprolactone) Nanoparticles in Lung and Breast Cancer Treatment: An in Vitro and in Vivo Study. *Eur. J. Pharm. Sci.* **2017**, *102*, 24–34. [CrossRef]
132. Akbari, E.; Mousazadeh, H.; Hanifehpour, Y.; Mostafavi, E.; Gorabi, A.M.; Nejati, K.; Peyman, keyhanvar; Pazoki-Toroudi, H.; Mohammadhosseini, M.; Akbarzadeh, A. Co-Loading of Cisplatin and Methotrexate in Nanoparticle-Based Pcl-Peg System Enhances Lung Cancer Chemotherapy Effects. *J. Clust. Sci.* **2022**, *33*, 1751–1762. [CrossRef]
133. Aorada, S.; Chittasupho, C.; Mangmool, S.; Angerhofer, A.; Imaram, W. Gallic Acid-Encapsulated Pamam Dendrimers as an Antioxidant Delivery System for Controlled Release and Reduced Cytotoxicity against Arpe-19 Cells. *Bioconjug. Chem.* **2024**, *35*, 1959–1969.
134. Guo, Z.; Li, S.; Liu, Z.; Xue, W. Tumor-Penetrating Peptide-Functionalized Redox-Responsive Hyperbranched Poly(Amido Amine) Delivering Sirna for Lung Cancer Therapy. *ACS Biomater. Sci. Eng.* **2018**, *4*, 988–996. [CrossRef]
135. Bai, S.-B.; Cheng, Y.; Liu, D.-Z.; Ji, Q.-F.; Liu, M.; Zhang, B.-L.; Mei, Q.-B.; Zhou, S.-Y. Bone-Targeted Pamam Nanoparticle to Treat Bone Metastases of Lung Cancer. *Nanomedicine* **2020**, *15*, 833–849. [CrossRef]
136. Yin, Y.; Li, Y.; Wang, S.; Dong, Z.; Liang, C.; Sun, J.; Wang, C.; Chai, R.; Fei, W.; Zhang, J.; et al. Mscs-Engineered Biomimetic Pmaa Nanomedicines for Multiple Bioimaging-Guided and Photothermal-Enhanced Radiotherapy of Nsclc. *J. Nanobiotechnol.* **2021**, *19*, 80. [CrossRef]
137. Zare, H.; Ahmadi, S.; Ghasemi, A.; Ghanbari, M.; Rabiee, N.; Bagherzadeh, M.; Karimi, M.; Webster, T.J.; Hamblin, M.R.; Mostafavi, E. Carbon Nanotubes: Smart Drug/Gene Delivery Carriers. *Int. J. Nanomed.* **2021**, *16*, 1681–1706. [CrossRef]
138. Hanafy, N.A.N. Optimally Designed Pegylatied Arabinoxylan Paclitaxel Nano-Micelles as Alternative Delivery for Abraxane®: A Potential Targeted Therapy against Breast and Lung Cancers. *Int. J. Biol. Macromol.* **2025**, *293*, 139355.
139. Bertino, E.M.; Williams, T.M.; Shilo, K.; Villalona-Calero, M.A.; Phillips, G.S.; Mo, X.; Otterson, G.A. Phase 2 Trial of Nab-Paclitaxel Plus Carboplatin for Advanced Nsclc in Patients at Risk of Bleeding from Vegf Directed Therapies: Metastatic Non-Small Cell Lung Cancer. *Int. J. Radiat. Oncol. Biol. Phys.* **2014**, *90*, S45. [CrossRef]
140. Subbiah, V.; Grilley-Olson, J.E.; Combest, A.J.; Sharma, N.; Tran, R.H.; Bobe, I.; Osada, A.; Takahashi, K.; Balkissoon, J.; Camp, A.; et al. Phase Ib/Ii Trial of Nc-6004 (Nanoparticle Cisplatin) Plus Gemcitabine in Patients with Advanced Solid Tumors. *Clin. Cancer Res.* **2018**, *24*, 43–51. [CrossRef]
141. Ibrahim, N.K.; Desai, N.; Legha, S.; Soon-Shiong, P.; Theriault, R.L.; Rivera, E.; Esmaeli, B.; Ring, S.E.; Bedikian, A.; Hortobagyi, G.N.; et al. Phase I and Pharmacokinetic Study of Abi-007, a Cremophor-Free, Protein-Stabilized, Nanoparticle Formulation of Paclitaxel. *Clin. Cancer Res.* **2002**, *8*, 1038–1044. [PubMed]
142. Lim, E.A.; Bendell, J.C.; Falchook, G.S.; Bauer, T.M.; Drake, C.G.; Choe, J.H.; George, D.J.; Karlix, J.L.; Ulahannan, S.; Sachsenmeier, K.F.; et al. Phase Ia/B, Open-LabelOpen-Label, Multicenter Study of Azd4635 (an Adenosine A2a Receptor Antagonist) as Monotherapy or Combined with Durvalumab, in Patients with Solid Tumors. *Clin. Cancer Res.* **2022**, *28*, 4871–4884. [CrossRef]
143. Ashique, S.; Garg, A.; Mishra, N.; Raina, N.; Ming, L.C.; Tulli, H.S.; Behl, T.; Rani, R.; Gupta, M. Nano-Mediated Strategy for Targeting and Treatment of Non-Small Cell Lung Cancer (Nsclc). *Naunyn-Schmiedeberg's Arch. Pharmacol.* **2023**, *396*, 2769–2792. [CrossRef]
144. Su, P.; Pei, W.; Wang, X.; Ma, Y.; Jiang, Q.; Liang, J.; Zhou, S.; Zhao, J.; Liu, J.; Lu, G.Q.M. Exceptional Electrochemical Her Performance with Enhanced Electron Transfer between Ru Nanoparticles and Single Atoms Dispersed on a Carbon Substrate. *Angew. Chem. Int. Ed. Engl.* **2021**, *60*, 16044–16050. [CrossRef]
145. Srinivasarao, D.A.; Shah, S.; Famta, P.; Vambhurkar, G.; Jain, N.; Pindiprolu, S.; Sharma, A.; Kumar, R.; Padhy, H.P.; Kumari, M.; et al. Unravelling the Role of Tumor Microenvironment Responsive Nanobiomaterials in Spatiotemporal Controlled Drug Delivery for Lung Cancer Therapy. *Drug Deliv. Transl. Res.* **2025**, *15*, 407–435. [CrossRef]

146. Xu, W.; Yang, S.; Lu, L.; Xu, Q.; Wu, S.; Zhou, J.; Lu, J.; Fan, X.; Meng, N.; Ding, Y. Influence of Lung Cancer Model Characteristics on Tumor Targeting Behavior of Nanodrugs. *J. Control. Release* **2023**, *354*, 538–553. [CrossRef] [PubMed]
147. Islam, R.; Maeda, H.; Fang, J. Factors Affecting the Dynamics and Heterogeneity of the Epr Effect: Pathophysiological and Pathoanatomic Features, Drug Formulations and Physicochemical Factors. *Expert. Opin. Drug Deliv.* **2022**, *19*, 199–212. [CrossRef] [PubMed]
148. Subhan, M.A. Satya Siva Kishan Yalamarty, Nina Filipczak, Farzana Parveen, Vladimir P Torchilin, Recent Advances in Tumor Targeting Via Epr Effect for Cancer Treatment. *J. Pers. Med.* **2021**, *11*, 571. [CrossRef]
149. Kim, J.; Cho, H.; Lim, D.-K.; Joo, M.K.; Kim, K. Perspectives for Improving the Tumor Targeting of Nanomedicine Via the Epr Effect in Clinical Tumors. *Int. J. Mol. Sci.* **2023**, *24*, 10082. [CrossRef] [PubMed]
150. Elmowafy, M.; Shalaby, K.; Elkomy, M.H.; Alsaidan, O.A.; Gomaa, H.A.M.; Abdelgawad, M.A.; Mostafa, E.M. Polymeric Nanoparticles for Delivery of Natural Bioactive Agents: Recent Advances and Challenges. *Polymers* **2023**, *15*, 1123. [CrossRef]
151. Sabir, F.; Qindeel, M.; Zeeshan, M.; Ain, Q.U.; Rahdar, A.; Barani, M.; González, E.; Aboudzadeh, M.A. Onco-Receptors Targeting in Lung Cancer Via Application of Surface-Modified and Hybrid Nanoparticles: A Cross-Disciplinary Review. *Processes* **2021**, *9*, 621. [CrossRef]
152. Yu, M.; Mathew, A.; Liu, D.; Chen, Y.; Wu, J.; Zhang, Y.; Zhang, N. Microfluidics for Formulation and Scale-up Production of Nanoparticles for Biopharma Industry. In *Microfluidics in Pharmaceutical Sciences: Formulation, Drug Delivery, Screening, and Diagnostics*; Lamprou, D.A., Weaver, E., Eds.; Springer Nature: Cham, Switzerland, 2024; pp. 395–420.
153. Worku, D.A.; Hewitt, V. The Role and Economics of Immunotherapy in Solid Tumour Management. *J. Oncol. Pharm. Pract.* **2020**, *26*, 2020–2024. [CrossRef]
154. Cangut, B.; Akinlusi, R.; Mohseny, A.; Ghesani, N.; Ghesani, M. Evolving Paradigms in Lung Cancer: Latest Trends in Diagnosis, Management, and Radiopharmaceuticals. *Semin. Nucl. Med.* **2025**, *55*, 264–276. [CrossRef]
155. Salaudeen, H.D.; Akinniranye, R.D. Precision Nanotechnology for Early Cancer Detection and Biomarker Identification. *Int. J. Res. Publ. Rev.* **2024**, *5*, 6313–6327. [CrossRef]
156. Desai, N.; Rana, D.; Patel, M.; Bajwa, N.; Prasad, R.; Vora, L.K. Nanoparticle Therapeutics in Clinical Perspective: Classification, Marketed Products, and Regulatory Landscape. *Small* **2025**, *21*, 2502315. [CrossRef]
157. Junyaprasert, V.B.; Thummarati, P. Innovative Design of Targeted Nanoparticles: Polymer–Drug Conjugates for Enhanced Cancer Therapy. *Pharmaceutics* **2023**, *15*, 2216. [CrossRef] [PubMed]
158. Yang, T.; Zhai, J.; Hu, D.; Yang, R.; Wang, G.; Li, Y.; Liang, G. "Targeting Design" of Nanoparticles in Tumor Therapy. *Pharmaceutics* **2022**, *14*, 1919. [CrossRef] [PubMed]
159. Kim, Y.; Kim, J.; Eom, S.; Jun, H.; Lee, H.B.; Jeong, D.; Kang, S. Protein Nanoparticles Simultaneously Displaying Trail and Egfr-Binding Ligands Effectively Induce Apoptotic Cancer Cell Death and Overcome Egfr-Tki Resistance in Lung Cancer. *ACS Appl. Mater. Interfaces* **2025**, *17*, 25139–25151. [CrossRef] [PubMed]
160. Nagime, P.V.; Singh, S.; Chidrawar, V.R.; Rajput, A.; Syukri, D.M.; Marwan, N.T.; Shafi, S. Moringa Oleifera: A Plethora of Bioactive Reservoirs with Tremendous Opportunity for Green Synthesis of Silver Nanoparticles Enabled with Multifaceted Applications. *Nano-Struct. Nano-Objects* **2024**, *40*, 101404. [CrossRef]
161. Rakaee, M.; Tafavvoghi, M.; Ricciuti, B.; Alessi, J.V.; Cortellini, A.; Citarella, F.; Nibid, L.; Perrone, G.; Adib, E.; Fulgenzi, C.A.M.; et al. Deep Learning Model for Predicting Immunotherapy Response in Advanced Non-Small Cell Lung Cancer. *JAMA Oncol.* **2025**, *11*, 109–118. [CrossRef]

pharmaceutics

MDPI

Article

Evaluation of Novel Nasal Mucoadhesive Nanoformulations Containing Lipid-Soluble EGCG for Long COVID Treatment

Nicolette Frank [1], Douglas Dickinson [2], Garrison Lovett [1], Yutao Liu [3], Hongfang Yu [3], Jingwen Cai [3], Bo Yao [4], Xiaocui Jiang [4] and Stephen Hsu [1,2,*]

1. Department of Oral Biology & Diagnostic Sciences, Dental College of Georgia, Augusta University, Augusta, GA 30912, USA; nicolettefrank6499@yahoo.com (N.F.); glovett@augusta.edu (G.L.)
2. Camellix Research Laboratory, Augusta, GA 30912, USA; gougdickinson4357@gmail.com
3. Department of Cellular Biology & Anatomy, Medical College of Georgia, Augusta University, Augusta, GA 30912, USA; yutliu@augusta.edu (Y.L.); hyu@augusta.edu (H.Y.); jincai@augusta.edu (J.C.)
4. Hangzhou Shanju Biotech Co., Ltd., Hangzhou 310030, China; yaobotea@126.com (B.Y.); jxc20057011222@126.com (X.J.)

* Correspondence: shsu@augusta.edu; Tel.: +1-706-721-4816

Abstract: Following recovery from the acute infection stage of the SARS-CoV-2 virus (COVID-19), survivors can experience a wide range of persistent Post-Acute Sequelae of COVID-19 (PASC), also referred to as long COVID. According to the US National Research Action Plan on Long COVID 2022, up to 23.7 million Americans suffer from long COVID, and approximately one million workers may be out of the workforce each day due to these symptoms, leading to a USD 50 billion annual loss of salary. Neurological symptoms associated with long COVID result from persistent infection with SARS-CoV-2 in the nasal neuroepithelial cells, leading to inflammation in the central nervous system (CNS). As of today, there is no evidence that vaccines or medications can clear the persistent viral infection in olfactory mucosa. Recently published clinical data demonstrate that only 5% of long COVID anosmia patients have fully recovered during the past 2 years, and 10.4% of COVID patients are still symptomatic 18 months post-infection. Our group demonstrated that epigallocatechin-3-gallate-monopalmitate (EC16m) nanoformulations possess strong antiviral activity against human coronavirus, suggesting that this green-tea-derived compound in nanoparticle formulations could be developed as an intranasally delivered new drug targeting the persistent SARS-CoV-2 infection, as well as inflammation and oxidative stress in the CNS, leading to restoration of neurologic functions. The objective of the current study was to evaluate the mucociliary safety of the EC16m nasal nanoformulations and their efficacy against human coronavirus. Methods: Nanoparticle size and Zeta potential were measured using the ZetaView Nanoparticle Tracking Analysis system; mucociliary safety was determined using the MucilAir human nasal model; contact antiviral activity and post-infection inhibition against the OC43 viral strain were assessed by the TCID50 assay for cytopathic effect on MRC-5 cells. Results: The saline-based EC16 mucoadhesive nanoformulations containing 0.005 to 0.02% *w*/*v* EC16m have no significant difference compared to saline (0.9% NaCl) with respect to tissue integrity, cytotoxicity, and cilia beat frequency. A 5 min contact resulted in 99.9% inactivation of β-coronavirus OC43. OC43 viral replication was inhibited by >90% after infected MRC-5 cells were treated with the formulations. Conclusion: The saline-based novel EC16m mucoadhesive nasal nanoformulations rapidly inactivated human coronavirus with mucociliary safety properties comparable to saline, a solution widely used for nasal applications.

Keywords: respiratory virus; long COVID; nasal drug; EC16; EGCG-palmitate; nanoformulations

1. Introduction

According to the Center of Disease Control and Prevention (CDC), long COVID symptoms can last for weeks, months, or years after the initial COVID-19 illness and can sometimes result in disability (CDC, long COVID or post-COVID conditions). A recent analysis of clinical data found that only 5% of long COVID anosmia patients have fully recovered during the past 2 years, and 10.4% of COVID patients are still symptomatic 18 months post-infection [1]. The neurologic symptoms associated with long COVID include fatigue; "brain fog"; cognitive impairment; headache; sleep, mood, smell, or taste disorders; myalgias; sensorimotor deficits; dizziness; anxiety; depression; earache; hearing loss and/or ringing in the ears; dysautonomia; and psychiatric manifestations [2–4]. The pathogenesis of these long COVID neurologic symptoms involves neuroinvasion of SARS-CoV-2 from the nasal neuroepithelium to the support and stem cells in the olfactory mucosa, causing persistent olfactory dysfunction (anosmia). This persistence of SARS-CoV-2 also induces dysregulation of innate and adaptive immunity with prolonged cytokine release, oxidative stress, and lymphocytic infiltration in the central nervous system (CNS), leading to stress, demyelination, and neurodegeneration [5]. Thus, therapeutic approaches targeting long COVID neurologic symptoms must address the persistent presence of SARS-CoV-2. Effort has been taken to reduce the neurologic symptoms using antiviral drugs such as Nirmatrelvir (oral), and steroids like fluticasone (nasal), mometasone (nasal), and naltrexone (oral). Among these clinical studies, the only randomized double-blind clinical trial showed that mometasone nasal spray did not lead to significant improvement in recovery rates or duration of anosmia [6]. A Phase II clinical trial using nasal irrigation with saline + 400 mg theophylline in long COVID patients did not generate satisfactory outcomes [7]. Despite induction of a robust intranasal antibody production by mRNA vaccines [8], there is presently no evidence for clearance of persistent SARS-CoV-2 from the olfactory mucosa by vaccination [9].

In order to address the "root cause" of long COVID neurologic symptoms, the best approach would be to simultaneously reduce the persistent SARS-CoV-2 viral presence (persistent infection), the inflammation of olfactory epithelia, and the oxidative stress and damage in the CNS [10]. Based on our recent studies, we hypothesized that epillagocatechin-3-gallate (EGCG)-mono-palmitate (EC16m), a compound with multiple mechanisms of antiviral activity, plus anti-inflammatory, antioxidant, and neuroprotective properties, has the potential to become a new nasal drug to minimize long-COVID-associated neurologic symptoms such as anosmia [11,12].

EGCG is a naturally formed hydrophilic major green tea polyphenol which has a wide spectrum of antiviral activity [13,14], including against SARS-CoV-2 [15]. EGCG can be esterified chemically to EC16m [16]. The mono-palmitoylated EGCG (EC16m), which is also a naturally occurring green tea catechin, is an amphipathic compound [17]. We have shown that EC16m is able to enter epithelial cells and is hydrolyzed by esterase in the cytoplasm, releasing free EGCG [17,18].

In addition to its antiviral properties, EGCG is able to reduce epithelial cell inflammation in vitro and in vivo [19]. In autoimmune animal studies, we found that EGCG significantly reduces lymphocyte infiltration and serum autoantibody levels and protects human cells from TNF-α-induced cytotoxicity [20,21]. Interestingly, our animal studies showed that EGCG modulates the antioxidant defense enzymes to protect cells from free-radical-induced damage [22], stabilizes p21 expression, and reduces DNA damage from inflammation-induced reactive oxygen species [23]. Also, it has been widely reported in preclinical studies and clinical trials that EGCG provides neuroprotective effects [24–26]. However, the poor bioavailability and instability of EGCG prevent its beneficial effects from being realized in new drug development [16,27–30].

In comparison to water-soluble EGCG, EC16m is significantly more potent against influenza virus, herpes simplex virus, and norovirus [29,31,32]. Our recent studies demonstrate that nanoparticles of EC16 (contains 50% EC16m) or EC16m in saline-based nasal formulations are able to rapidly inactivate human coronavirus [33,34].

The preparation of green tea catechin (polyphenol) NPs has been explored previously with different methods. For example, lipid-based NPs of EGCG were produced with different lipids and a surfactant and evaluated for potential use in cancer treatment through oral administration [34–36]. EGCG can be encapsulated in hordein NPs [37] and further coated with chitosan [38] to improve the bioavailability of EGCG [39].

Our invention of a "facilitated self-assembling" method to generate nanoparticles of the amphipathic compound EC16/EC16m is a milestone in nanotechnology (U.S. Application No. 63/490,712) which enables us to formulate aqueous suspensions of the nanoparticles for various purposes, especially for disease control and prevention using highly effective and natural compounds. The EC16 nanoparticles (NPs) do not belong to any current NP classifications such as incidental, bioinspired, anthropogenic, or engineered NPs [33].

Unlike other engineered EGCG NPs, the EC16/EC16m NPs were produced through a "facilitated self-assembling" method (proprietary, patent pending), which does not involve association with metals, monomers, oil, or encapsulation. In addition to the high efficacy of rapid inactivation of human coronaviruses OC43 and 229E, the water-based nanoformulations are not associated with cytotoxicity [11,12]. In contrast, recently published data showed that engineered EGCG-AgNPs are cytotoxic to human skin cells (CC50 = 30 μg/mL), and the efficacy is poor against human herpes simplex virus type 1 and type 2, with less than $\log_{10}$ 2 reduction (<99%) after 60 min of incubation [40]. Another study indicated that EGCG-AgNPs become cytotoxic at nM levels [41], while EC16 NPs at 1.4 mM are not associated with cytotoxicity [12].

For intranasally applied formulations, nasal mucociliary clearance transit time (MCCTT) and mucociliary toxicity must be considered. The previously tested, saline-based, aqueous EC16m nanoformulations required further formulation to create a mucoadhesive formulation because the human MCCT is under 20 min [42]. These novel EC6m mucoadhesive nanoformulations should possess a rapid antiviral effect without mucociliary toxicity. The objectives of the current study were to evaluate the mucociliary safety of the EC16m nasal nanoformulations using a 3D Human Nasal Epithelium Model (MucilAir) and test the efficacy against human β-coronavirus OC43.

2. Materials and Methods

2.1. Virus and Cell Lines

OC43 human β-coronavirus (ATCC VR-1558), HCT-8 human epithelial cells (ATCC HRT-18), and MRC-5 human respiratory fibroblast cells (ATCC CCL-171) were purchased from ATCC (Manassas, VA, USA). HCT-8 cells were used in cell viability (MTT) assays. MRC-5 cells were used for antiviral assays on OC43 virus.

2.2. EC16m and Other Supplies

Epigallocatechin-3-gallate-4′ mono-palmitate (EC16m, CAS# 507453-56-7) was provided by Camellix, LLC (Evans, GA, USA). Dulbecco's Modified Eagle's Medium (DMEM) was purchased from ATCC (30-2002). Trypsin-EDTA solution was purchased from ATCC (30-2101). Fetal bovine serum (FBS) was obtained from Neuromics (Edina, MN, USA). Penicillin, streptomycin, and amphotericin B solution (100×) was obtained from Corning (Glendale, AR, USA). Carboxymethylcellulose (CMC) was obtained from Fisher Scientific (Waltham, MA, USA). Plasticwares were purchased from Southern Labware (Cumming, GA, USA).

2.3. EC16 Mucoadhesive Nanoformulations

EC16m (formula weight 697) nanoparticles were initially dispersed in 90% glycerol as stable stocks at 1% *w/v* using a facilitated self-assembling method (proprietary) that does not involve specialized equipment. Four formulations, A-D (FA-FD), were prepared by diluting this EC16m nanoparticle stock with normal saline (0.9% NaCl) containing carboxymethylcellulose (CMC) for a final concentration of 0.5% CMC. Formulation C contained 0.05% (700 μM) EC16m. Formulations A, B, and D each contained the same amount of a food-grade dispersing agent (proprietary information) and EC16m at 0.02% (280 μM), 0.01% (140 μM), and 0.002% (28 μM), respectively. The nanoformulations with the dispersing agent had a pH of 6.22 and viscosity of 19 mPa. The FC nanoformulation without the dispersing agent had a pH of 6.21 and viscosity of 14 mPa.

2.4. Evaluation of Particle Size Distribution

ZetaView nanoparticle tracking analysis was performed according to a method described previously [12,43]. The particle size distribution and concentration were measured using the Zetaview ×20 (Particle Metrix, Meerbusch, Germany) and corresponding software. The measuring range for particle diameter is 10–2000 nm. These samples were diluted by the same volume of 1× PBS and then loaded into the cell. Particle information was collected from the instrument at 11 different positions across the cell, with two cycles of readings. Standard operating procedure was set to a temperature of 23 °C, a sensitivity of 70, a frame rate of 30 frames per second, and a shutter speed of 100. The post-acquisition parameters were set to a minimum brightness of 20, a maximum area of 1000, a minimum area of 10, and a trace length of 15 [43].

2.5. Evaluation of Cytotoxicity by Cell Viability (MTT) Assay

HCT-8 cells were seeded in 48-well plates at 5×10^4 cells/well and allowed to form a monolayer prior to incubation with the nanoformulations (Formulations A, B, C, and D) and control formulation (normal saline) in a series of dilutions for 60 min. The tested nanoformulations were then replaced with DMEM medium with 10% FBS and incubated overnight. The MTT assay was performed the next day using CytoSelect MTT Cell Proliferation Assay kit (Cell Biolabs, Inc., San Diego, CA, USA) according to the method provided by the manufacturer.

2.6. Evaluation of Mucociliary Toxicities of EC16m Nanoformulations by 3D MucilAir Human Nasal Epithelium Model (Performed by Epithelix Sàrl, Plan-les-Ouates, Switzerland)

The aim of this evaluation was to study the acute mucociliary toxicological effect of EC16m nasal spray nanoformulations using fully differentiated human nasal epithelial cells cultured at the air–liquid interface. Human nasal epithelia (MucilAir™-Pool, Epithelix Sàrl, Plan-les-Ouates, Switzerland) were reconstituted with a mixture of cells isolated from 14 different normal nasal donors. Formulations A, B, C, and D (FA, FB, FC, and FD) were exposed apically for 2 days. At time 0 (day 1), 10 µL of the nanoformulation was applied apically on MucilAir™-Pool inserts (Figure 1) twice a day over a 30 min period with a 6 h interval. This was repeated on day 2. Subsequently, tissue integrity (TEER), lactate dehydrogenase (LDH) release (cytotoxicity), and cilia beating frequency (CBF) were measured at the end of the experiment. Detailed protocols are presented as appendices. The experiments were repeated three times in the Epithelix laboratories. Normal saline (0.9% NaCl) was used as the vehicle control, and 10% Triton X in the culture medium was used as positive control.

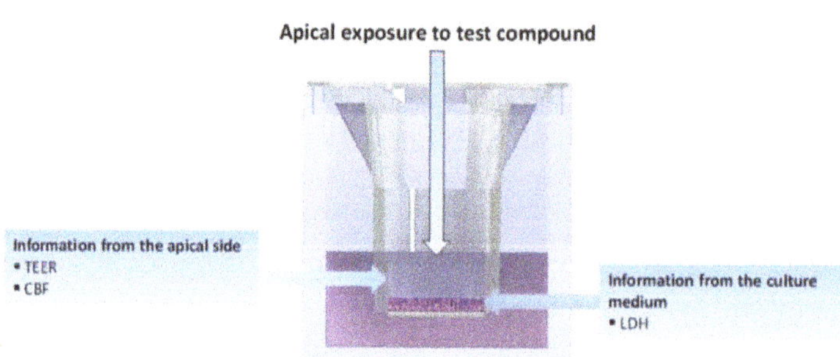

Figure 1. Schematic illustration of sample application and end-point measurements of MucilAire 3D human nasal model.

2.7. Direct Contact Antiviral Activity Tests

Infection of cells by OC43 virus and viral titer: MRC-5 cells were cultured in DMEM Medium supplemented with 10% FBS and 1% penicillin, streptomycin, and amphotericin

B. The viral infection assay and viral titering were performed in 96-well cell culture plates when the cells had reached 90% confluency. A 10-fold series dilution of OC43 virus in DMEM containing 2% FBS (MM) was loaded into wells in quadruplicates per dilution. After a one-hour adsorption, the viral dilutions were removed, and 100 µL MM was added, followed by incubation at 33 °C with 5% CO_2 for >4 days to allow a CPE (cytopathic effect) to become visible. Viral titer was calculated by TCID50 Excel software (November 2013, Research Gate) based on the Reed–Muench method [44]. A minimum of three independent experiments were performed and results recorded.

2.8. Post-Infection Test

To test whether EC16m nasal nanoformulations possess a post-infection effect, MRC-5 cells were allowed to form a monolayer (90% confluent) in a 96-well cell culture plate prior to a one-hour infection of OC43 virus in a series dilution to 10–9 before removal of the virus. Then, 50 µL of EC16m nanoformulation was applied to the designated wells for 5 min before being replaced by MM. The vehicle control wells were treated with the vehicles after viral infection for 5 min before medium change. The cytopathic effect (CPE) was captured after incubation for at least 6 days.

2.9. Statistical Analysis

The primary statistical tests were parametric one-way ANOVAs based on three or more repeated test points. Alpha was 0.05. GraphPad Prism version 6.0 software (www.graphpad.com) was used for most analyses. Reported errors are given as standard deviation (SD).

3. Results

Based on the results from cell viability assays and 3D MucilAire Human Nasal Model evaluation, only Formulations C and D were selected for the rest of the experiments.

3.1. Size Distribution and Zeta Potentials of Particles

As shown in Figure 2A the FC nanoformulation showed a polydisperse particle size distribution, with a median size of 252.6 ± 109 nm (SD, $n = 2$). More than 90% of the particles were within 222 to 296 nm range, while 6.2% were in the 100 nm range. The concentration of particles was 2.2×10^9/mL (2.2 billion particles/mL). For the FD nanoformulation (Figure 2B), the particle size distribution ranged from 45.4 to 331 nm, with a broader left (smaller particle) tail than FC. The median size for FD was 257 ± 134 nm (SD, $n = 2$). About 78% of the particles were in the 265 to 331 nm range, and rest were within 45 and 128 nm range. The particle concentration was 6.5×10^9 (6.5 billion/mL). (Figure 2C). The Zeta Potential of the FC nanoformulation diluted 30× with water at 25 °C was −29.49 ± 1.02 mV. The Zeta Potential Distribution was 29.49 mV FWHM 6.05 (SL1/2). (Figure 2D). The Zeta Potential of the FD nanoformulation similarly diluted was −51.31 ± 1.22 mV. The Zeta Potential Distribution was 51.31 mV FWHM 4.47 (SL1/2). Thus, particles appeared to be more evenly distributed in FD compared to FC.

3.2. Cell Viability after 1 h Incubation with HCT-8 Cells

To assess the initial toxicity of the novel EC16m mucoadhesive nanoformulations prior to the 3D Human Nasal Epithelium Model safety tests, HCT-8 human intestinal epithelial cells were exposed to the mucoadhesive nasal nanoformulations for 1 h. After replacement of the formulations with media and overnight incubation, an MTT assay was performed. One-way ANOVA showed a significant difference between the groups ($p = 0.009$). As shown in Figure 3, the untreated control cell viability measured by absorption at 450 nm was 0.844 ± 0.105, and the value for the vehicle (saline) was 0.774 ± 0.014. There is no significant difference between these controls (Tukey's multiple comparisons test, $p = 0.63$, $n = 3$).

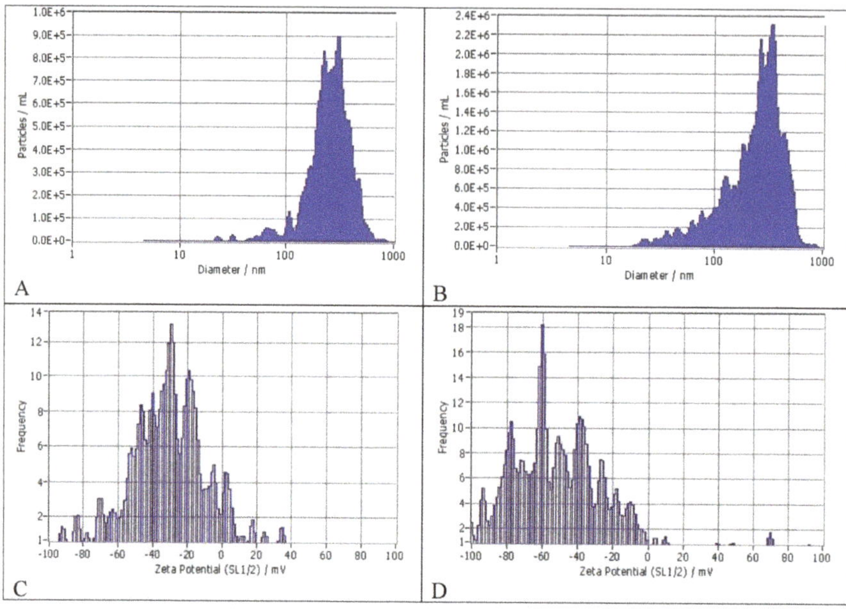

Figure 2. (**A**) Size distribution of particles in FC. (**B**) Size distribution of particles in FD The size distribution profile for one representative sample/formulation determined by NTA is shown. (**C**) Zeta potential and distribution of FC. (**D**) Zeta potential and distribution of FD.

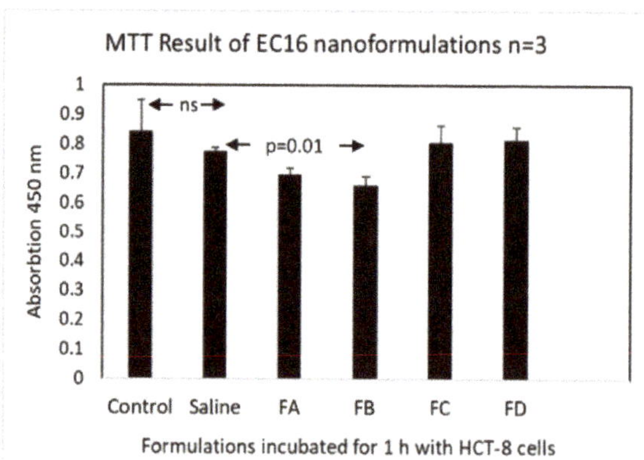

Figure 3. Cell viability (MTT) assay results for the four nanoformulations in comparison to saline as vehicle control and untreated cell control. The assay was conducted in 48-well tissue culture plates with confluent HCT-8 cells in each well ($n = 3$). Select p values are shown; ns: not significant ($p >$ 0.05). Arrows point to columns with differences.

In comparison to the saline vehicle control, only FB (140 µM EC16m in combination with a dispersing agent), showed a significant reduction in MTT value (0.66 ± 0.03; Dunnett's multiple comparisons test, $p = 0.01$). Formulations A, C and D were not significant different from saline treated cells (0.70± 0.02, 0.81 ± 0.06, $p = 0.44$ and 0.82 ± 0.04, respectively, $p \geq 0.08$). Thus, there was no indication for a cytotoxic effect for CMC in combination with EC16m (FC, $p = 0.69$ vs saline), and there was a weak trend towards a cytotoxic effect at the higher concentrations of EC16m tested in combination with a dispersing agent.

3.3. Mucociliary Toxicity

3.3.1. Tissue Integrity

The normal range of tissue integrity (TEER) for MucilAir Human Nasal Epithelium Model are 200–800 Ω.cm^2. Figure 4 shows that the saline vehicle control, and FB-FD were all in the normal range (with saline and FB at the lower end of the range), while FA was below the normal range. One-way ANOVA showed a significant difference between the groups (saline and FA-FD; $p = 0.003$). Formulation A gave a TEER value (167 ± 24.4) significantly lower than FB-FD (Tukey's multiple comparisons test, $p \leq 0.026$), but there was no significant difference between FB-FD ($p > 0.50$) and the controls. In comparison to the saline vehicle control, there was no significant difference with FA-FD (Dunnett's multiple comparisons test, $p \geq 0.052$) with FA showing the borderline significance. Thus, FB, FC, and FD did not impair tissue integrity in this model.

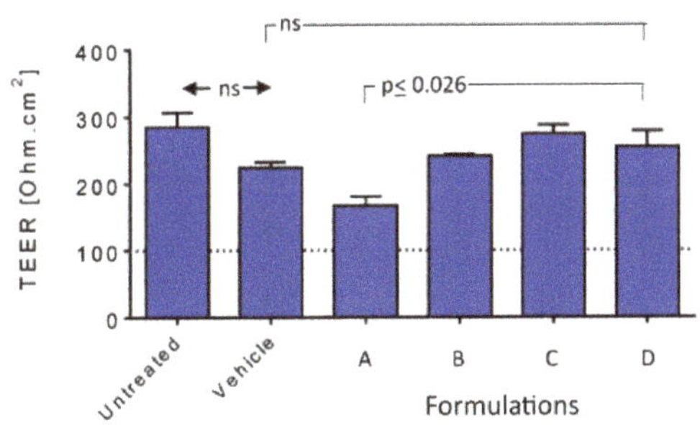

Figure 4. Impact on tissue integrity of the nanoformulations in comparison to saline (vehicle). Select p values are shown; ns: not significant ($p > 0.05$). Arrows indicate significant difference between pairs of columns. Brackets show groups of columns with no significant difference to indicated column to adjacent columns.

3.3.2. Cytotoxicity Measured by LDH Release

Results shown in Figure 5 indicated that the untreated control and the vehicle gave similar, modest levels of LDH release (8.03 and 8.40%, respectively; $p = 1.0$, Tukey's multiple comparisons test) based on 100% release with 10% Triton-X100. For these two controls and the formulations, one-way ANOVA showed a significant difference between the groups ($p < 0.0001$). Formulation A exhibited a significant increase in release (16.0 ± 1.7%) relative to the two controls ($p \leq 0.0002$) and in comparison to FB-FD ($p \leq 0.011$). There were no significant differences between FB-FD and the controls ($p \geq 0.13$). Therefore, FA induced a significant increase in FA, but FB-FD did not and were comparable to saline.

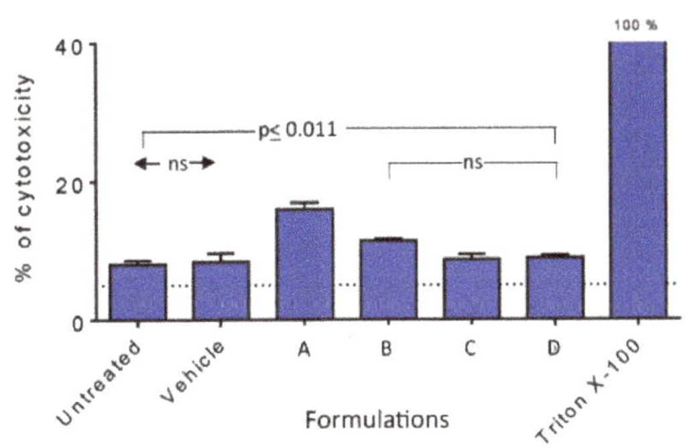

Figure 5. Cytotoxicity induced by the nanoformulations in comparison to untreated control, saline (vehicle), and positive control (10% Triton X-100). Select p values are shown; ns: not significant ($p > 0.05$). Arrows indicate significant difference between pairs of columns. Brackets show groups of columns with no significant difference to indicated column to adjacent columns.

3.3.3. Cilia Beating Frequency (CBF)

After application of the nanoformulations or controls twice daily for two consecutive days, one-way ANOVA showed significant differences between the groups ($p = 0.001$) (Figure 6). The untreated and vehicle controls displayed CBF values of 4.96 and 4.01 Hz that were not significantly different (Tukey's multiple comparisons test, $p = 0.073$). Formulation A gave a CBF of 3.15 ± 0.37 Hz, significantly lower than the CBF for the vehicle (Dunnett's multiple comparisons test, $p = 0.023$), whilst FB-FD were not significantly different from the vehicle ($p \geq 0.24$). There were no significant differences among FB-FD (Tukey's multiple comparisons test, $p \geq 0.16$). The three formulations containing the dispersal agent (FA, FB, FD) all gave CBF values significantly lower than the untreated control ($p \leq 0.028$), whilst FC was not significantly different ($p = 0.38$). Formulation A was also significantly lower than FC ($p = 0.019$). Therefore, the dispersal agent had a modest effect on CBF, with a trend to a dose dependence for F18m, but at the lower levels of EC16m there was no significant difference compared to the saline control.

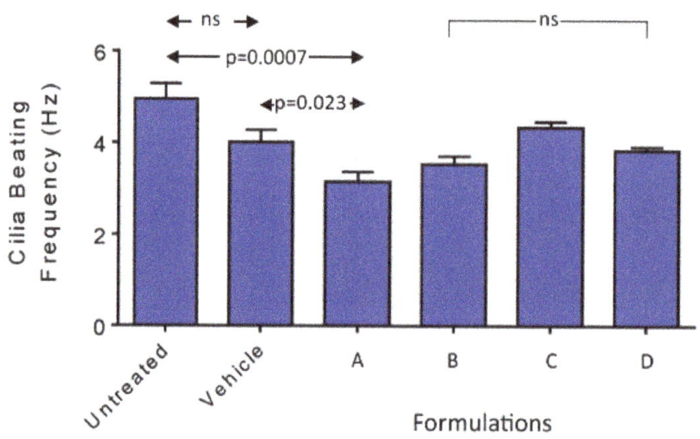

Figure 6. Cilia beating frequency measurements for the four nanoformulations in comparison to saline (vehicle) after two days of applications based on twice daily 30 min application/each schedule. FA was the only nanoformulation associated with significantly reduced CBF among the formulations. Select p values are shown; ns: not significant ($p > 0.05$). Arrows indicate significant difference between pairs of columns. Brackets show groups of columns with no significant difference to indicated column to adjacent columns.

3.4. Contact Inhibition of OC43 Viral Infection

According to the results from cell viability and the 3D Human Nasal Epithelium Model safety assessment, FA and FB demonstrated viability and mucociliary safety levels consistently lower than the vehicle (either as a trend or statistically significant), while both FC and FD showed similar levels for these safety parameters, comparable to the vehicle. Therefore, only FC and FD were used in follow-up in vitro efficacy tests to validate the antiviral activity of the novel mucoadhesive nasal nanoformulations.

As shown in Figure 7, both FC and FD were able to reduce OC43 infectivity by approximately 99.9% at the times tested (2.83–3.42 $\log_{10}$), FC at both times and FD at 5 min. This reduction was significantly greater than zero (one-sample t-test, $p \leq 0.0076$, Bonferroni correction to alpha = 0.0125 ($n = 4$)). For FD at 15 min (2.83 $\log_{10}$), the difference was not significant after Bonferroni correction ($p = 0.023$). The vehicle control did not reduce the infectivity of the virus ($p > 0.05$). For FC, 5 min and 15 min incubations with the virus led to titer reductions of $\log_{10}$ 3.42 ± 0.52 and $\log_{10}$ 3.33 ± 0.14, respectively. For FD, 5 min and 15 min incubations with the virus led to $\log_{10}$ 2.92 ± 0.38 and $\log_{10}$ 2.83 ± 0.76 reduction, respectively. There were no statistical differences among pairwise comparisons of incubation times or formulations (repeat measures one-way ANOVA, $p = 0.26$), which were all significantly greater than the vehicle controls ($p \leq 0.024$).

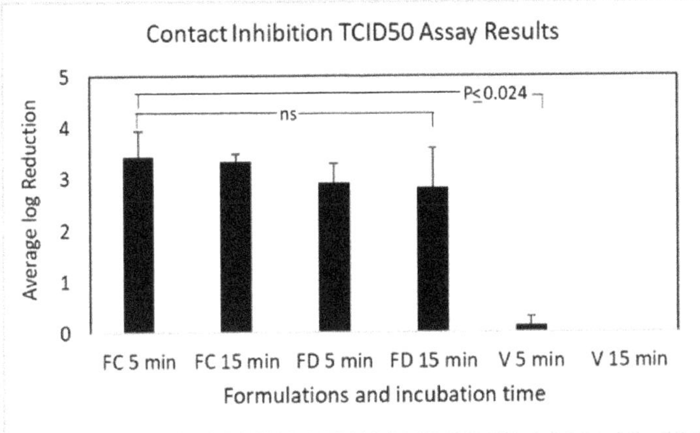

Figure 7. Mean $\log_{10}$ reduction in OC43 infectivity after incubation with FC and FD for 5 and 10 min (V: vehicle control). The results are from three independent TCID50 assays for the nanoformulations. Select p values are shown; ns: not significant ($p > 0.05$).

3.5. *Post-Infection Inhibition of OC43 Viral Replication*

As shown in Figure 8, the vehicle controls showed a minimal effect on viral titer ($\log_{10}$ 0.50 ± 0.25 reduction) that was not significantly different from zero (one sample t-test $p = 0.07$). There was no statistical difference between the vehicle controls. The FC nanoformulation without the dispersal agent exhibited only a small inhibitory effect against OC43 replication in MRC-5 cells ($\log_{10}$ 0.83 ± 0.14 vs. vehicle control) that was significantly greater than zero (one sample t-test, $p = 0.01$) but without a statistical difference compared to the vehicle control (one-way ANOVA, Tukey's multiple comparisons test, $p = 0.26$). In contrast, post-treatment with the FD nanoformulation reduced OC43 viral replication in infected cells by more than 99% ($\log_{10}$ 2.33 ± 0.14), which was significantly different from the vehicle control ($p = 0.005$) and from FC ($p = 0.023$).

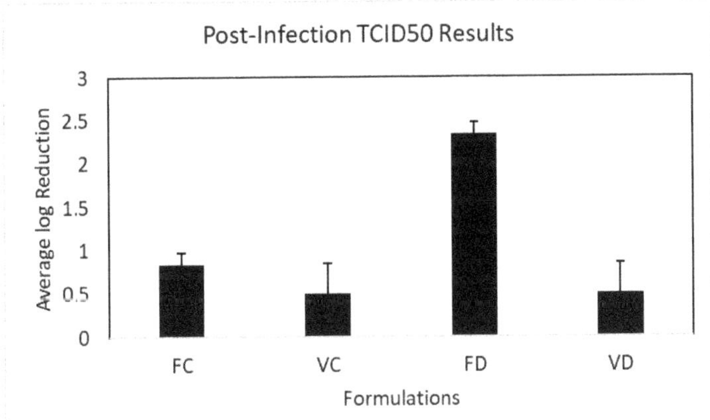

Figure 8. Mean $\log_{10}$ reduction in OC43 infectivity after post-infection incubation with FC and FD for 5 min (VC, VD: vehicle controls for FC and FD, respectively). The results are from three independent TCID50 assays for the nanoformulations.

4. Discussion

Our understanding of the etiology of long COVID is still evolving. Currently, it is thought that the persistence of SARS-CoV-2 in the neuroepithelium leads to invasion of the support and stem cells in the olfactory mucosa, causing persistent olfactory dysfunction (anosmia) and induction of dysregulation of innate and adaptive immunity with prolonged cytokine release, oxidative stress, and lymphocytic infiltration in the central nervous system (CNS), leading to stress, and demyelination, and neurodegeneration [2–5].

The current study is the first attempt to validate the suitability of EC16m nanoparticle/saline based mucoadhesive nasal formulations for use in humans using the 3D MucilAir Human Nasal Epithelium Model (MucilAir) and in vitro toxicity/efficacy methods. We previously reported a series of test results from EC16 and EC16m nanoformulations that were saline-based aqueous suspensions of the nanoparticles designated as F18, F18D, F18m, and F18Dm that showed high antiviral efficacy [11,12]. However, these high-efficacy nanoformulations lack a mucoadhesive nature, and therefore they are not suitable for nasal application due to the mucociliary clearance transit time of 20 min [42]. Accordingly, the novel nanoformulations tested in the current study were adjusted with the addition of CMC to increase the viscosity up to 19 mPa, which increased the mucoadhesive capability and the mucociliary clearance time. With a pH of approximately 6.20, the FC and FD formulations are comparable to the pH of the human nasal cavity.

Compared to EC16 nanoparticles such as those in F18D [12], the EC16m nanoparticles in FC and FD have a slightly larger particle size (Figure 2). Nanoparticles in FC have a median size of 252.6 + 109 nm (SD, n = 2), and nanoparticles in FD have a median size is 257 + 134 nm (SD, n = 2). In contrast, the F18D particles have a median particle size of 186.6 + 20.62 nm. Differences in the particle size distributions could be due to the chemical and physical differences between EC16m (EGCG-mono-palmitate) and EC16, which is a mixture of EGCG-mono-palmitate, EGCG-di-palmitate, and EGCG-tri-palmitates (EGCG-palmitates containing 50% EC16m) [45]. Comparing FC and FD, the particle density of FD is 3 times greater than FC, which is reflected in the more polydisperse particle size in FD (with more smaller particles) (Figure 2). The addition of the food-grade dispersing agent in FD also resulted in a greater Zeta potential (−51.31 ± 1.22 mV) than the FC nanoformulation (−29.49 ± 1.02 mV), suggesting that the FD nanoformulation is more stable (Figure 2). Thus, the dispersing agent plays an important role in FD formulation's stability by preventing particle aggregation. On the other hand, the direct contact antiviral activities between the two nanoformulations are comparable without a statistical difference (Figure 7).

The results from MTT assays and the 3D Human Nasal Epithelium Model demonstrated that both the FC and FD nanoformulations showed levels of cytotoxicity and mucociliary toxicity comparable to the vehicle (saline) control (Figures 3–6). Normal saline (0.9% NaCl) is used widely in nasal formulations with a consistent safety record [46]. The current toxicity data for FC and FD confirmed that these mucoadhesive nasal nanoformulations can be expected to be tolerable by human nasal epithelium without acute toxicity, pending clinical studies.

As shown in Figure 7, the saline-based vehicles had no antiviral activity with regard to OC43 virus, while both the FC and FD nanoformulations possessed potent antiviral activity in a concentration range from 0.005 (70 μM) to 0.02% (280 μM). This effective concentration range is significantly lower than for the F18D nanoformulation (0.1% or 1.25 mm) [12]. The F18D nanoformulation is extremely active against OC43 with a 1 min exposure time–kill rate of 99.9999% [12]. However, a 3D Human Nasal Epithelium Model test indicated that the F18D nanoformulation is toxic to the nasal epithelium. It is interesting to observe that all of the EC16 or EC16m nanoformulations tested so far demonstrated a rapid action against human coronavirus [11,12]. This rapid antiviral action was also seen for FC and FD, where 5 min and 15 min incubations with OC43 virus did not have significant difference in efficacy (Figure 7), suggesting that the damage to the viral structure occurred immediately upon contact with the formulations, as we reported previously [12].

Whilst it would have been ideal to use SARS-CoV-2 for these studies, that would have required a BSL-3 facility, and since our goal was to test the formulations by exclusion (i.e., whether they had been cytotoxic or shown low or no antiviral activity), the use of a lower-risk model beta-coronavirus was appropriate [47]. As the formulations have now been shown to be suitable, future studies using the SARS-CoV-2 virus are warranted.

Unlike the results from the contact time–kill assays, where both FC and FD exhibited comparable potent antiviral activity, a surprising result from the post-infection inhibition assay was observed, which indicated that unlike FD, FC has little effectiveness against OC43 viral replication after the virus has already entered the cells (Figure 8). One explanation could be the difference in particle size distribution between the two nanoformulations. Only 6.2% of FC nanoparticles are in 100 nm range, while 93.8% of particles are in the 200 to 300 nm range (Figure 2A). In contrast, 78.3% particles in FD are in the 200 to 330 nm range, and the rest (21.7%) are in the 45 to 124 nm range (Figure 2B). It could be postulated that smaller particles enter the cells with higher efficiency than larger particles, and therefore the FD nanoformulation shows a significantly higher antiviral effect (>99%) (Figure 8). This particle-size-associated intracellular antiviral activity should be further explored, as it would be important for new drug designs. Another factor that cannot be ruled out is the presence of CMC, which increased the viscosity of the nanoformulation. That is, the large particles in the FC formulation may have a slower release from a thick formulation onto the cell membrane within 5 min.

According to the pathogenesis of long COVID, an ideal intranasal intervention targeting the cause of long COVID neurologic symptoms should contain an agent (drug) possessing a potent antiviral activity to clear the persistent viral presence; a strong anti-inflammatory property to reduce inflammation in the affected tissues; and powerful antioxidant activity. In addition, the formulation must be safe and stable. Based on the overall considerations, Formulation D (FD) is the best EC16m nanoformulation for this purpose due to the efficacy (>99% reduction in viral infectivity in direct contact or post-infection application in 5 min), mucociliary safety (similar to normal saline), and stability (Zeta potential at about −50 mV). There are several advantages to the patent-pending "facilitated self-assembled" EGCG-mono-palmitate nanoparticles (EC16m) over other reported nanoparticles containing EGCG. One of the advantages is the amphipathic chemical nature of EC16m, which not only significantly increased the antiviral activity against human pathogenic viruses, but also increased the bioavailability in reaching target tissues. Quantitatively, the particle density of FD is 6.5 billion particles/mL. In general, a full nasal spray volume is approximately 0.07 mL. Therefore, each spray of the FD nanoformulation delivers about 455 million EC16m nanoparticles. Since the persistent infection of the olfactory mucosa is associated with >0.5 million RNA copies [48] (not the number of viable viral particles), on a proportional basis, 0.07 mL/nostril (0.14) of the FD nanoformulation (910 million nanoparticles) would be sufficient to inactivate (by structural change and other mechanisms) the coronavirus in the olfactory mucosa within a short period of time. In addition, a small quantity of the EC16m nanoparticles will enter the CNS prior to cilia clearance and release free EGCG to perform anti-inflammatory, antioxidant, and neuroprotective activities.

To date, we have no information on the nature of the physicochemical interaction of the EC16m nanoparticles with the virus or the interaction of the nanoparticles with mucin and other mucosa secretions or how this interaction could affect mucociliary clearance rates. These important issues will be addressed in future studies.

In summary, the results of the current study demonstrated that the saline-based EC16m mucoadhesive nasal Formulation D was highly effective against human β-coronavirus OC43, a strain with high genome homology with SARS-CoV-2 [47], and in reducing viral replication after a single 5 min post-infection treatment, without mucociliary toxicity. With the known anti-inflammatory, antioxidant, and neuroprotective properties, intranasally delivered EC16m by Formulation D could not only terminate the "persistent infection" in the olfactory epithelium but also inhibit local inflammation and apoptosis, thereby restoring the olfactory function and reducing free radical levels and inflammation in the CNS.

5. Conclusions

In conclusion, the EC16m (drug grade) intranasal Nanoformulation D is suitable for a new intranasal drug to minimize long-COVID-associated anosmia and other neurologic symptoms, pending chronic mucociliary safety and human studies.

6. Patents

PCT/US23/74377 Pending: Compositions and methods of minimizing Long COVID. Inventor: Stephen Hsu (2023).

Author Contributions: Conceptualization, S.H. and D.D.; methodology, N.F., G.L. and X.J.; ZetaView validation, Y.L., H.Y., J.C. and D.D.; formulation data analysis, S.H., D.D., H.Y. and J.C.; investigation, N.F., D.D. and S.H.; resources, Y.L.; data curation, S.H.; EC16m purification and analysis, B.Y. and X.J.; writing—original draft preparation, S.H.; writing—review and editing, D.D.; visualization, N.F.; supervision, S.H. and D.D.; project administration, S.H.; funding acquisition, S.H. and D.D. All authors have read and agreed to the published version of the manuscript.

Funding: This work was funded by a grant from the National Institute on Deafness and Other Communication Disorders (NIDCD) (1R41DC020678-01). The content is solely the responsibility of the authors and does not necessarily represent the official views of the National Institutes of Health.

Institutional Review Board Statement: Not applicable.

Informed Consent Statement: Not applicable.

Data Availability Statement: The original contributions presented in the study are included in the article, further inquiries can be directed to the corresponding authors for the Protocol of Epithelix Sarl's 3D Human Nasal Epithelium Model evaluation of the nanoformulation samples and detailed original ZetaView results shown in Figure 2.

Acknowledgments: The authors want to thank Epithelix Sàrl for mucociliary toxicity evaluations and support from Augusta University Research Institute and Office of Innovation Commercialization.

Conflicts of Interest: B.Y. and X. J. are associated with the manufacturer that supplies EC16m to Camellix, LLC. The remaining authors declare that the research was conducted in the absence of any commercial or financial relationships that could be construed as a potential conflict of interest.

References

1. Hastie, C.E.; Lowe, D.J.; Mcauley, A.; Mills, N.L.; Winter, A.J.; Black, C.; Scott, J.T.; O'Donnell, C.A.; Blane, D.N.; Browne, S.; et al. True prevalence of long-COVID in a nationwide, population cohort study. *Nat. Commun.* **2023**, *14*, 7892. [CrossRef] [PubMed]
2. Leonel, J.W.; Ciurleo, G.C.V.; Formiga, A.M.; Vasconcelos, T.d.M.F.; de Andrade, M.H.; Feitosa, W.L.Q.; Sobreira-Neto, A.A.; Portugal, C.G.; Morais, L.M.; Marinho, S.C.; et al. Long COVID: Neurological manifestations—An updated narrative review. *Dement. Neuropsychol.* **2024**, *18*, e20230076. [CrossRef] [PubMed]
3. Stefanou, M.-I.; Palaiodimou, L.; Bakola, E.; Smyrnis, N.; Papadopoulou, M.; Paraskevas, G.P.; Rizos, E.; Boutati, E.; Grigoriadis, N.; Krogias, C.; et al. Neurological manifestations of long-COVID syndrome: A narrative review. *Ther. Adv. Chronic Dis.* **2022**, *13*, 20406223221076890. [CrossRef] [PubMed]
4. Groff, D.; Sun, A.; Ssentongo, A.E.; Ba, D.M.; Parsons, N.; Poudel, G.R.; Lekoubou, A.; Oh, J.S.; Ericson, J.E.; Ssentongo, P.; et al. Short-term and Long-term Rates of Postacute Sequelae of SARS-CoV-2 Infection: A Systematic Review. *JAMA Netw. Open* **2021**, *4*, e2128568. [CrossRef] [PubMed]
5. Leng, A.; Shah, M.; Ahmad, S.A.; Premraj, L.; Wildi, K.; Bassi, G.L.; Pardo, C.A.; Choi, A.; Cho, S.-M. Pathogenesis Underlying Neurological Manifestations of Long COVID Syndrome and Potential Therapeutics. *Cells* **2023**, *12*, 816. [CrossRef] [PubMed]
6. Abdelalim, A.A.; Mohamady, A.A.; Elsayed, R.A.; Elawady, M.A.; Ghallab, A.F. Corticosteroid nasal spray for recovery of smell sensation in COVID-19 patients: A randomized controlled trial. *Am. J. Otolaryngol.* **2021**, *42*, 102884. [CrossRef] [PubMed]
7. Gupta, S.; Lee, J.J.; Perrin, A.; Khan, A.; Smith, H.J.; Farrell, N.; Kallogjeri, D.; Piccirillo, J.F. Efficacy and Safety of Saline Nasal Irrigation Plus Theophylline for Treatment of COVID-19-Related Olfactory Dysfunction: The SCENT2 Phase 2 Randomized Clinical Trial. *JAMA Otolaryngol. Head. Neck Surg.* **2022**, *148*, 830–837. [CrossRef] [PubMed]
8. Cao, K.T.; Cobos-Uribe, C.; Knight, N.; Jonnalagadda, R.; Robinette, C.; Jaspers, I.; Rebuli, M.E. SARS-CoV-2 mRNA vaccination induces an intranasal mucosal response characterized by neutralizing antibodies. *J. Allergy Clin. Immunol. Glob.* **2023**, *2*, 100129. [CrossRef] [PubMed]

9. Bleier, B.S.; Ramanathan, M., Jr.; Lane, A.P. COVID-19 Vaccines May Not Prevent Nasal SARS-CoV-2 Infection and Asymptomatic Transmission. *Otolaryngol. Head. Neck Surg.* **2021**, *164*, 305–307. [CrossRef]
10. Najafloo, R.; Majidi, J.; Asghari, A.; Aleemardani, M.; Kamrava, S.K.; Simorgh, S.; Seifalian, A.; Bagher, Z.; Seifalian, A.M. Mechanism of Anosmia Caused by Symptoms of COVID-19 and Emerging Treatments. *ACS Chem. Neurosci.* **2021**, *12*, 3795–3805. [CrossRef]
11. Frank, N.; Dickinson, D.; Garcia, W.; Xiao, L.; Xayaraj, A.; Lee, L.H.; Chu, T.; Kumar, M.; Stone, S.; Liu, Y.; et al. Evaluation of Aqueous Nanoformulations of Epigallocatechin-3-Gallate-Palmitate (EC16) Against Human Coronavirus as a Potential Intervention Drug. *Biomed. J. Sci. Tech. Res.* **2023**, *50*, 2023.
12. Frank, N.; Dickinson, D.; Garcia, W.; Liu, Y.; Yu, H.; Cai, J.; Patel, S.; Yao, B.; Jiang, X.; Hsu, S. Feasibility Study of Developing a Saline-Based Antiviral Nanoformulation Containing Lipid-Soluble EGCG: A Potential Nasal Drug to Treat Long COVID. *Viruses* **2024**, *16*, 196. [CrossRef] [PubMed]
13. Dinda, B.; Dinda, S.; Dinda, M. Therapeutic potential of green tea catechin, (-)-epigallocatechin-3-O-gallate (EGCG) in SARS-CoV-2 infection: Major interactions with host/virus proteases. *Phytomed. Plus* **2023**, *3*, 100402. [CrossRef] [PubMed]
14. Hsu, S. Compounds Derived from Epigallocatechin-3-Gallate (EGCG) as a Novel Approach to the Prevention of Viral Infections. *Inflamm. Allergy Drug Targets* **2015**, *14*, 13–18. [CrossRef] [PubMed]
15. Hurst, B.L.; Dickinson, D.; Hsu, S. Epigallocatechin-3-Gallate (EGCG) Inhibits SARS-CoV-2 Infection in Primate Epithelial Cells: (A Short Communication). *Microbiol. Infect. Dis.* **2021**, *5*, 1116. [CrossRef] [PubMed]
16. Liu, B.; Kang, Z.; Yan, W. Synthesis, Stability, and Antidiabetic Activity Evaluation of (−)-Epigallo-catechin Gallate (EGCG) Palmitate Derived from Natural Tea Polyphenols. *Molecules* **2021**, *26*, 393. [CrossRef]
17. Chen, P.; Dickinson, D.; Hsu, S. Lipid-soluble Green Tea Polyphenols: Stabilized for Effective Formulation. In *Handbook of Green Tea and Health Research*; McKinley, H., Jamieson, M., Eds.; Nova Science Publishers, Inc.: New York, NY, USA, 2009; pp. 45–61.
18. Hsu, S.; Dickinson, D. Green tea and skin protection: Mechanism of action and practical applications. *Househ. Pers. Care Today* **2009**, *2*, 33–36.
19. Hsu, S.; Dickinson, D.; Borke, J.; Walsh, D.S.; Wood, J.; Qin, H.; Winger, J.; Pearl, H.; Schuster, G.; Bollag, W.B. Green tea polyphenol induces caspase 14 in epidermal keratinocytes via MAPK pathways and reduces psoriasiform lesions in the flaky skin mouse model. *Exp. Dermatol.* **2007**, *16*, 678–684. [CrossRef] [PubMed]
20. Gillespie, K.; Kodani, I.; Dickinson, D.P.; Ogbureke, K.U.E.; Camba, A.M.; Wu, M.; Looney, S.; Chu, T.-C.; Qin, H.; Bisch, F.; et al. Effects of oral consumption of the green tea polyphenol EGCG in a murine model for human Sjogren's syndrome, an autoimmune disease. *Life Sci.* **2008**, *83*, 581–588. [CrossRef]
21. Hsu, S.D.; Dickinson, D.P.; Qin, H.; Borke, J.; Ogbureke, K.U.E.; Winger, J.N.; Camba, A.M.; Bollag, W.B.; Stöppler, H.J.; Sharawy, M.M.; et al. Green tea polyphenols reduce autoimmune symptoms in a murine model for human Sjogren's syndrome and protect human salivary acinar cells from TNF-alpha-induced cytotoxicity. *Autoimmunity* **2007**, *40*, 138–147. [CrossRef]
22. Dickinson, D.; DeRossi, S.; Yu, H.; Thomas, C.; Kragor, C.; Paquin, B.; Hahn, E.; Ohno, S.; Yamamoto, T.; Hsu, S. Epigallocatechin-3-gallate modulates antioxidant defense enzyme expression in murine submandibular and pancreatic exocrine gland cells and human HSG cells. *Autoimmunity* **2014**, *47*, 177–184. [CrossRef] [PubMed]
23. Dickinson, D.; Yu, H.; Ohno, S.; Thomas, C.; DeRossi, S.; Ma, Y.-H.; Yates, N.; Hahn, E.; Bisch, F.; Yamamoto, T.; et al. Epigallocatechin-3-gallate prevents autoimmune-associated down- regulation of p21 in salivary gland cells through a p53-independent pathway. *Inflamm. Allergy-Drug Targets* **2014**, *13*, 15–24. [CrossRef] [PubMed]
24. de la Torre, R.; de Sola, S.; Farre, M.; Xicota, L.; Cuenca-Royo, A.; Rodriguez, J.; Leon, A.; Langohr, K.; Gomis-Gonzalez, M.; Hernandez, G.; et al. A phase 1, randomized double-blind, placebo controlled trial to evaluate safety and efficacy of epigallocatechin-3-gallate and cognitive training in adults with Fragile X syndrome. *Clin. Nutr.* **2020**, *39*, 378–387. [CrossRef] [PubMed]
25. Singh, N.A.; Mandal, A.K.; Khan, Z.A. Potential neuroprotective properties of epigallocatechin-3-gallate (EGCG). *Nutr. J.* **2016**, *15*, 60. [CrossRef] [PubMed]
26. Cano, A.; Ettcheto, M.; Espina, M.; Auladell, C.; Calpena, A.C.; Folch, J.; Barenys, M.; Sánchez-López, E.; Camins, A.; García, M.L. Epigallocatechin-3-gallate loaded PEGylated-PLGA nanoparticles: A new anti-seizure strategy for temporal lobe epilepsy. *Nanomed. Nanotechnol. Biol. Med.* **2018**, *14*, 1073–1085. [CrossRef] [PubMed]
27. Cai, Z.Y.; Li, X.M.; Liang, J.P.; Xiang, L.P.; Wang, K.R.; Shi, Y.L.; Yang, R.; Shi, M.; Ye, J.H.; Lu, J.L.; et al. Bioavailability of Tea Catechins and Its Improvement. *Molecules* **2018**, *23*, 2346. [CrossRef] [PubMed]
28. Yang, C.S.; Chen, L.; Lee, M.J.; Balentine, D.; Kuo, M.C.; Schantz, S.P. Blood and urine levels of tea catechins after ingestion of different amounts of green tea by human volunteers. *Cancer Epidemiol. Biomark. Prev.* **1998**, *7*, 351–354.
29. Zhong, J.; Dickinson, D.; Sampath, L.; Hsu, S. Effects of Epigallocatechin-3-Gallate-Palmitate (EC16) on In Vitro Norovirus Infection. *Microbiol. Infect. Dis.* **2021**, *5*, 1–7. [CrossRef]
30. Wei, Y.; Chen, P.; Ling, T.; Wang, Y.; Dong, R.; Zhang, C.; Zhang, L.; Han, M.; Wang, D.; Wan, X.; et al. Certain (-)-epigallocatechin-3-gallate (EGCG) auto-oxidation products (EAOPs) retain the cytotoxic activities of EGCG. *Food Chem.* **2016**, *204*, 218–226. [CrossRef]
31. de Oliveira, A.; Adams, S.D.; Lee, L.H.; Murray, S.R.; Hsu, S.D.; Hammond, J.R.; Dickinson, D.; Chen, P.; Chu, T.C. Inhibition of herpes simplex virus type 1 with the modified green tea polyphenol palmitoyl-epigallocatechin gallate. *Food Chem. Toxicol.* **2013**, *52*, 207–215. [CrossRef]

32. Mori, S.; Miyake, S.; Kobe, T.; Nakaya, T.; Fuller, S.D.; Kato, N.; Kaihatsu, K. Enhanced anti-influenza A virus activity of (-)-epigallocatechin-3-O-gallate fatty acid monoester derivatives: Effect of alkyl chain length. *Bioorg Med. Chem. Lett.* **2008**, *18*, 4249–4252. [CrossRef] [PubMed]
33. Barhoum, A.; García-Betancourt, M.L.; Jeevanandam, J.; Hussien, E.A.; Mekkawy, S.A.; Mostafa, M.; Omran, M.M.; Abdalla, M.S.; Bechelany, M. Review on Natural, Incidental, Bioinspired, and Engineered Nanomaterials: History, Definitions, Classifications, Synthesis, Properties, Market, Toxicities, Risks, and Regulations. *Nanomaterials* **2022**, *12*, 177. [CrossRef]
34. Farabegoli, F.; Granja, A.; Magalhães, J.; Purgato, S.; Voltattorni, M.; Pinheiro, M. Epigallocatechin-3-gallate Delivered in Nanoparticles Increases Cytotoxicity in Three Breast Carcinoma Cell Lines. *ACS Omega* **2022**, *7*, 41872–41881. [CrossRef]
35. Farabegoli, F.; Pinheiro, M. Epigallocatechin-3-Gallate Delivery in Lipid-Based Nanoparticles: Potentiality and Perspectives for Future Applications in Cancer Chemoprevention and Therapy. *Front. Pharmacol.* **2022**, *13*, 809706. [CrossRef] [PubMed]
36. Chen, B.-H.; Hsieh, C.-H.; Tsai, S.-Y.; Wang, C.-Y.; Wang, C.-C. Anticancer effects of epigallocatechin-3-gallate nanoemulsion on lung cancer cells through the activation of AMP-activated protein kinase signaling pathway. *Sci. Rep.* **2020**, *10*, 5163. [CrossRef] [PubMed]
37. He, A.; Guan, X.; Song, H.; Li, S.; Huang, K. Encapsulation of (−)-epigallocatechin-gallate (EGCG) in hordein nanoparticles. *Food Biosci.* **2020**, *37*, 100727. [CrossRef]
38. Song, H.; He, A.; Guan, X.; Chen, Z.; Bao, Y.; Huang, K. Fabrication of chitosan-coated epigallocatechin-3-gallate (EGCG)-hordein nanoparticles and their transcellular permeability in Caco-2/HT29 cocultures. *Int. J. Biol. Macromol.* **2022**, *196*, 144–150. [CrossRef]
39. Jiang, Y.; Jiang, Z.; Ma, L.; Huang, Q. Advances in Nanodelivery of Green Tea Catechins to Enhance the Anticancer Activity. *Molecules* **2021**, *26*, 3301. [CrossRef] [PubMed]
40. Krzyzowska, M.; Janicka, M.; Chodkowski, M.; Patrycy, M.; Obuch-Woszczatyńska, O.; Tomaszewska, E.; Ranoszek-Soliwoda, K.; Celichowski, G.; Grobelny, J. Epigallocatechin Gallate-Modified Silver Nanoparticles Show Antiviral Activity against Herpes Simplex Type 1 and 2. *Viruses* **2023**, *15*, 2024. [CrossRef]
41. Meesaragandla, B.; Hayet, S.; Fine, T.; Janke, U.; Chai, L.; Delcea, M. Inhibitory Effect of Epigallocatechin Gallate-Silver Nanoparticles and Their Lysozyme Bioconjugates on Biofilm Formation and Cytotoxicity. *ACS Appl. Bio Mater.* **2022**, *5*, 4213–4221. [CrossRef]
42. Sakakura, Y.; Ukai, K.; Majima, Y.; Murai, S.; Harada, T.; Miyoshi, Y. Nasal mucociliary clearance under various conditions. *Acta Otolaryngol.* **1983**, *96*, 167–173. [CrossRef] [PubMed]
43. Helwa, I.; Cai, J.; Drewry, M.D.; Zimmerman, A.; Dinkins, M.B.; Khaled, M.L.; Seremwe, M.; Dismuke, W.M.; Bieberich, E.; Stamer, W.D.; et al. A Comparative Study of Serum Exosome Isolation Using Differential Ultracentrifugation and Three Commercial Reagents. *PLoS ONE* **2017**, *12*, e0170628. [CrossRef] [PubMed]
44. Muench, H.R. A simple method of estimating 50 per cent end points. *Am. J. Hyg.* **1938**, *27*, 493–497.
45. US FDA GRAS NOTICE 772, U.G.N. GRAS Notice for Oil-Soluble Green Tea Extract (Green Tea Catechin Palmitate). 2018. Available online: https://www.fda.gov/media/126906/download (accessed on 28 April 2024).
46. Rabago, D.; Zgierska, A. Saline nasal irrigation for upper respiratory conditions. *Am. Fam. Physician* **2009**, *80*, 1117–1119. [PubMed]
47. Kim, M.I.; Lee, C. Human Coronavirus OC43 as a Low-Risk Model to Study COVID-19. *Viruses* **2023**, *15*, 578. [CrossRef]
48. de Melo, G.D.; Lazarini, F.; Levallois, S.; Hautefort, C.; Michel, V.; Larrous, F.; Verillaud, B.; Aparicio, C.; Wagner, S.; Gheusi, G.; et al. COVID-19-related anosmia is associated with viral persistence and inflammation in human olfactory epithelium and brain infection in hamsters. *Sci. Transl. Med.* **2021**, *13*, abf8396. [CrossRef]

pharmaceutics

MDPI

Article

Peptide Functionalization of Emulsion-Based Nanocarrier to Improve Uptake across Blood–Brain Barrier

Alberta De Capua [1,2], Raffaele Vecchione [1,*], Cinzia Sgambato [1,2], Marco Chino [3], Elena Lagreca [1], Angela Lombardi [3] and Paolo Antonio Netti [1,2,4]

[1] Center for Advanced Biomaterials for Health Care (CABHC), Istituto Italiano di Tecnologia, Largo Barsanti e Matteucci 53, 80125 Napoli, Italy; elena.lagreca@iit.it (E.L.); paolo.netti@iit.it (P.A.N.)
[2] Department of Chemical Materials and Industrial Production (DICMaPI), University of Naples Federico II, P.le Tecchio 80, 80125 Naples, Italy
[3] Department of Chemical Sciences, University of Naples Federico II, Complesso Universitario Monte S. Angelo, Via Cintia 21, 80126 Naples, Italy; marco.chino@unina.it (M.C.)
[4] Interdisciplinary Research Centre on Biomaterials (CRIB), University of Naples Federico II, P.le Tecchio 80, 80125 Naples, Italy
* Correspondence: raffaele.vecchione@iit.it

Abstract: New strategies for enhancing drug delivery to the blood–brain barrier (BBB) represent a major challenge in treating cerebral diseases. Nanoemulsion-based nanocarriers represent an ideal candidate to improve drug delivery thanks to their versatility in functionalization and cargo protection. In this work, a paclitaxel-loaded nano-emulsion has been firstly functionalized and stabilized with two layers constituted of chitosan and hyaluronic acid, and, secondly, the latter has been conjugated to the CRT peptide. CRT is a bioactive peptide that selectively recognizes bEnd.3 cells, a model of the BBB, thanks to its interactions with transferrin (Tf) and its receptor (TfR). Cytotoxic results showed a 41.5% higher uptake of CRT functionalized nano-emulsion than the negative control, demonstrating the ability of this novel tool to be accumulated in brain endothelium tissue. Based upon these results, our approach can be fully generalizable to the design of multifunctional nanocarriers for delivery of therapeutic agents to the central nervous systems.

Keywords: blood–brain barrier; peptide targeting; nanoemulsions; drug delivery; transferrin receptor; glioblastoma

1. Introduction

Effective cancer therapy for the treatment of brain tumors and central nervous system (CNS) diseases remains one of the most challenging areas in drug delivery research. One of the major issues is represented by the inability to cross the physical obstacle of the blood–brain barrier (BBB) [1,2]. Identifying routes for non-invasive drug delivery to the brain and developing targeting strategies to transport biologics into the brain represent a research area of growing importance [3]. It is known from the literature that enhancing lipophilicity and positive charge is a possible strategy to increase passive diffusion, like glucose, water, amino acids and small lipophilic molecules that are crucial to neural function [4]. This is, for instance, the penetration mechanism of cell-penetrating peptides possessing multiple positive charges. However, these modifications generally lead to higher unspecific uptake in many tissues, often resulting in off-target effects since they are not selective. A promising strategy for overcoming the BBB to deliver biologics is targeting endogenous receptor-mediated transport (RMT) systems that engage vesicular trafficking to transport ligands across the BBB endothelium. Drug delivery systems (DDS), modified with appropriate targeting ligands, could improve access to the brain via RMT and release their cargo [5].

The transferrin receptor (TfR) is one of the first RMT systems studied for BBB drug delivery applications [6]. TfR is ubiquitously overexpressed on brain capillary endothelial cells because it mediates iron delivery to the brain via binding and intracellular trafficking of the iron-binding protein transferrin (Tf). Most importantly, compared to healthy brain cells, TfR has much higher expression levels in human glioblastoma because it is required for cancer cell proliferation [7,8]. The use of Tf as a targeting ligand has been demonstrated [9]. Unfortunately, there is competitive binding to TfR between the endogenous Tf and the Tf-modified DDS in vivo, thus inducing insufficient delivery to the tumor site [10]. One approach to overcome this issue is the use of targeting moieties whose TfR recognition is mediated by a different molecular pathway. A recently published disulfide-bridged cyclic peptide, CRTIGPSVC (CRT), was discovered by selection of a phage display peptide library in vivo [11]. CRT functionally mimics iron by binding to *apo*-Tf and causes the adoption of the iron-bound *holo*-Tf conformation, thereby gaining access to the brain through the Tf-TfR interaction. A CRT-labeled DDS would then be able to overcome the limitations of Tf-modified DDS by directly activating endogenous Tf towards TfR binding and internalization. This peptide exhibited promising results for the treatment of brain tumors by delivering the herpes simplex virus thymidine kinase gene to a mouse model of human glioma [11,12]. The delivery was accomplished via intravenous administration of a CRT-targeted adeno-associated virus and phage hybrid vector and resulted in significant tumor shrinkage. Another example of the therapeutic potential of CRT for glioblastoma involved the treatment with paclitaxel-loaded CRT–nanoparticles (NPs) in diseased mice, resulting in a remarkably prolonged median survival was observed [13]. Chitosan-based NPs have also been covalently modified with the widely used T7 peptide, and with the CRT-NPs have 1.9-fold higher availability by avoiding the competitive inhibition of endogenous Tf [14]. More recently, CRT functionalized core-shell nanoparticles, featuring a midazolam coating, enabled transferrin receptor (TfR)-mediated brain-targeting in mice [15].

However, although biodegradable, solid nanoparticles are classified as nanomaterials and there are some concerns regarding their use with heavy regulatory paths. Here, we propose an alternative liquid-based nanocarrier, which is an oil in water nanoemulsion (O/W NE). An O/W NE is a versatile tool in the drug delivery field, which can be easily functionalized with several targeting moieties while protecting drugs solubilized in their lipophilic phase, as in the case of solid NPs [16]. In this work, we prepare O/W NEs that are stabilized by two polymeric layers of chitosan and biotinylated hyaluronic acid, the latter being functionalized with the CRT bioactive peptide to promote the NEs' accumulation on the BBB. An easy additive decoration strategy that exploits biotin−streptavidin physical interaction [12] is adopted to conjugate the peptide outside the system. CRT is linked to a biotinylated poly(ethylene glycol) (PEG) chain in order to inhibit the NEs' clearance by reticuloendothelial system (RES) and expose the peptide on the external side. To verify its specificity toward cells over-expressing TfR receptor, biological tests of peptide functionalized O/W NEs have been carried out. We chose a mouse brain cell line (bEnd.3) as our model of the blood–brain barrier, and paclitaxel (PTX), a well-known cytotoxic drug, to assess peptide-mediated accumulation on bEnd.3 cells by the induced cytotoxicity. Even though PTX is an anticancer drug, its purpose in this context is mainly to assess the ability of NEs to accumulate a harmful substance toward healthy BBB cells thanks to the CRT peptide. Our results suggest that the designed vector is ready for targeted pharmacophore delivery and that further integration on the surface of a cell-penetrating peptide [17] will lead to safe crossing of the BBB by the carrier to reach the tumor site.

2. Materials and Methods

2.1. Materials

Surfactant Lipoid E80 (egg lecithin powder 80–85% enriched with phosphatidylcholine and 7–9.5% content in phosphatidylethanolamine) was purchased from Lipoid GmbH and used without further purification. Soybean oil (dietary source of long-chain triglycerides and other lipids, $n_{20/D}$ 1.4743), ethanol (99%) and acetic acid (99.7%) were purchased from Merck. CO_2 ($\geq$99.998%) was purchased from Nippon gases (Italy). For preparation of all nanoemulsions and solutions, Millipore Milli-Q water was used. Chitosan

(CT, LMW 90–150 kDa, degree of deacetylation (DDA) 84% determined via 1H-NMR), 1-hydroxybenzotriazole hydrate (HOBt, ≥95%), N,N′-Diisopropylcarbodiimide (DIC, 99%), N,N-Diisopropylethylamine (DIEA, 99.5%), trifluoroacetic acid (TFA, ≥99.5%, Romil), dimethyl sulfoxide (DMSO, ≥99.9%), dichloromethane (DCM, ≥99.5%), anhydrous N,N-dimethyl-formamide (DMF, 99.8%), 1,2-Ethanedithiol (EDT, 99%), Triisopropylsilane (TIS, ≥98%), piperidine (≥99%), acetone (reagent grade) and diethyl ether (≥99%) were purchased from Sigma Aldrich and used without further purification. Hyaluronic acid 250 kDa and Biotin-PEG-COOH 2kDa were purchased from Creative PEGWorks. N-α-Fmoc amino acids and coupling reagents were provided by NovaBiochem (>98%). Paclitaxel (99% purity) was purchased from Discovery Fine Chemicals Ltd, Wimborne, UK.

2.2. Peptide Synthesis and Purification

CRT (βA-CRTIGPSVC-βA-K) peptides were synthesized using the standard solid-phase-9-fluorenyl methoxy carbonyl (Fmoc) procedure and were obtained with good overall yields (50–60%). The syntheses were performed using a Biotage®Syro WaveTM peptide synthesizer (Biotage, Uppsala, Sweden). The peptide scale synthesis was 0.1 mmol. It was assembled on Rink amide resin with a substitution level of 0.71 mmol/g. The following protected amino acids were used to synthesize the peptide:

Fmoc-Lys(Boc)-OH; Fmoc-Ile-OH; Fmoc-Gly-OH; Fmoc-Ser(tBu)-OH; Fmoc-Arg(Pbf)-OH; Fmoc-Pro-OH; Fmoc-Cys(Trt)-OH; Fmoc-Ala-OH; Fmoc-Thr(tBu)-OH; Fmoc-Val-OH

The synthetic procedure can be summarized as follow:

1. Deprotection: Fmoc group was removed at the beginning of cycle with a 20% piperidine solution in DMF. After deprotection, the resin was washed with DMF to remove the residual piperidine. The peptide resin was then ready for coupling.
2. Activation: The carboxyl group of each Fmoc-amino acid was activated by addition of HBTU (2 eq.)/Oxyma Pure (2 eq.)/DIEA (4 eq.).
3. Coupling: The pre-activated Fmoc-amino acid reacted with the free amino-terminal group of the growing peptide chain on the resin using DMF as the reaction solvent
4. Capping: This reaction was performed after each coupling step, using a solution of Ac_2O 20% and DIEA 5% in DMF. Capping cycle was introduced to prevent deletion byproducts.

Deprotection, coupling and capping steps were repeated for each subsequent amino acid, until the chain assembly was completed. When the coupling was complete, the resin was washed with DMF. At completion of the synthesis, the resin was washed several times with DMF and finally dried. The peptide was cleaved from the resin by treating it with 94% TFA/2.5% EDT/2.5% water/1% TIS for 2 h at room temperature. The mixture was then concentrated and transferred to glass centrifugal tubes for compound precipitation using ice-cold diethyl ether, which was performed repeatedly. Purified CRT peptide was obtained by preparative RP-HPLC with a Vydac C18 column (Grace, Columbia, MD, USA, 22 mm × 250 cm; 10 μm), eluted with a linear gradient (solvent A, H_2O 0.1% TFA; solvent B, Acetonitrile, ACN 0.1% TFA) from 20 to 70% B over 58 min at a flow rate of 23 mL·min^{-1}. All analyses were performed at detection wavelength of 220 nm and reported after blank chromatogram subtraction.

2.3. Peptide Cyclization

CRT peptide was dissolved in an aqueous solution at a concentration of 0.1 mM. Then, DMSO was added dropwise until its final concentration was 5% (10 mL final volume). The reaction mixture was kept open to atmosphere under vigorous magnetic stirring overnight. The product was monitored by LC-MS analysis. When the reaction was completed, the water was evaporated; the peptide was precipitated in cold ethyl acetate and lyophilized.

2.4. Biotin-PEG-COOH Peptide Conjugation

The peptide was conjugated at the N-term with Biotin-PEG-COOH directly on the resin. Firstly, the Fmoc protecting group was removed with a 20% piperidine solution in

DMF, followed by several washing steps. Then, the coupling reaction with Biotin-PEG-COOH (2 equivalents) was conducted directly on the resin (10 mg) with DIC/HOBt/DIEA (1:1:2) 0.1 M, using DMF as solvent, overnight under nitrogen flow. At completion of the synthesis, the resin was washed several times with DMF, NMP, DCM, isopropanol and methanol, and finally dried. Biotin-PEG-peptide was cleaved from the resin by treating with 94% TFA/2.5% EDT/2.5% water/1% TIS for 2 h, precipitated in ice-cold diethyl ether and lyophilized.

2.5. Peptides Analysis and Purification

The identity of crude peptides was analyzed by analytical RP-HPLC–ESI-MS. The LC-MS was performed with a Shimadzu LC-10ADvp equipped with an SPDM10Avp diode-array detector. ESI-MS spectra were recorded on a Shimadzu LC-MS-2010EV system with ESI interface and Shimadzu LC-MS solution Workstation software (ver. 3.41) was used for the data processing. A Q-array-Octapole-Quadrupole mass analyzer was used as the detector. Argon was used as ion gas in the CID cell and data were analyzed by Shimadzu LC-MS solution Workstation software. The optimized MS parameters were selected as followed: curved desolvation line (CDL) temperature 200 °C; block temperature 200 °C; probe temperature 200 °C; detector gain 1.6 kV; probe voltage +4.5 kV; CDL voltage −15 V. Nitrogen served as nebulizer gas (flow rate: 1.5 L·min^{-1}). All analyses were performed with a Vydac C18 column (4.6 mm × 150 mm; 5 µm), eluted with a linear elution gradient from 1% to 70% B over 35 min at a flow rate 1 mL·min^{-1}. The running eluents were: solvent A, H_2O 0.1% TFA and solvent B, ACN 0.1% TFA.

The crude non-cyclic peptide was further purified by preparative RP-HPLC with a Vydac C18 column (22 mm × 250 cm; 10 µm), eluted with a linear gradient (solvent A, H_2O 0.1% TFA; solvent B, ACN 0.1% TFA) from 20 to 80% B over 58 min at flow rate of 23 mL·min^{-1}. All analyses were performed at detection wavelength of 220 nm. The pooled fractions, containing the desired products, were lyophilized. The peptides homogeneity was assessed by analytical HPLC and by ESI mass spectrometry. The crude Biotin-PEG-peptide was purified by preparative flash chromatography, using a Biotage ISOLERA flash purification system, ISO-1SW model, equipped with a diode-array detector. The product was eluted with a linear gradient (solvent A, H_2O 0.1% TFA; solvent B, ACN 0.1% TFA) from 0% to 95% B over 20 column volumes, using SNAP C18 12 g as column. The pooled fractions, containing the desired products, were analyzed by analytical RP-HPLC–ESI-MS.

2.6. MALDI-TOF Analysis of PEGylated Peptides

PEGylated peptide was characterized by matrix-assisted laser desorption/ionization mass spectrometry coupled to two times of flight analyzers (MALDI-TOF-TOF). The sample was prepared with a final concentration of ~2 pmol/µL in the matrix by mixing the peptide with a solution 60% of α-cyano-4-hydroxycinnamic acid (CHCA) and 40% of 5-Dihydroxybenzoic acid (DHB).

The two matrix solutions were prepared as follows:

1. 20 mg/mL of CHCA in a solution of H_2O 5% formic acid in ACN (30/70 *v*/*v*);
2. 20 mg/mL of DHB in a solution of H_2O 0.1% TFA in ACN (30/70 *v*/*v*).

Approximately 0.25 µL of the sample was deposited on the MALDI plate after a layer deposition of a saturated solution of CHCA in acetone and allowed to dry prior to analysis. The mass spectra were recorded on an AB SCIEX TOF/TOF 5800 instrument operated in the reflector positive mode. MALDI-TOF MS analyses were conducted at a laser intensity of 4287 units and laser pulse rate of 400 Hz with a set mass range of 1000 to 6000 Da. A continuous stage motion set in a random pattern at 600 µm/s was used for sampling. Calibration was performed using Cal mix 5 from AB SCIEX as calibrants, which contained des-Arg^1-Bradykinin, Angiotensin I, Glu^1-Fibrinopeptide B, adrenocorticotropic hormone ACTH (1–17 clip), ACTH (18–39 clip) and ACTH (7–38 clip), resulting in a mass accuracy of 50 ppm. Each spectrum represents the sum of 2040 laser pulses from randomly chosen spots per sample position. Raw data were analyzed using TOF/TOF Series Explorer software (ver. 4.1) provided by the manufacturer and are reported as monoisotopic masses.

2.7. Paclitaxel-Loaded Oil in Water Nanoemulsion

Firstly, a 20 wt% oil in water pre-emulsion was prepared. A mass of 5.8 g of lecithin Lipoid E 80 (egg lecithin powder 80–85% enriched with phosphatidyl choline and 7–9.5% content in phosphatidyl ethanolamine) was dissolved in 24 mL of soybean oil (density at 20 °C of 0.922 g·mL^{-1}) at 60 °C using the immersion sonicator (Ultrasonic Processor VCX500 Sonic and Materials, Hielscher Ultrasonics GmbH Teltow, Germany), performing runs of 10 s for 1 min at 10% of sonication amplitude (microtip screwed). Then, 1 mL of ethanol solution of PTX (5 mg/mL) was added to the oil phase and kept for 1 h at 70 °C to evaporate the ethanol. Subsequently, the oil phase was added to the aqueous phase (Milli-Q water), and mixed using the immersion sonicator with runs of 10 s for 8 min at 70% of amplitude (a pulse-on and a pulse-off respectively of 10 s). The pre-emulsion was finally homogenized for 3 single cycles and 200 steps at a pressure of 2000 bar by a high-pressure homogenizer (110P series microfluidizer) to obtain the final nanoemulsion.

2.8. Polymers Multilayer Deposition above Paclitaxel-Loaded O/W NEs

Firstly, a layer of chitosan was deposited around the oil template with a final concentration of oil and chitosan of 10 wt% and 0.1 wt%, respectively. A 0.1 M acetic acid solution of chitosan (0.125 wt%) was prepared with a final pH = 4. Nanoemulsion 20 wt% oil was added quickly to the chitosan solution under vigorous stirring and kept under stirring for 15 min to allow uniform chitosan deposition. The nanoemulsion with the first positive layer of chitosan was passed through a high-pressure valve homogenizer at 700 bars for 100 continuous steps. The next hyaluronic acid layer was prepared by aid of two syringe pumps (HARVARD APPARATUS 11 PLUS) and an ultrasonic bath (FALC INSTRUMENTS). Starting from the secondary nanoemulsion 10 wt% oil–0.1 wt% CT, a negative charged polymer layer was deposited by mixing 1:1 (*v*:*v*) of a 0.24 wt% aqueous solution of biotinylated hyaluronic acid with the secondary nanoemulsion suspension. The two liquid phases were injected at the same flow rate (0.4 mL min^{-1}) through two Polymicro flexible fused silica micrometric capillaries (inner diameter of 200 μm) interfaced at their extremities (Molex). Each drop was then collected inside a glass tube immersed in the ultrasonic bath at room temperature, 59 kHz and 100% power for 15 min. The NCs were characterized at each step of preparation by dynamic light scattering (DLS) analysis.

2.9. Nanocarrier Assembly

The streptavidin solution was prepared by dissolving 1 mg in 1 mL of Milli-Q water (16.6 μM). It was added to the HA-Biotin 0.12 wt%-CT 0.05 wt%-NEs 5 wt% oil, under sonication for 15 min and T = 20 °C, at a final concentration of 5.69 μM. In the same way the compound CRT-PEG-Biotin was added under sonication for 15 min and T = 20 °C to the streptavidin-HA-Biotin-CT-NEs (SAV-HA-Biotin-CT-NEs) at a molar ratio 2:1 between CRT-PEG-biotin and the streptavidin. The final concentrations were 3.2 μM and 6.4 μM for streptavidin and CRT, respectively, while the final oil weight percentage was 2.78 wt%. The NCs were characterized at each step of preparation measuring the size and Z-potential by DLS, as described next.

2.10. Particle Size and Z-Potential Measurements

All nanoemulsions and their successive functionalization were characterized at each step of preparation by measuring size and polydispersity index (PdI), using a Zetasizer Nano ZS device (Malvern Instruments) with a 4 mW He-Ne ion laser at the wavelength of 633 nm and a photodiode detector at an angle of 173°. All the samples were diluted to a droplet concentration of 0.025 wt% using 20 mM acetic acid at pH 4 for monolayer, and Milli-Q water for emulsions and bilayer suspensions. The calculation of the particle size distribution was performed using a default refractive index ratio (1.59) and 5 runs for each measurement (1 run lasting 100 s), at least 3 times for each sample. A particle electrophoresis instrument (Zetasizer zs nano series ZEN 3600, Malvern Instruments Ltd., Malvern, UK) was used for the Z-potential determinations. Samples were diluted as for the particle size analysis. Setting 50 runs for each measurement carried out the Z-potential analysis. Samples were collected into polystyrene cuvettes and measured three times, and the results presented are the averages of these measurements. Experiments were carried out at 25 °C. Zetasizer software 7.11 (Malvern Instruments) was used to obtain the data. Cumulate analysis was used to give the Z-average value, hydrodynamic diameter, polydispersity index and the intensity size distribution graphs.

2.11. Cryo-TEM Characterization

For the preparation of the frozen-hydrated sample the plunge freezing method was performed. Briefly, a drop of 3 µL of the samples were put on a previously glow-discharged 200 mesh holey carbon grids (Ted Pella, Redding, CA, USA) after that the grid was inserted in the chamber of a FEI Vitrobot Mark IV (FEI company, Eindhoven, The Netherlands) at 4 °C and 90% of humidity. The droplet of sample was blotted with filter paper for 1 s, (blot force 1, drain time 0.5 s) and then the grid was plunged into the liquid propane. The grid was then stored in liquid nitrogen in a grid box until it was finally transferred to a cryo-specimen 626 holder (Gatan, Inc., Pleasanton, CA, USA) and loaded into the cryo-transmission electron microscope for imaging. To obtain the image of the nanoparticles we used a Tecnai G2 20, a cryo-tomo transmission electron microscope (FEI company, The Netherland) equipped with LaB6 emitter (acceleration voltage of 200 kV) and recorded at with a 2 k × 2 k CCD-Eagle 2HS camera. The frozen-hydrated sample is radiation-sensitive material, so to avoid damaging; the observation was carried out in Low Dose Mode.

2.12. Cell Culture

bEnd.3 cells were grown in DMEM (10% FBS, 1% L-Glu, 1% streptomycin–penicillin). Cell cultures were always performed at 37 °C in 5% CO_2 and 100% relative humidity (RH). Cells were used from passages 23 to 30.

2.13. Cytotoxicity Analysis

Cell viability was quantified by the PrestoBlue Assay (Invitrogen, Waltham, MA, USA) and compared to non-treated cells, which were used as a control. Briefly, 1×10^4 bEnd.3 cells were seeded in a 96-well and incubated several times (30 min, 2 h and 4 h) with PTX-loaded CRT-PEG-NEs, PEG-NEs and free PTX, diluted 1:5 in cells, at a final PTX concentration of 1.4 µM. PrestoBlue Assay was performed according to the manufacturer's procedure, after 24 h. Fluorescence of PrestoBlue reagent solution (excitation 535 nm) was read at 615 nm using a spectrofluorometer (Wallac 1420 Victor2, Perkin–Elmer, Waltham, MA, USA). All experiments were performed in triplicate. We conducted one-way ANOVA tests using MATLAB® (MathWorks, Natick, MA, USA) to quantitatively evaluate cell viability at 30 min, 2 h, 4 h and 24 h.

2.14. Uptake of PTX-Loaded NCs

bEnd.3 cells were cultured in DMEM supplemented with 10% FBS, 1% L-Glutamine and 1% streptomycin–penicillin. After seeding 1×10^4 cells, they were incubated overnight to allow for attachment. The cells were then treated with PTX-loaded CRT-PEG-NEs and PEG-NEs, incorporating rhodaminated streptavidin during the nanoparticle assembly for 4 h at 37 °C in cell-specific medium. Post-treatment, the cells were washed twice with PBS and fixed with 4% paraformaldehyde (PAF) for 20 min. To label the nuclei and cell membranes, DRAQ5 (excitation at 633 nm) and WGA 555 were used, respectively. Fluorescence intensity was analyzed using a Zeiss LSM 700 confocal microscope equipped with a 20×/0.8NA objective lens.

Image reconstruction was performed with ImageJ. For each sample, at least 5 confocal digital images were randomly collected and analyzed. Images were acquired with a resolution of 512 × 512 pixels (949 × 949 μm). The experiment was performed in triplicates. Significant differences of data obtained from image analysis were assessed by ANOVA using MATLAB® (MathWorks, Natick, MA, USA).

3. Results and Discussion

3.1. Synthesis, Purification and Characterization of CRT

One of the most promising peptides that can recognize the TfR receptor is the cyclic 9-mer peptide CRT, with the sequence CRTIGPSVC. The amino acid sequence is slightly modified here by introducing a non-α-amino acid (β-alanine) at the N-term and C-term, and a lysine at the C-term as the last amino acid (βA-CRTIGPSVC-βA-K). β-alanine acts as a spacer to reduce the influence of the nanocapsule surface on the peptide conformation, while the Lys was inserted to exploit its side chain (-NH_2) for further functionalization (i.e., fluorophore labelling). The peptide synthesis was performed using solid-phase protocols (SPPS), utilizing 9-fluorenylmethoxycarbonyl (Fmoc) chemistry and a super acid labile resin. The resulting peptide was deprotected and cleaved from the resin. The crude peptide purity was assessed by analytical RP-LC-MS, with the chromatogram and mass spectra reported in Figure 1A. The purified CRT peptide was cyclized by dissolving it in an aqueous solution containing 5% DMSO to the recommended dilute concentration of the thiol moieties (0.01–0.1 mM), thus favoring the intra-chain disulfide bond formation between the two cysteine residues over dimer formation, which is a frequent side reaction during cyclization [18,19]. Unambiguous indication of peptide cyclization was accomplished by LC-MS analysis (Figure 1B). A slight shift in the retention time of non-cyclic and cyclic peptides can be observed, followed by a two unit difference in the $[M+H]^+$ values corresponding to the expected loss of two protons upon disulfide bridge formation.

Peptide Biotin/PEGylation was performed to anchor the cyclic peptide to our nanocapsules by exploiting biotin–streptavidin affinity (Figure 2).

A biotin functionalized PEG linker (Biotin-PEG-COOH) was conjugated at the N-term of the peptide sequence, directly on the resin, via amide bond formation with the carboxyl group.

Solid-phase synthesis was preferred to simplify the removal of impurities from PEGylated peptide by several wash steps with a set of solvents (NMP-DCM-iPrOH-Et_2O). Once the peptide coupling with the Biotin-PEG-COOH linker was complete, the Biotin-PEG-CRT peptide (CRT-PEG) was deprotected and cleaved from the resin. The crude CRT-PEG was purified by preparative flash chromatography, obtaining a pure product with a final yield of approximately 5%. The pooled fractions containing the desired products were analyzed by analytical RP-HPLC, with the chromatogram reported in Figure 3.

Identification of the product at R_t = 23.78 min was performed by matrix-assisted laser desorption/ionization mass spectrometry coupled to two time of flight analyzers (MALDI-TOF-TOF). Comparison of the centroid mass peaks of Biotin-PEG-COOH linker acquired before and after peptide conjugation confirms the reaction outcome (Figure 4).

The intense polydisperse Gaussian distribution and the typical expected ethylene oxide repeat units of 44 Da were observed, corroborating the presence of PEG within the samples. The mass increment for CRT-PEG perfectly corresponds to the CRT peptide, clear evidence that the reaction took place. The observed mass increment (2589.11 $\pm n \times$ 44 Da) with respect to the expected theoretical isotopic mass of CRT-PEG (2566.22 $\pm n \times$ 44 Da) is due to the +23 m/z sodiated species $(MNa)^+$. Peptide cyclization was then performed as described before directly on the obtained pegylated peptide.

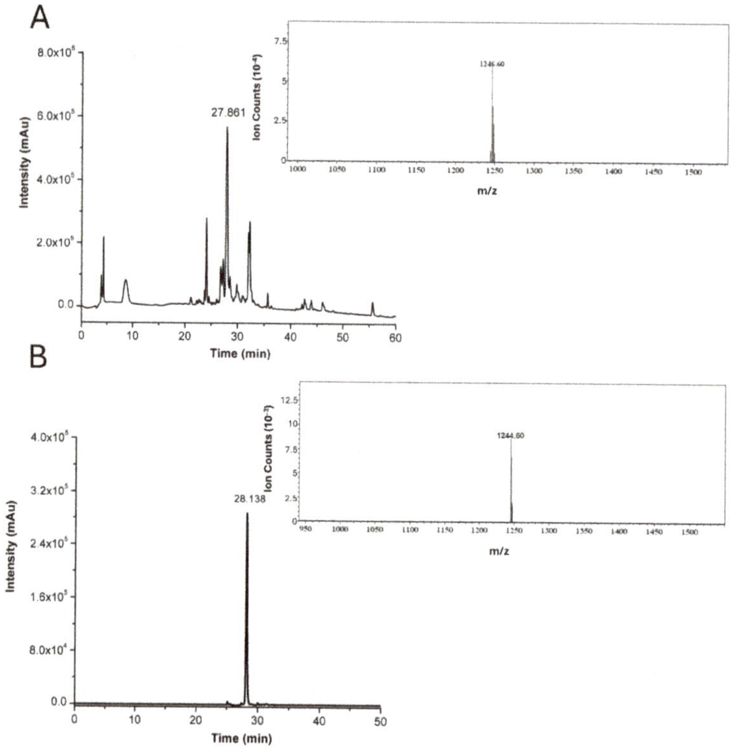

Figure 1. (**A**) RP-HPLC chromatogram of crude CRT. ESI-MS spectrum relative to the peak at R_t = 27.86 min that corresponds to $[M+H]^+$ of CRT (theoretical mass: 1246.50 Da; observed mass: 1246.65 Da). (**B**) RP-HPLC chromatogram of pure cyclic CRT. ESI-MS spectrum relative to the peak at R_t = 28.14 min that corresponds to $[M+H]^+$ of cyclic-CRT (theoretical mass: 1244.49 Da; observed mass: 1244.60 Da). Detector counts of the photodiode array are reported in arbitrary units (Au).

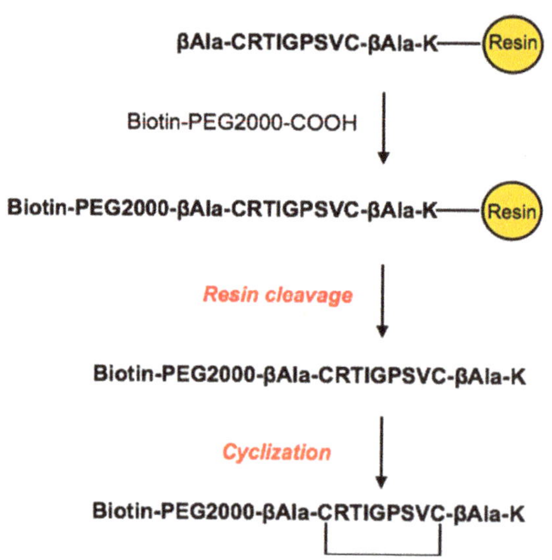

Figure 2. Schematic representation of the solid-phase synthetic strategy of CRT peptide PEGylation.

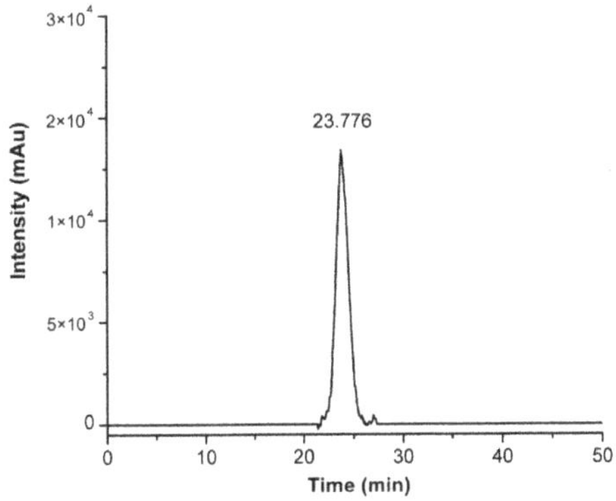

Figure 3. Analytical RP-HPLC chromatogram of purified CRT-PEG, detected at λ = 220 nm. Detector counts of the photodiode array are reported in arbitrary units (Au).

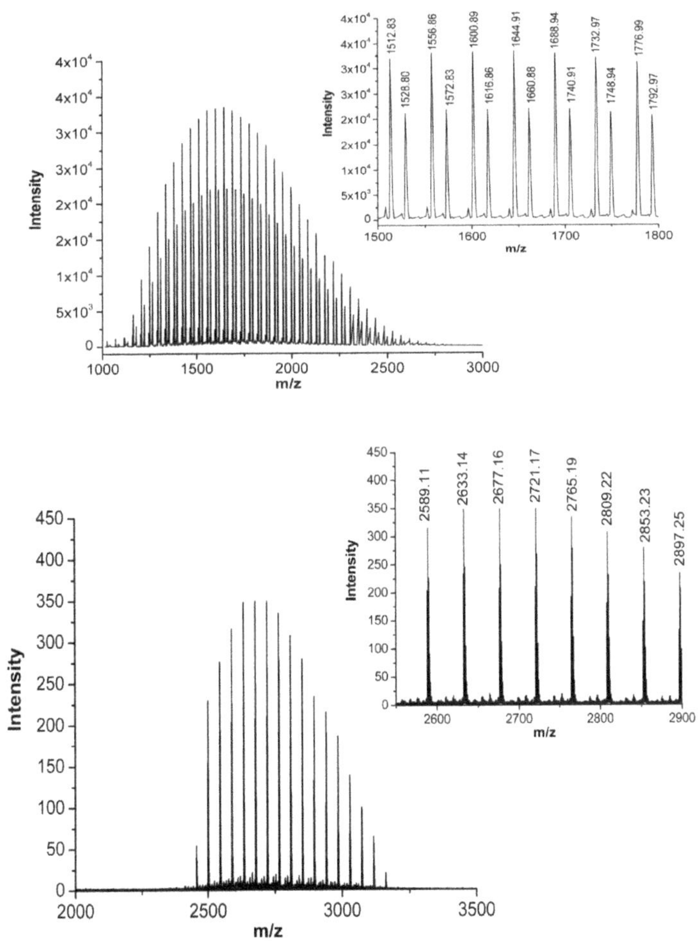

Figure 4. MALDI mass spectra (centroid) of Biotin-PEG-COOH linker on the **top** and CRT-PEG peptide at the **bottom**, showing the expected mass shift upon peptide conjugation. Insets report the zoomed centroid section of the spectra.

3.2. Nanocarrier Assembly

The CRT-PEG was then linked to the multi-layered NEs (CRT-PEG-NEs) to exploit its recognition of overexpressed TfR on the brain endothelium. We used a previously optimized decoration strategy [16], featuring an outer shell of biotinylated hyaluronic acid (HA-Biotin-NEs), with the only difference being that the oil core of the nanoemulsion is loaded with paclitaxel. Briefly, the streptavidin (SAV) was added to HA-Biotin-NEs under sonication, followed by the addition of CRT-PEG. The system was characterized at all stages by DLS (Table 1 and Figure 5). At all stages, narrowly monodispersed functionalized nanocapsules were obtained with a net negative charge according to the Z-potential measurements.

Table 1. Size, PdI and Z-potential measurement of each NCs component, loaded with paclitaxel, during the several steps of assembly. Data are reported as mean (n = 3) ± SD.

	Size (nm)	PdI	Z-Potential (mV)
NEs	89 ± 1.	0.080 ± 0.015	−27 ± 4
CT-NEs	93 ± 2	0.086 ± 0.003	+25 ± 1
HA-Biotin-CT-NEs	113 ± 9	0.094 ± 0.015	−31 ± 2
SAV-HA-Biotin-CT-NEs	131 ± 7	0.145 ± 0.017	−28 ± 3
CRT-PEG-SAV-HA-Biotin-CT-NEs	137 ± 7	0.152 ± 0.002	−30 ± 1
PEG-SAV-HA-Biotin CT-NEs	138 ± 5	0.139 ± 0.002	−30 ± 1

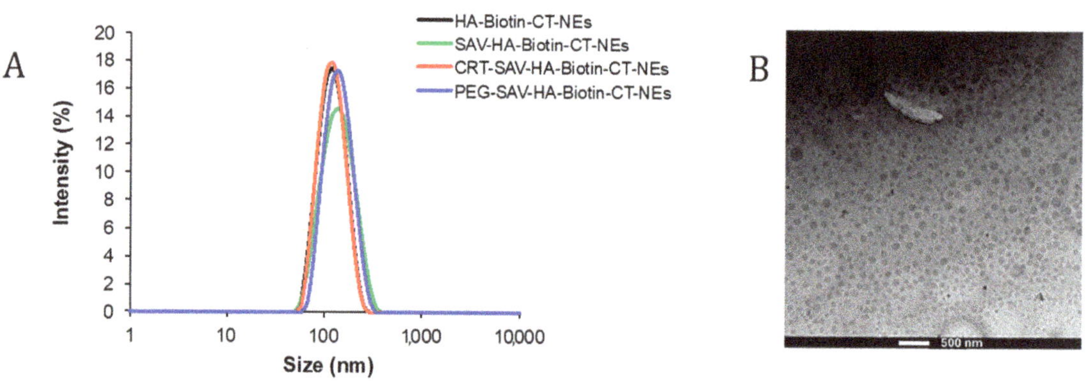

Figure 5. (**A**) Overlapping of hydrodynamic size of each component deposited around NEs. (**B**) Cryo-TEM projection image of paclitaxel-loaded HA-CT-NEs. Scale bar corresponds to 500 nm.

The morphological characterization of the paclitaxel-loaded nanocapsule was performed by cryo-TEM analysis (Figure 5B). The microscopy clearly shows a homogeneous sample fixed in its frozen hydrated state confirming the size distribution observed by DLS. The image displays a well-defined electron-dense core, corresponding to the PTX-loaded in the oil core that makes it difficult to observe the polymer layers around the O/W NEs.

DLS periodical measurements were performed to evaluate the size evolution over time. Figure 6 shows that the hydrodynamic diameter of CRT-PEG-NEs is quite stable within the one-month detection window, with only minimal coalescence in the last 5 days (~20% radius increase).

3.3. Preliminary Biological Tests

bEnd.3 cells, an immortalized mouse brain endothelial cell line, represent an attractive candidate as model of the BBB due to their rapid growth, maintenance of BBB characteristics and formation of functional barriers [20]. It was demonstrated in our group that the number of TfRs per bEnd3 cells was 100-fold higher than in HUVEC cells, confirming a high value expression of TfRs [21]. To understand the best conditions to appreciate the selective activity of CRT-PEG-NEs toward bEnd.3 cells, thanks to CRT-TfRs interaction, we investigated the PTX effect on this cell line using the same procedure of Falanga et al. [22].

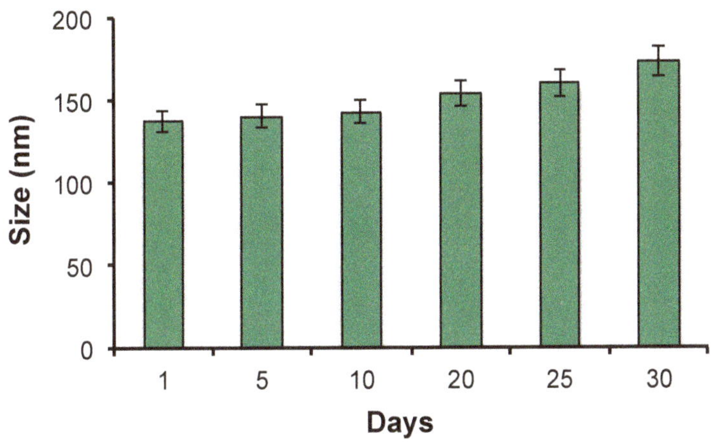

Figure 6. Dimensional behavior over time for PTX-loaded CRT-NEs measured by DLS analysis.

A confluent monolayer of bEnd.3 cells was incubated with PTX-loaded CRT-PEG-NEs, diluted 1:5 in cells, at a final PTX concentration of 1.4 μM for 30 min, 2 h and 4 h. Moreover, cells were treated with cell medium alone as positive control, free PTX as negative control and un-functionalized PEG-NEs that act as blanks. After incubation, the cells were washed and a quantitative evaluation of cell viability (normalized to positive control, which is set to 100%) was obtained by PrestoBlue® assay after 48 h (Figure 7).

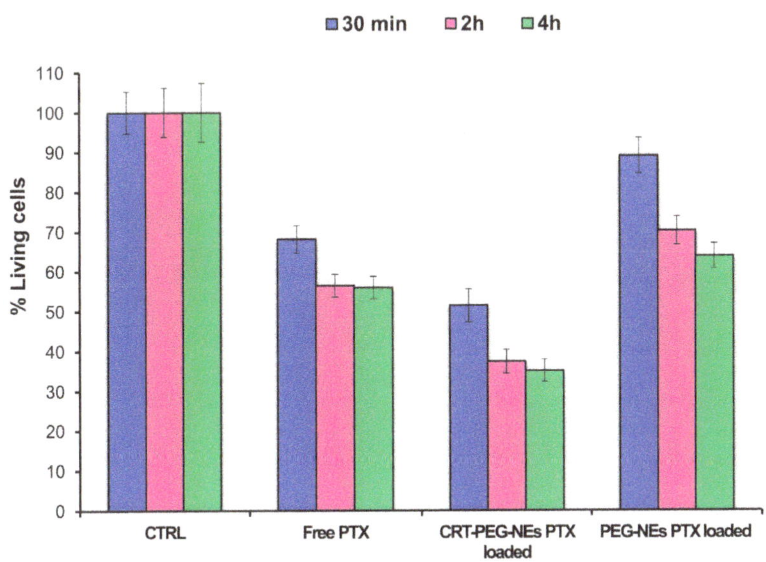

Figure 7. Cytotoxicity assay of PTX-loaded CRT-PEG-NEs, PEG-NEs and free PTX. bEnd.3 cells were treated at several incubation times (30 min, 2 h and 4 h) and cell viability was evaluated after 48 h. Data are reported as mean of three independent experiments (n = 3 ± SD) and expressed as percentage compared to control cells.

We conducted one-way ANOVA tests using MATLAB® (MathWorks, Natick, MA, USA) to quantitatively evaluate cell viability at 30 min, 2 h, 4 h and 24 h. Data are reported as mean of three independent experiments and expressed as percentage compared to control cells. The column Prob ≥ F shows the *p*-values for the treatments (CTRL, Free drug, CRT-PEG-PXT and PEG-PXT, Prob ≥ F of 0) and time of treatments (30 min, 2 h, 4 h; Prob ≥ F of 0.0004) and the interaction between them (Prob ≥ F of 0.2525) (Table 2). These values indicate that the type and time of treatments affect the percentage yield of living cells, but there is no evidence of interaction effect of the two (Table 2).

Table 2. Two-way ANOVA for cytotoxicity analysis.

Two-Way ANOVA					
Source	**SS**	**Df**	**MS**	**F**	**Prob ≥ F**
Columns	570.513	3	190.171	236.64	0
Rows	14.9	2	7.45	9.27	0.0004
Interaction	6.527	6	1.088	1.35	0.2525
Error	38.575	48	0.804		
Total	630.515	59			

The data showed an increase of cell mortality both for free PTX and for CRT-PEG-NEs over time, with no significant difference between two and four hours. Very interestingly, it was possible to observe a significant cytotoxicity effect of CRT-PEG-NEs compared to blank (PEG-NEs). This is an evident consequence of peptide capability to accumulate the nanocarrier on the cells surface, and allow its internalization, thanks to ligand–receptor interaction. For all the incubation times, CRT-PEG-NEs exhibit an increase of cell death relative to PEG-NEs and free PTX. For instance, at 30 min of incubation, CRT-PEG-NEs exhibit an increase of 38% of cell death relative to PEG-NEs and almost 17% relative to free PTX. Considering an average over all times, relative to un-functionalized PEG-NEs, CRT-PEG-NEs induced increased cell death of 33.05 ± 4.42%.

To confirm the effect due to the targeted accumulation, cellular uptake assays by confocal microscopy were performed. A confluent monolayer of bEnd.3 cells were incubated with CRT-PEG-NEs and PEG-NEs for 4 h at the same experimental condition of cytotoxicity test described before. Rhodaminated streptavidin was used during the nanocapsules assembling to detect their fluorescence. In addition, the control (CTRL) consisted in cells treated with cell medium alone. Figure 8 shows confocal microscopic images of the bEnd.3 cell monolayer after NCs uptakes, which sheds light on the previous cytotoxicity results. The presence of functionalized nanocapsules is shown in red, while the cell cytoplasm and the nuclei are in green and blue respectively. In Figure 9, the plot of rhodaminated NCs 'mean' fluorescence intensity normalized for the bEnd.3 cell number is reported. Particularly, three replicates for sample and at least five images for replicate are considered and statistical results are reported in Table 3.

Table 3. One-way ANOVA for uptake analysis.

One Way-ANOVA					
Source	**SS**	**Df**	**MS**	**F**	**Prob ≥ F**
Columns	$2.28946e^{+10}$	1	$2.28946e^{+10}$	0.3	0.6156
Error	$3.09993e^{+11}$	4	$7.74983e^{+10}$		
Total	$3.32888e^{+11}$	5			

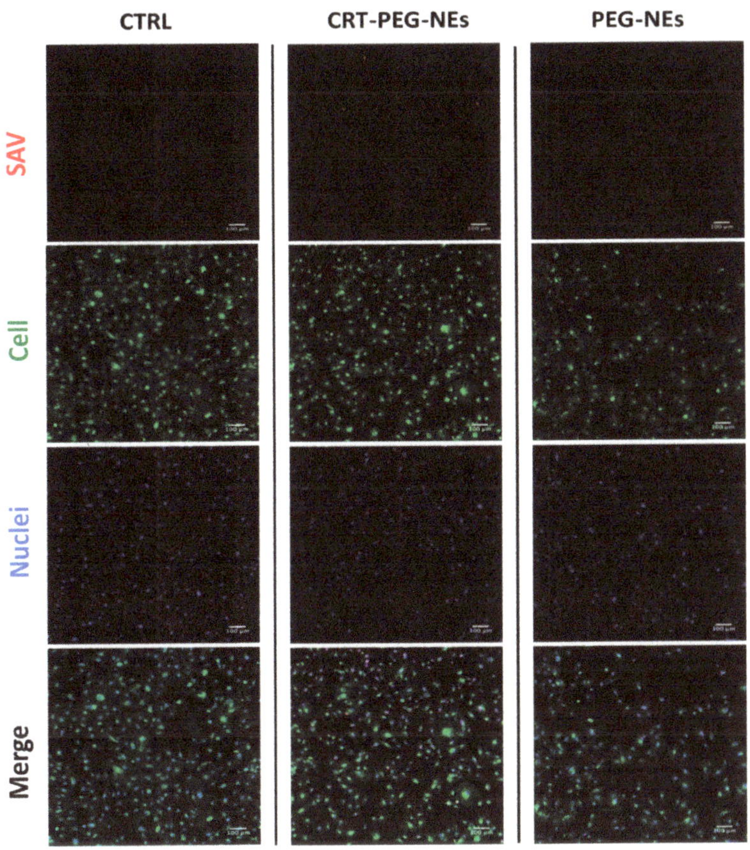

Figure 8. Confocal images of bEnd.3 cell. Untreated, rhodaminated CRT-PEG-NEs and PEG-NEs interactions with a confluent monolayer of bEnd.3 cells. Nuclei (blue) and cellular membrane (green) of the cells were stained with DAPI and WGA 555, respectively, while the red color represents rhodamine uptake. Scale bar is 100 µm.

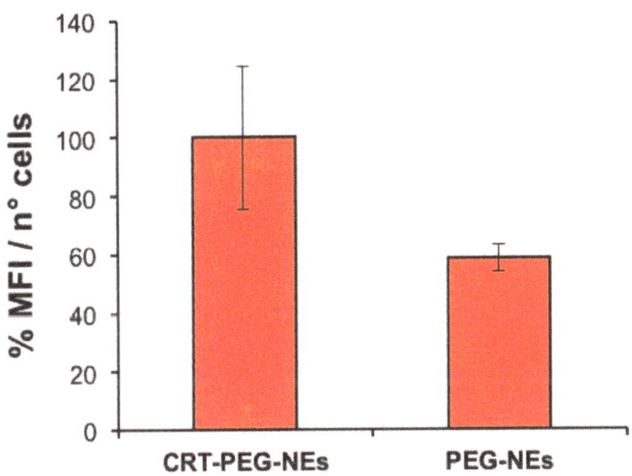

Figure 9. Plot of mean fluorescence intensity of rhodaminated nanocapsules normalized to cells number. bEnd.3 cells were treated with CRT-PEG-NEs and PEG-NEs. Data are reported as mean (n = 3) ± SD, p-value = 0.6150.

The uptake data are consistent with the previous cytotoxicity results, showing a clear fluorescence difference between CRT-PEG-NEs and PEG-NEs. No fluorescence was detected in the images of the control cells. Specifically, the fluorescence intensity of peptide-functionalized NCs was approximately 40% higher compared to the negative control. Furthermore, a small amount of PEG-NEs was detected in the cells.

4. Conclusions

In this study, we demonstrate the versatility of engineered layer by layer nanocapsules in active targeting. To develop a novel nanocapsule able to target the BBB, our O/W NEs have been functionalized with the CRT peptide, which is able to recognize the BBB cells thanks to the Tf/TfR mechanism. O/W NEs are stabilized by a double layer of chitosan and biotinylated hyaluronic acid, and, thanks to their long shelf-life, represent an ideal candidate for further decoration of a bioactive peptide on the outer layer by an additive strategy, without purification steps. CRT-functionalized NEs preserve a narrowly distributed hydrodynamic diameter below 150 nm and are stable over time. Preliminary cytotoxic tests were carried out using a mouse brain cell line (bEnd.3) as a model of endothelial brain tissue. Results show an increased cellular death of 33.05 ± 4.42% compared to undecorated PEG-NEs that acted as a control, while the uptake of CRT-PEG-NEs increased 41.5 ± 24% with respect to the negative control. We proved in vitro the ability of the CRT peptide to target brain endothelium tissue while conjugated to the proposed O/W NE-based carrier. These in vitro results are in agreement with the current literature on CRT peptide decorate NCs. Previous strategies have utilized various nanotechnology-based delivery systems such as liposomes, polymeric nanoparticles (like PLGA NPs) and micelles. These systems often incorporate surface modifications to exploit receptors overexpressed at the BBB. For instance, PLGA NPs have been modified with transferrin receptor-targeting peptides, demonstrating improved brain targeting and reduced cytotoxicity in vitro and in vivo [23,24].

Recent advancements also include the use of receptor-mediated transcytosis to enhance drug delivery to the brain. Strategies such as the SGT-53 liposome, which targets the transferrin receptor, have shown promising results in glioblastoma treatment when combined with immunotherapy. Peptide-based targeting and advanced nanocarrier systems offer significant potential for improving drug delivery across the BBB, targeting tumors, and enhancing therapeutic efficacy. By leveraging passive and active targeting strategies and employing novel nanotechnologies, these approaches aim to optimize drug delivery while minimizing adverse effects. It is remarkable that here we were able to develop such an advanced functional nanocarrier starting with a simple basic nanocarrier, namely O/W NEs, thanks to the stability we can confer to it, bringing the intrinsic advantage of oil to encapsulate powerful lipophilic drugs. The proposed platform paves the way to the design of novel multifunctional nanocarriers for the delivery of therapeutic agents to the CNS with the additional advantage that the vegetable oil core is easily degraded and absorbed with no concerns in terms of accumulation in the body, in contrast with many solid-based nanomaterials. Further development will concern the integration of a cell-penetrating peptide, like gH625, able to cross the BBB. Moreover, for anticancer therapeutic purposes, potentially new therapeutics [25] or the well-established curcumin instead of PTX can be used, the latter being used in this case only as a cytotoxic molecule to test NCs' accumulation ability. Indeed, curcumin not only has been demonstrated to be a good anticancer substance, but it is also specific for tumor cells [16]. Specifically, curcumin delivery systems that exploit the unique properties of nanoparticles and surface modifications demonstrate enhanced brain targeting and therapeutic outcomes for neurological diseases and brain tumors. The integration of these innovative delivery methods represents a promising frontier in the treatment of CNS disorders, potentially leading to more effective and safer therapies [23]. Therefore, it may be an ideal candidate for blood–brain barrier treatments, because, although a fraction of NCs will break and release curcumin in the endothelial cells instead of crossing them, such a molecule has no activity toward these cells, avoiding side effects to healthy sites.

Author Contributions: Conceptualization, R.V. and P.A.N.; Methodology, A.D.C. and M.C.; Validation, A.D.C.; Formal analysis, R.V., M.C., A.L. and E.L.; Data curation, A.D.C. and C.S.; Writing—original draft, A.D.C., R.V. and M.C.; Supervision, R.V., M.C., A.L. and P.A.N.; Funding acquisition, A.L. and P.A.N. All authors have read and agreed to the published version of the manuscript.

Funding: This research was supported by EU funding within the MUR PNRR "National Center for Gene Therapy and Drugs based on RNA Technology" (Project no. CN00000041 CN3 Spoke #8 "Platforms for RNA/DNA delivery", CUP E63C22000940007, CUP J33C22001130001).

Institutional Review Board Statement: Not applicable.

Informed Consent Statement: Not applicable.

Data Availability Statement: Data are contained within the article.

Acknowledgments: The authors thank Valentina Mollo for the TEM analyses and Denise Pagliara for the support on the statistical analysis.

Conflicts of Interest: The authors declare no conflict of interest.

References

1. Pardridge, W.M. Drug Transport across the Blood–Brain Barrier. *J. Cereb. Blood Flow Metab.* **2012**, *32*, 1959–1972. [CrossRef]
2. Oller-Salvia, B.; Sánchez-Navarro, M.; Giralt, E.; Teixidó, M. Blood–Brain Barrier Shuttle Peptides: An Emerging Paradigm for Brain Delivery. *Chem. Soc. Rev.* **2016**, *45*, 4690–4707. [CrossRef]
3. Del Genio, V.; Bellavita, R.; Falanga, A.; Hervé-Aubert, K.; Chourpa, I.; Galdiero, S. Peptides to Overcome the Limitations of Current Anticancer and Antimicrobial Nanotherapies. *Pharmaceutics* **2022**, *14*, 1235. [CrossRef]
4. Walter, F.R.; Santa-Maria, A.R.; Mészáros, M.; Veszelka, S.; Dér, A.; Deli, M.A. Surface Charge, Glycocalyx, and Blood-Brain Barrier Function. *Tissue Barriers* **2021**, *9*, 1904773. [CrossRef]
5. Barra, T.; Falanga, A.; Bellavita, R.; Laforgia, V.; Prisco, M.; Galdiero, S.; Valiante, S. gH625-Liposomes Deliver PACAP through a Dynamic in Vitro Model of the Blood–Brain Barrier. *Front. Physiol.* **2022**, *13*, 932099. [CrossRef]
6. Friden, P.M.; Walus, L.R.; Musso, G.F.; Taylor, M.A.; Malfroy, B.; Starzyk, R.M. Anti-Transferrin Receptor Antibody and Antibody-Drug Conjugates Cross the Blood-Brain Barrier. *Proc. Natl. Acad. Sci. USA* **1991**, *88*, 4771–4775. [CrossRef]
7. Recht, L.; Torres, C.O.; Smith, T.W.; Raso, V.; Griffin, T.W. Transferrin Receptor in Normal and Neoplastic Brain Tissue: Implications for Brain-Tumor Immunotherapy. *J. Neurosurg.* **1990**, *72*, 941–945. [CrossRef]
8. Gomme, P.T.; McCann, K.B.; Bertolini, J. Transferrin: Structure, Function and Potential Therapeutic Actions. *Drug Discov. Today* **2005**, *10*, 267–273. [CrossRef]
9. Silva, M.C.; Torres, J.A.; Castro, A.A.; da Cunha, E.F.F.; Alves de Oliveira, L.C.; Corrêa, A.D.; Ramalho, T.C. Combined Experimental and Theoretical Study on the Removal of Pollutant Compounds by Peroxidases: Affinity and Reactivity toward a Bioremediation Catalyst. *J. Biomol. Struct. Dyn.* **2016**, *34*, 1839–1848. [CrossRef]
10. Salvati, A.; Pitek, A.S.; Monopoli, M.P.; Prapainop, K.; Bombelli, F.B.; Hristov, D.R.; Kelly, P.M.; Åberg, C.; Mahon, E.; Dawson, K.A. Transferrin-Functionalized Nanoparticles Lose Their Targeting Capabilities When a Biomolecule Corona Adsorbs on the Surface. *Nat. Nanotechnol.* **2013**, *8*, 137–143. [CrossRef]
11. Staquicini, F.I.; Ozawa, M.G.; Moya, C.A.; Driessen, W.H.P.; Barbu, E.M.; Nishimori, H.; Soghomonyan, S.; Flores, L.G.; Liang, X.; Paolillo, V.; et al. Systemic Combinatorial Peptide Selection Yields a Non-Canonical Iron-Mimicry Mechanism for Targeting Tumors in a Mouse Model of Human Glioblastoma. *J. Clin. Investig.* **2011**, *121*, 161–173. [CrossRef]
12. Hajitou, A.; Trepel, M.; Lilley, C.E.; Soghomonyan, S.; Alauddin, M.M.; Marini, F.C.; Restel, B.H.; Ozawa, M.G.; Moya, C.A.; Rangel, R.; et al. A Hybrid Vector for Ligand-Directed Tumor Targeting and Molecular Imaging. *Cell* **2006**, *125*, 385–398. [CrossRef]
13. Kang, T.; Jiang, M.; Jiang, D.; Feng, X.; Yao, J.; Song, Q.; Chen, H.; Gao, X.; Chen, J. Enhancing Glioblastoma-Specific Penetration by Functionalization of Nanoparticles with an Iron-Mimic Peptide Targeting Transferrin/Transferrin Receptor Complex. *Mol. Pharm.* **2015**, *12*, 2947–2961. [CrossRef]
14. Liu, M.; Wu, L.; Shan, W.; Cui, Y.; Huang, Y. Iron-Mimic Peptide Converts Transferrin from Foe to Friend for Orally Targeting Insulin Delivery. *J. Mater. Chem. B* **2018**, *6*, 593–601. [CrossRef]
15. Du, Y.; Gao, J.; Zhang, H.; Meng, X.; Qiu, D.; Gao, X.; Zheng, A. Brain-Targeting Delivery of MMB4 DMS Using Carrier-Free Nanomedicine CRT-MMB4@MDZ. *Drug Deliv.* **2021**, *28*, 1822–1835. [CrossRef]
16. De Capua, A.; Palladino, A.; Chino, M.; Attanasio, C.; Lombardi, A.; Vecchione, R.; Netti, P.A. Active Targeting of Cancer Cells by CD44 Binding Peptide-Functionalized Oil Core-Based Nanocapsules. *RSC Adv.* **2021**, *11*, 24487–24499. [CrossRef]
17. Fotticchia, T.; Vecchione, R.; Scognamiglio, P.L.; Guarnieri, D.; Calcagno, V.; Di Natale, C.; Attanasio, C.; De Gregorio, M.; Di Cicco, C.; Quagliariello, V.; et al. Enhanced Drug Delivery into Cell Cytosol via Glycoprotein H-Derived Peptide Conjugated Nanoemulsions. *ACS Nano* **2017**, *11*, 9802–9813. [CrossRef]

18. Chan, W.; White, P. (Eds.) *Fmoc Solid Phase Peptide Synthesis: A Practical Approach;* Oxford University Press: Oxford, UK, 1999; ISBN 978-0-19-963725-6.
19. Chen, L.; Annis, I.; Barany, G. Disulfide Bond Formation in Peptides. *Curr. Protoc. Protein Sci.* **2001**, *23*, 18.6.1–18.6.19. [CrossRef]
20. Brown, R.C.; Morris, A.P.; O'Neil, R.G. Tight Junction Protein Expression and Barrier Properties of Immortalized Mouse Brain Microvessel Endothelial Cells. *Brain Res.* **2007**, *1130*, 17–30. [CrossRef]
21. Fotticchia, I.; Guarnieri, D.; Fotticchia, T.; Falanga, A.P.; Vecchione, R.; Giancola, C.; Netti, P.A. Energetics of Ligand-Receptor Binding Affinity on Endothelial Cells: An in Vitro Model. *Colloids Surf. B Biointerfaces* **2016**, *144*, 250–256. [CrossRef]
22. Falanga, A.P.; Melone, P.; Cagliani, R.; Borbone, N.; D'Errico, S.; Piccialli, G.; Netti, P.A.; Guarnieri, D. Design, Synthesis and Characterization of Novel Co-Polymers Decorated with Peptides for the Selective Nanoparticle Transport across the Cerebral Endothelium. *Molecules* **2018**, *23*, 1655. [CrossRef] [PubMed]
23. Askarizadeh, A.; Barreto, G.E.; Henney, N.C.; Majeed, M.; Sahebkar, A. Neuroprotection by Curcumin: A Review on Brain Delivery Strategies. *Int. J. Pharm.* **2020**, *585*, 119476. [CrossRef] [PubMed]
24. Na, Y.; Zhang, N.; Zhong, X.; Gu, J.; Yan, C.; Yin, S.; Lei, X.; Zhao, J.; Geng, F. Polylactic-Co-Glycolic Acid-Based Nanoparticles Modified with Peptides and Other Linkers Cross The Blood–Brain Barrier for Targeted Drug Delivery. *Nanomedicine* **2023**, *18*, 125–143. [CrossRef]
25. De Fenza, M.; Esposito, A.; D'Alonzo, D.; Guaragna, A. Synthesis of Piperidine Nucleosides as Conformationally Restricted Immucillin Mimics. *Molecules* **2021**, *26*, 1652. [CrossRef] [PubMed]

pharmaceutics

MDPI

Article

Development of a Water-Soluble Nanomicellar Formulation Loaded with Trans-Resveratrol Using Polyethylene Glycol Monostearate for the Treatment of Intracerebral Hemorrhage

Kengo Banshoya [1,*,†], Aoi Machida [1,2,†], Saki Kawamura [1], Tetsuhiro Yamada [1], Riko Okada [1], Yui Kawamoto [1], Hikaru Kimura [1,3], Sachi Shibata [4], Yuhzo Hieda [1], Yoshiharu Kaneo [1], Tetsuro Tanaka [1] and Masatoshi Ohnishi [1,2,*]

1 Faculty of Pharmacy and Pharmaceutical Sciences, Fukuyama University, Fukuyama 729-0292, Hiroshima, Japan; machida@fukuyama-u.ac.jp (A.M.); p7119028@fukuyama-u.ac.jp (S.K.); p7120097@fukuyama-u.ac.jp (T.Y.); p7120020@fukuyama-u.ac.jp (R.O.); p7120029@fukuyama-u.ac.jp (Y.K.); kimurahikaru@yamaguchi-u.ac.jp (H.K.); hieda@fukuyama-u.ac.jp (Y.H.); kaneoyoshiharu@gmail.com (Y.K.); tanaka@fukuyama-u.ac.jp (T.T.)

2 Graduate School of Pharmacy and Pharmaceutical Sciences, Fukuyama University, Fukuyama 729-0292, Hiroshima, Japan

3 Pharmacy Department, Yamaguchi University Hospital, Ube 755-8505, Yamaguchi, Japan

4 Faculty of Health and Welfare Science, Okayama Prefectural University, Soja 719-1197, Okayama, Japan; sachi_shibata@fhw.oka-pu.ac.jp

* Correspondence: kban@fukuyama-u.ac.jp (K.B.); ohnishi@fukuyama-u.ac.jp (M.O.); Tel.: +81-84-936-2111 (K.B. & M.O.)

† These authors contributed equally to this work.

Abstract: Background/Objectives: Trans-resveratrol (Res) has been reported to possess many biological activities, including neuroprotective effects, owing to its anti-inflammatory and antioxidant properties. However, Res has very low water solubility, which limits its therapeutic application. In this work, we formulated water-soluble micellar formulations incorporating Res using polyethylene glycol monostearate (stPEG). **Methods:** These formulations (stPEG/Res) were developed using five types of stPEG containing 10, 25, 40, 55 and 140 PEG repeat units. The formulations were characterized for Res content, water solubility, particle size, zeta potential, precipitation, biodistribution, and efficacy against neuronal and motor dysfunction in intracerebral hemorrhage (ICH). **Results:** Intravenous administration of stPEG40/Res, which demonstrated particle size, water solubility, and biodistribution properties suitable for intravenous administration, suppressed neurological and motor dysfunction following in a collagenase-induced ICH mouse model. These effects were inhibited by zinc protoporphyrin-9, an inhibitor of the antioxidant enzyme heme oxygenase-1, suggesting that Res contributes to antioxidant enzyme expression and anti-inflammatory activity. **Conclusions:** The stPEG/Res micellar formulation developed in this study may offer a promising therapeutic approach for ICH treatment.

Keywords: trans-resveratrol; micellar formulation; water-soluble; polyethylene glycol monostearate; intracerebral hemorrhage

1. Introduction

Intracerebral hemorrhage (ICH) is a neurological disease caused by nontraumatic bursting of blood vessels in the brain. It is the second most common stroke subtype after ischemic stroke [1]. The mortality rate within the first month after ICH onset exceeds 30%, with only 14% of survivors regaining functional independence after 12 months [2]. The primary treatment for ICH is surgical evacuation of the hematoma, which itself can cause secondary damage. Due to the lack of pharmacological treatment options, ICH remains the most challenging subtype of stroke [3]. Therefore, there is an urgent need to develop safe and effective treatment strategies. Numerous studies have demonstrated that factors such as oxidative stress caused by red blood cell lysis products and plasma components lead to brain damage after ICH [4]. Oxidative stress is involved in the pathophysiology of numerous brain diseases, including neurodegenerative diseases, depression, and ischemic stroke [5]. Importantly, oxidative stress is also a major mediator of secondary brain injury after ICH [5]. Thus, oxidative stress plays an important role in brain injury following ICH [6]. Due to the lack of effective pharmacological strategies for the treatment of ICH, researchers have increasingly turned to traditional and alternative medicines, particularly natural products [7].

One natural product that is an effective antioxidant for the treatment of ICH is trans-resveratrol (Res). Res is the most useful, common, and abundantly available polyphenol in pharmaceutical and dietary supplements [8]. Res is a safe compound that has been proven to exhibit many neuroprotective effects in various neurological diseases, such as stroke, Alzheimer's disease, and Parkinson's disease [9–12]. Importantly, frequent oral administration of Res has been shown to reduce brain injury in experimental ICH models [13]. However, when Res is administered orally, most of it is rapidly metabolized in the liver and excreted in urine and feces [14]. In addition, the need for pre-administration of Res and the fact that some patients with ICH are comatose makes oral administration of Res unsuitable for the treatment of ICH. Res is highly lipophilic, with a low water solubility of approximately 0.03 mg/mL [15]. Therefore, a study has been reported in which Res was dissolved in dimethyl sulfoxide and administered intravenously to investigate its therapeutic effect in ICH model animals [16]. However, intravenous administration of organic solvents, such as dimethyl sulfoxide, is clinically problematic due to their toxicity. Water-insoluble compounds are usually difficult to formulate as injectable drugs, and Res is no exception. As a solubilized formulation of Res that can be safely administered intravenously, there is a report on nano-formulation using lipids [17]. Lipid-based nano-formulations loaded with compounds other than Res have also been reported to be effective against ICH [18,19]. However, lipid-based nano-formulations are expensive and require complex manufacturing procedures. Therefore, a water-soluble Res injection formulation that overcomes these issues is required for the treatment of ICH.

One way to improve the performance of hydrophobic compounds is to use polymeric micelle systems [20–23]. Polymeric micelles, composed of amphiphilic polymers containing polyethylene glycol (PEG), have several advantages, including improved drug solubilization, stability, and ease of preparation and storage [24]. Polyethylene glycol monostearate (stPEG) is generally regarded as an essentially nontoxic and nonirritating material, and this amphiphilic block polymer containing PEG can be used in topical, oral, and intravenous formulations [25]. Furthermore, stPEG has various HLB values depending on the PEG length [25]. This makes it possible to use stPEG with optimal properties for solubilization of hydrophobic compounds.

We have successfully developed water-soluble stPEG micellar formulations of α-tocopherol and coenzyme Q10 [26,27]. Similar to these, Res is a poorly water-soluble antioxidant. In the present study, we developed novel intravenous Res micelle formulations using five different PEG lengths of stPEG, encapsulating Res in the inner hydrophobic part of the micelle. These micelles were investigated for their Res content, particle size, zeta potential, precipitation, saturation solubility, toxicity, and therapeutic effects in ICH models.

2. Materials and Methods

2.1. Materials

Res was purchased from Tokyo Chemical Industry (Tokyo, Japan). Medetomidine and pentobarbital were obtained from Kyoritsu Seiyaku (Tokyo, Japan). MuLV reverse transcriptase was purchased from Thermo Fisher Scientific (Waltham, MA, USA). Random primers were obtained from Takara Bio (Shiga, Japan). Other primers were obtained from Eurofins Genomics (Tokyo, Japan). The LightCycler FastStart DNA Master SYBER Green I was purchased from Roche Diagnostics Japan (Tokyo, Japan). Collagenase type IV was purchased from Sigma-Aldrich (Tokyo, Japan). The MTX-LDH was sourced from Kyokuto Seiyaku (Tokyo, Japan). Zinc protoporphyrin-9 (ZnPPIX) was purchased from Cayman Chemicals (Ann Arbor, MI, USA), and paraformaldehyde (PFA) was purchased from Nacalai Tesque (Kyoto, Japan). Mouse anti-neuronal nuclei (NeuN) antibody was purchased from Merck Millipore (Burlington, MA, USA). Biotinylated anti-mouse IgG (H + L) and an avidin-biotinylated horseradish peroxidase complex (Vectastain Elite ABC Kit) were obtained from Vector Laboratories (Burlingame, CA, USA). stPEG140 was kindly provided by Kao (Tokyo, Japan), and other stPEGs, reagents, and solvents were obtained from FUJIFILM Wako Pure Chemical (Osaka, Japan). All other chemicals and reagents used were of the highest grade commercially available and were used without further purification.

2.2. Cell Culture and Animals

Dulbecco's Modified Eagle Medium (DMEM), penicillin/streptomycin, and trypsin/EDTA solution were purchased from FUJIFILM Wako Pure Chemical (Osaka, Japan). Fetal bovine serum (FBS) and fetal horse serum (HS) were purchased from Japan Bioserum (Fukuyama, Japan). PC-12 cells (established by Greene L.; JCRB0733), a rat pheochromocytoma cell line commonly used as a neuroblast model, were provided by JCRB Cell Bank (Osaka, Japan). The cells were maintained in DMEM supplemented with 5% FBS, 10% HS, 100 U/mL penicillin, and 100 μg/mL streptomycin. They were cultured at 37 °C in a humidified atmosphere containing 5% CO_2. Four-week-old male ddY mice were purchased from Shimizu Laboratory Supplies (Kyoto, Japan) and housed in cages with sawdust bedding and free access to commercial food and distilled water.

2.3. Preparation of stPEG/Res

The stPEG/Res micelles were prepared by dissolving stPEG (400 mg) in 80 mL of distilled water under magnetic stirring. Subsequently, a solution of Res (40 mg) dissolved in 2 mL of acetone was slowly added to each stPEG solution using a syringe while stirring. These solutions were stirred for 3 h at room temperature to volatilize the acetone and then lyophilized with light shielding to obtain stPEG/Res micelle powder. Samples were stored at 4 °C, protected from light. Distilled water or 5% glucose was added to the micelle powder before use, followed by sonication.

2.4. High-Performance Liquid Chromatography (HPLC) Analysis

Res was quantified using HPLC. The HPLC system consisted of an LC-20AD pump (Shimadzu, Kyoto, Japan) and a variable-wavelength ultraviolet detector (SPD-20A, Shimadzu, Kyoto, Japan). The detection wavelength was set at 305 nm, and a 4.6 × 150 mm C18 reversed-phase column (TSKgel ODS 80TM, Tosoh, Tokyo, Japan) was maintained at 40 °C. The mobile phase comprised methanol, distilled water, and acetic acid (200:199:1, *v*/*v*/*v*), with a flow rate of 1.0 mL/min. The injection volume was 20 μL. The Res content (*w*/*w*)% of each micelle was calculated as follows:

Res content (*w*/*w*)% = (mass of Res loaded in micelles / mass of micelles) × 100

2.5. Measurement of Particle Size and Zeta Potential

The average particle size and zeta potential of each stPEG/Res micelle were measured using a Malvern Zetasizer Nano ZS (Malvern Instruments, Worcestershire, UK), as described previously [26]. Each micelle type was prepared at a concentration of 1 mg/mL.

2.6. Measurement of Saturation Solubility

The saturation solubility of each stPEG/Res sample was measured using the aforementioned HPLC system. Specifically, 50 mg of the Res-equivalent of stPEG/Res was mixed with 2 mL of distilled water and centrifuged at 10,000 rpm for 30 min. The supernatant was filtered through a 0.45 μm filter (Millex®-LH filter, Merck Millipore, Burlington, MA, USA) and diluted appropriately with the mobile phase for HPLC measurement.

2.7. Precipitation of stPEG/Res

The formation of stPEG/Res precipitates in the aqueous solution was evaluated. Briefly, 5 mg/mL solution of each stPEG/Res micelle formulation was prepared, filtered through a 0.45 μm filter, and stored in the dark at room temperature for 1 week. The formation of precipitates was visually confirmed.

2.8. In Vitro and in Vivo Toxicity

In vitro toxicity was evaluated by measuring the cytotoxicity of PC-12 cells using the lactate dehydrogenase (LDH) assay. Cells were seeded in 96-well plastic plates (CORNING, New York, NY, USA) at a density of 5×10^5 cells/well. After 24 h of culture, the cells were treated with the test sample for 24 h. Subsequently, 25 μL of culture supernatant was transferred to each well of a 96-well plate pre-loaded with 75 μL/well of the chromogenic substrate for MTX-LDH assay. The plates were then incubated in the dark for 15 min at room temperature. The reaction was terminated with 1 M HCl, and the absorbance at 540 nm was measured using a microplate reader (Bio-Rad, Hercules, CA, USA).

The in vivo toxicity and the LD_{50} of stPEG40/Res were determined by limiting studies in five male ddY mice (4 weeks old). The mice received intravenous injections of 100 mg/kg Res-equivalent stPEG40/Res via the tail vein every 24 h. Mice that became moribund were euthanized for humane reasons (none of them exhibited morbidity in this study). Surviving animals were monitored for delayed mortality over a 14-day period.

2.9. RNA Preparation and Quantitative Real-Time PCR

Changes in the expression of antioxidant genes were measured in PC-12 cells. The cells were seeded at a density of 1×10^6 cells/well in 35 mm dishes and treated with the test samples. After incubation for 6 h, the cells were washed with phosphate-buffered saline (PBS), and total RNA was extracted using the acid guanidinium thiocyanate-phenol-chloroform method. Complementary DNA (cDNA) was synthesized using MuLV reverse transcriptase and 50 pmol of random primers. Real-time reverse transcriptase-polymerase chain reaction (RT-PCR) was performed using a LightCycler 2.0 system (Roche Diagnostics Japan, Tokyo, Japan). In brief, reverse-transcribed cDNA, equivalent to 0.1 μg of total RNA, was used per reaction with LightCycler FastStart DNA Master SYBER Green I and 1 μM of primers. For the heme oxygenase-1 (HO-1) gene, the forward primer was 5′-TCTATCGTGCTCGCATG-3′ and the reverse primer was 5′-TTCCTCTGTCAGCAGTG-3′. For the glyceraldehyde 3-phosphate dehydrogenase (GAPDH) gene, the forward primer was 5′-TCTTCACCACCATGGAGA-3′ and the reverse primer was 5′-TGTCATGGATGACCTTGG-3′. Relative expression was calculated using the ΔΔCt method, with normalized to GAPDH as the endogenous control.

2.10. In Vivo ICH Mouse Model and Administration of stPEG/Res

To evaluate the therapeutic effects of stPEG/Res, a collagenase-induced ICH mouse model [28] was established. Mice were placed in a stereotactic frame (Narishige Scientific Instrument Lab, Tokyo, Japan) after an intraperitoneal injection (i.p.) of a mixture of medetomidine (0.3 mg/kg) and pentobarbital (40 mg/kg). The scalp was incised and the skull drilled. Collagenase type IV (0.03 U; volume: 3 μL/mouse; rate: 1 μL/min) or an equal volume of saline was injected into the right striatum (3.5 mm ventral from the skull surface, 2.2 mm lateral to the bregma midline suture, and 0.2 mm anterior to the coronal plane) using a Hamilton syringe (outer diameter: 0.019 inches; Hamilton, Reno, NV, USA). The Hamilton syringe was left for 5 min after injection and then gradually withdrawn. The scalp incision was closed with sutures and the mice were placed in a cage with ad libitum access to water and food. One hour prior to the collagenase injection, ZnPPIX, an inhibitor of HO-1 activity, was administered intraperitoneally at a dose of 5 mg/kg. ZnPPIX was dissolved in 0.1 M NaOH, neutralized with an equal volume of 0.1 M HCl, diluted 25-fold with 0.9% NaCl to a concentration of 1 mg/mL, and administered to the mice. Twenty minutes after the collagenase injection, stPEG40/Res (20 mg/kg Res-equivalent) or stPEG40 was administered via the tail vein. Both stPEG40/Res and stPEG40 were dissolved in 5% glucose before being administered.

2.11. Immunohistochemical Examinations

After 24 h of ICH induction, mice were reanesthetized with a mixture of medetomidine/pentobarbital i.p., and transcardial perfusion was performed with 10 mL of cold PBS and 4% PFA. Whole brains were excised, fixed in 4% PFA for 1 h, and then immersed in 15% sucrose overnight at 4 °C. The cerebellum was removed from the whole brain, frozen on dry ice, and stored at −80 °C until sectioning. Subsequently, 12 μm thick sections were prepared using a cryostat (Leica Biosystems, Wetzlar, Germany) and mounted on CREST-coated glass slides (Matsunami Glass Ind., Ltd., Osaka, Japan). NeuN antigen revival was performed by autoclaving the sections in 10 mM citric acid (pH 6.0) at 121 °C for 15 min (HIRAYAMA Manufacturing, Saitama, Japan). Sections were incubated with mouse anti-NeuN antibody (1:200) overnight at 4 °C following immersion in PBS containing 5% HS and 0.1% Triton X-100 for 1 h at room temperature. After washing with PBS, sections were incubated with biotinylated anti-mouse IgG (H + L) (1:200) for 1 h at room temperature. Finally, the sections were treated with an avidin-biotinylated horseradish peroxidase complex, and peroxidase activity was visualized using diaminobenzidine and H_2O_2.

Bright-field images were captured randomly using an inverted microscope (ECLIPSE TE300; Nikon, Tokyo, Japan) with a camera (DS-Ri2; Nikon). The NIS-Elements D software 4.30 (Nikon) was used for image acquisition and processing. Hematoma size was measured using coronal section images, including a syringe mask for collagenase injection. The area inside the hematoma was quantified using the ImageJ software 1.53t [29].

2.12. Behavioral Tests

The motor function of mice one day after ICH induction was assessed using corner-turn [30], modified beam-walking [29], and pole tests [31]. All animals were pretested, and investigators were blinded to group allocation.

In the beam-walking test, a 90 cm long beam, 1.5 cm wide and elevated 50 cm, was placed with dark home cages containing food on either end. The mice were placed at the center of the beam, and their behavior was observed and scored according to the criteria listed in Supplementary Table S2. Three trials of 2 min each were conducted, and the total score from all trials was calculated.

In the corner-turn test, the mice were directed toward a 30° corner, with the choice to turn left or right based on whisker contact with the wall. Ten trials were performed, and the percentage of turns to the right was calculated.

The pole test used a horizontal, rough-surfaced pole (8 mm diameter). The mice were placed on this pole with their forelimbs. Their behavior was observed and scored according to the criteria shown in Supplementary Table S3. Three trials of 30 s each were conducted, and the total score was calculated from all trials.

2.13. Statistical Analysis

Data are presented as mean ± standard deviation or standard error of the mean. ANOVA followed by Tukey's multiple comparison test was used to analyze statistical differences between multiple groups. A p-value < 0.05 was considered statistically significant. R version 4.3.0 (Vienna, Austria) was used for all statistical analyses.

3. Results and Discussion

3.1. Characterization of stPEG/Res

stPEG forms nanoparticles at a concentration of approximately 1–10 μg/mL in an aqueous environment due to aggregation of stearyl groups [27]. At concentrations well above the CMC of stPEG, most stPEG molecules are thought to form micelles in water. Therefore, to efficiently incorporate Res into the hydrophobic core of the micelles, we used an stPEG aqueous solution at 5 mg/mL in the formulation, significantly exceeding the CMC. The Res content in the stPEG/Res formulation was determined using HPLC and ranged from 8.1% to 10.3% *w*/*w* (Table 1), which was approximately consistent with the initial weight ratio of Res during preparation. Table 1 summarizes the results of the DLS analysis and zeta potential measurements. DLS analysis of each formulation revealed that the diameter of stPEG10/Res was 85.4 nm, while the other stPEG/Res formulations had diameters of approximately 10–20 nm. The zeta potential values of all formulations were approximately −20 mV. Nanoparticles with diameters smaller than 100 nm are known to increase blood circulation to avoid rapid trapping by the reticuloendothelial system in the liver, spleen, and lungs [32]. The size of stPEG/Res was within this range, suggesting that it was suitable for intravenous formulations. The particle size and zeta potential of the PEGylated nanoparticles depend on the surface density, length, and structure of the PEG chains. The areal density of PEG on nanoparticles can affect its conformation, with a high surface density resulting in a brush-like appearance and a low surface density producing a mushroom-like appearance [33]. Owing to these factors, stPEG10/Res may have a larger particle size than the other stPEG/Res formulations.

Table 1. Properties of the stPEG/Res.

Sample	Res Content (% *w/w*)	Diameter (nm)	Zeta Potential (mV)	Saturation Solubility (mg/mL) †
stPEG10/Res	10.3 ± 1.8	85.4 ± 7.2	−20.7 ± 0.5	0.93 ± 0.03
stPEG25/Res	8.8 ± 0.2	10.6 ± 0.6	−16.0 ± 4.2	12.42 ± 0.24
stPEG40/Res	8.6 ± 0.6	11.8 ± 0.2	−17.6 ± 4.6	12.30 ± 0.35
stPEG55/Res	9.8 ± 0.6	11.4 ± 0.3	−18.9 ± 2.0	12.19 ± 0.48
stPEG140/Res	8.1 ± 1.2	16.6 ± 1.2	−21.3 ± 1.4	0.43 ± 0.01

Values are expressed as mean ± standard deviation (n = 3). † The saturation solubility immediately after dissolution is expressed as Res-equivalent.

Saturation solubility measurements revealed that the Res formulation with stPEG10 and 140 had low saturation solubility, less than 1 mg/mL Res equivalent, while the formulations consisting of stPEG25, 40, and 55 demonstrated very high saturation solubility, approximately 12 mg/mL Res equivalent (Table 1). The emulsifying and dispersing abilities of a nonionic surfactant can be estimated by calculating the hydrophile–lipophile balance (HLB) value from its chemical structure. The HLB values of stPEG10–140 were calculated using the formula proposed by Griffin [34]: HLB = PEG (wt. %)/5. The calculations resulted in HLB values of 12.2, 15.9, 17.2, 17.9, and 19.1 for stPEG 10–140, respectively. Griffin classified surfactants in the HLB range of 10–13 as o/w emulsifiers, 13–15 as detergents, and 15–18 as solubilizers [34]. The saturation solubility results suggest that a solubilizers with HLB values of 15–18 are suitable for solubilizing Res using nonionic surfactant stPEG.

We further evaluated the precipitation properties of the three stPEG/Res samples with high saturation solubility after one week of storage. At a micellar concentration of 5 mg/mL, the stPEG40/Res solution remained clear, whereas the stPEG25/Res and stPEG55/Res solutions appeared cloudy (Figure 1). In addition, stPEG10/Res and stPEG140/Res at 5 mg/mL could not be dissolved without sonication. However, stPEG25-55/Res remained soluble without sonication under gentle agitation. It was suggested that stPEG10/Res and stPEG140/Res are unstable particles that tend to form aggregates and do not disperse unless energy such as sonication is applied.

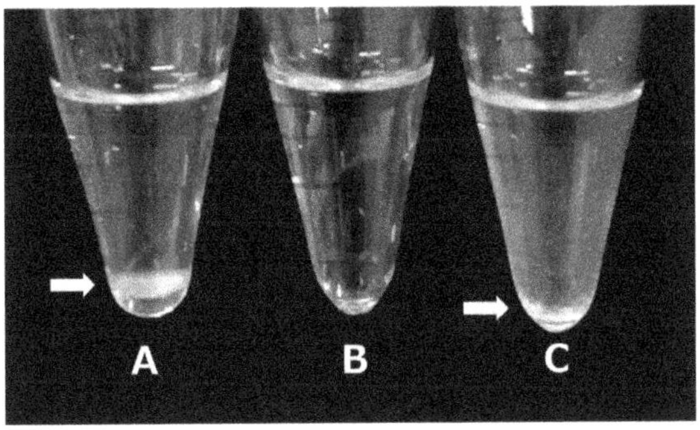

Figure 1. Precipitation of stPEG25/Res (**A**), stPEG40/Res (**B**), and stPEG55/Res (**C**) at a formulation concentration of 5 mg/mL in water after 1 week.

3.2. Neuroprotective Effect of stPEG/Res in ICH

Res exhibits protective activity against central nervous system damage by increasing the expression of antioxidant enzymes, including HO-1 [35]. Therefore, we focused on stPEG40/Res, which demonstrated no precipitation (Figure 1) and had the highest amount of Res concentration in the brain 10 min after administration (Supplementary Table S1). We inferred HO-1-mediated neuroprotective activity by measuring the increase in HO-1 gene expression following stPEG40/Res exposure in PC-12 cells (Figure 2). Exposure of 15 μM Res, a non-toxic concentration (Figure 3), significantly increased HO-1 mRNA expression compared to the control group in PC-12 cells. Furthermore, exposure of stPEG40/Res at an equivalent Res concentration of 15 μM significantly increased HO-1 mRNA expression compared to the control and the Res-exposed groups. The results showed that the Res formulation with stPEG may have a higher drug efficacy. stPEG promotes the intracellular translocation of loaded drugs [36]. The greater increase in HO-1 gene expression after exposure to this formulation may have been due to the enhanced internalization of Res by stPEG.

Following administration of stPEG40/Res at 20 mg/kg Res-equivalent dose, the average Res concentration in the brain was 0.75 μg/brain (Supplementary Table S1), approximately 11 μM. Given the results shown in Figures 2 and 3, this concentration was expected to increase HO-1 expression with low toxicity. Therefore, we evaluated the in vivo toxicity of stPEG40/Res and the in vivo efficacy of stPEG40/Res following ICH. To evaluate the in vivo toxicity of stPEG40/Res, we performed LD_{50} measurements, which indicated that the LD_{50} for stPEG40/Res was higher than that of the 100 mg/kg Res-equivalent (Table 2).

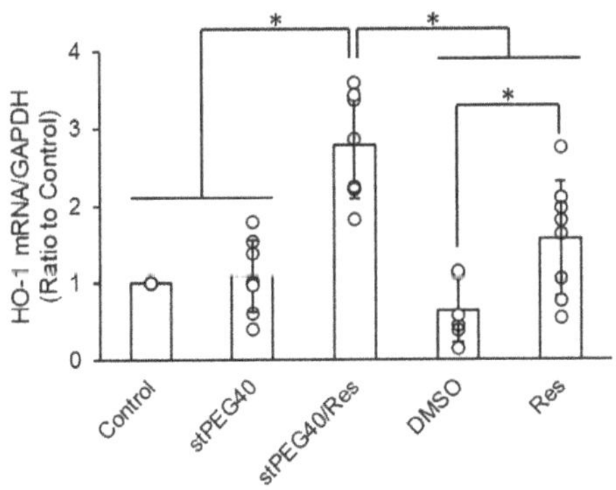

Figure 2. Effect of treatment with stPEG40/Res on the mRNA expression of HO-1 in PC-12 cells (n = 6–8). The values are presented as mean ± standard deviation. *—$p < 0.05$.

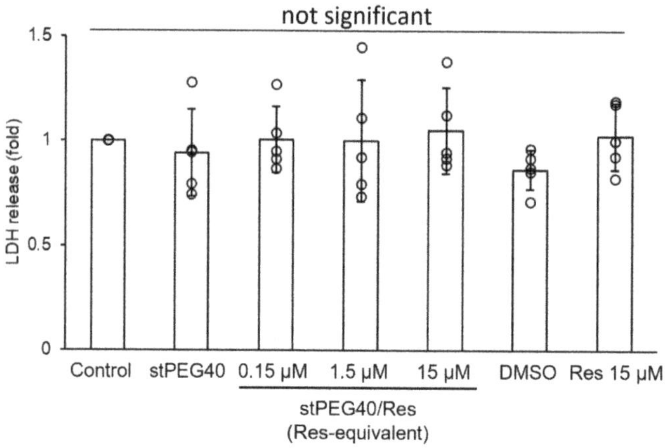

Figure 3. Cell injury in PC-12 cells induced by Res and stPEG40/Res, as assessed by the LDH assay. Cells were exposed to the samples for 24 h (n = 5). The values are presented as mean ± standard deviation.

Table 2. In vivo acute toxicity of stPEG40/Res in ddY mice.

Sample	LD_{50} (mg/kg)	95% Confidence Interval (mg/kg)
stPEG40/Res	>100 †	–

†—Res-equivalent.

Based on previous results, we evaluated the efficacy of stPEG40/Res at a 20 mg/kg Res-equivalent dose in an ICH model (Figure 4A), which was sufficiently lower than the LD_{50} value and was expected to achieve pharmacologically effective Res concentrations in the brain. The in vivo efficacy of stPEG/Res against ICH was evaluated using a collagenase-induced ICH mouse model, a widely used animal experimental model [37]. The neuroprotective effects of stPEG40/Res were assessed by measuring nerve damage and hematoma size through immunohistochemical staining with an anti-NeuN antibody. The hematoma size 24 h after ICH induction was not significantly different between the solvent and stPEG40/Res groups (Figure 4E). In contrast, the number of NeuN-positive neurons around the hematoma was significantly higher in the stPEG40/Res group than that in the solvent group (Figure 4B–D). Furthermore, administration of the HO-1 inhibitor ZnPPIX prior to stPEG40/Res treatment reversed the increase in NeuN-positive neurons caused by stPEG40/Res (Figure 4B–D). Thus, the neuroprotective effects of stPEG40/Res were mediated by HO-1.

Given the neuroprotective effects of stPEG40/Res, we expected that this formulation would be effective against motor dysfunction caused by ICH. We investigated its effect in suppressing motor dysfunction (Figure 5). Intravenous administration of stPEG40/Res at 20 mg/kg Res-equivalent to an ICH mouse model significantly improved performance on the beam walking test and corner turn test compared to the vehicle-administered group (Figure 5A,B). In addition, pole test scores tended to improve with stPEG40/Res administration (Figure 5C). Similar to the immunohistochemical staining evaluation, we performed a previous study in which ZnPPIX was administered. Regarding the beam-walking and pole test scores, the groups administered with both ZnPPIX and stPEG40/Res showed scores comparable to the vehicle-administered group (Figure 5A,C). In other words, similarly to the examination of the neuroprotective effect, ZnPPIX inactivated the effect of stPEG40/Res in improving motor dysfunction. These results indicate that the neuroprotective and motor dysfunction-improving effects of stPEG40/Res are mediated via HO-1.

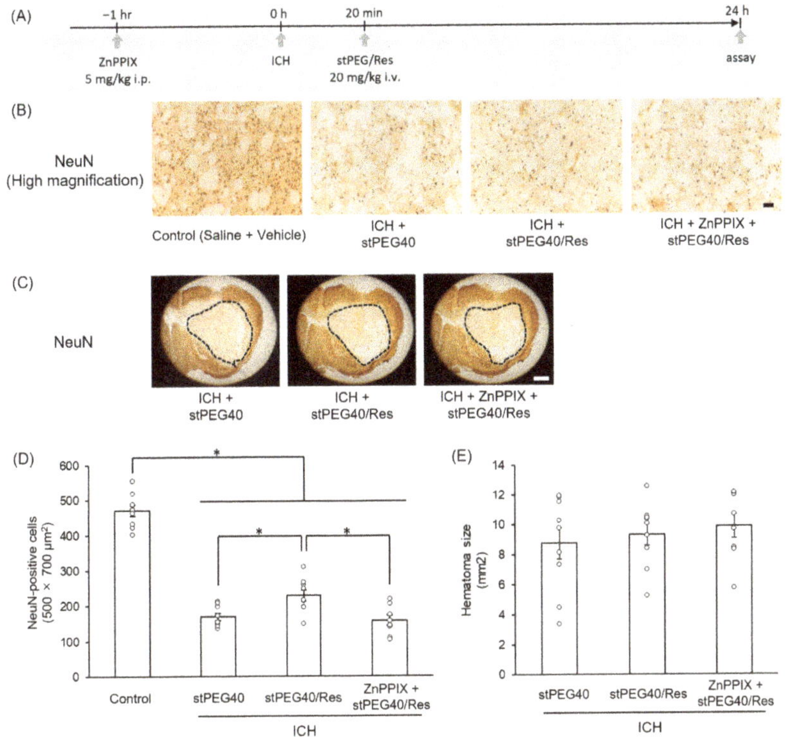

Figure 4. Neuroprotective effect of stPEG40/Res against intracerebral hemorrhage (ICH). (**A**) Experimental schedule using ICH model. (**B**) Magnified images of NeuN-immunopositive signals in the perihematomal area 24 h after ICH. Scale bar = 50 mm. (**C**) Representative images of NeuN immunostaining in the hematoma 24 h after ICH. Scale bar = 1 mm. Weak signals inside the dotted line are considered as a hematoma. (**D**) Number of NeuN-positive cells in the periphery of the hematoma (n = 8–9). (**E**) Summary of the hematoma size (n = 8–9). The values are presented as mean ± standard error. *—$p < 0.05$.

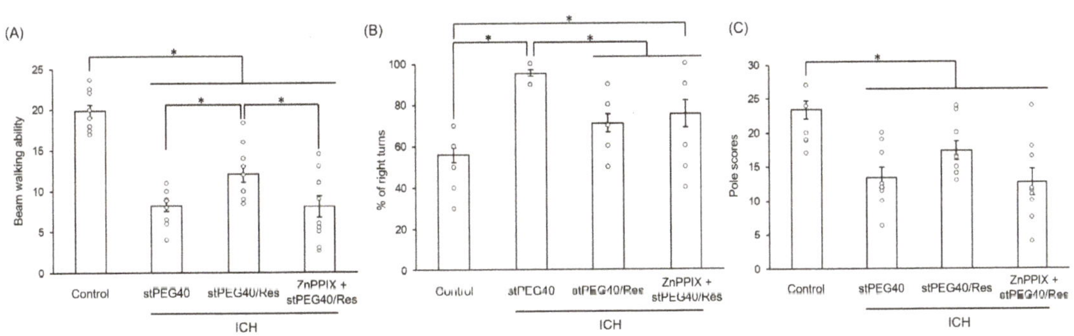

Figure 5. Efficacy of stPEG40/Res in alleviating motor dysfunction caused by intracerebral hemorrhage (ICH). (**A**) Effects of ICH and stPEG40/Res on beam walking ability (n = 10–11). (**B**) Effects of ICH and stPEG40/Res on corner turn test scores (n = 11–12). (**C**) Effects of ICH and stPEG40/Res on pole test scores (n = 9–10). The values are presented as mean ± standard error. *—$p < 0.05$.

Based on these findings, we successfully developed stPEG/Res, a water-soluble Res formulation that is effective for treating ICH and can be easily administered intravenously, even to unconscious patients. Furthermore, Res has been reported to be effective against various diseases, such as cardiovascular disease [38], kidney disease [39], malignant tumors [40], and neurological diseases [9–12]. Thus, stPEG/Res may have potential applications as therapeutic agents for these diseases. In this study, we used ZnPPIX to show that the neuroprotective effects and motor dysfunction improvement effects of Res in ICH were mostly mediated through HO-1. To the best of our knowledge, this is the first study to show that the mechanism of action of Res in ICH is primarily mediated through HO-1. This research is an example showing that poorly soluble compounds that could not previously be used as therapeutic agents because of bioavailability issues can be used as therapeutic agents by incorporating them into nanoparticles such as nanomicelles using stPEG or other materials. We hope that this research will have an impact on the development of new therapeutic agents that solubilize poorly soluble compounds.

4. Conclusions

In this study, we developed a novel intravenous micelle formulation composed of Res and stPEG, which exhibited suitable properties for intravenous administration. This formulation is inexpensive and easy to prepare compared with other nano-formulations. Furthermore, in vitro experiments using PC-12 cells demonstrated that stPEG40/Res induced the gene expression of enzymes that protect against oxidative stress and inflammation. In vivo experiments using a collagenase-induced ICH mouse model showed that intravenous administration of stPEG40/Res provided neuroprotective effects and suppressed motor dysfunction following ICH. These effects may be due to the upregulation of antioxidant and anti-inflammatory enzyme expression by Res. Consequently, these results suggest that stPEG/Res represents a novel and effective therapeutic formulation for ICH treatment.

Supplementary Materials: The following supporting information can be downloaded at: https://www.mdpi.com/article/10.3390/pharmaceutics16111462/s1, File S1: Biodistribution to the brain. Table S1: Distribution of Res in the brain; File S2: Scoring criteria for behavioral tests. Table S2: Score criteria for modified beam-walking test; Table S3: Score criteria for pole test.

Author Contributions: Conceptualization, K.B., Y.K. (Yoshiharu Kaneo), T.T. and M.O.; methodology, K.B., A.M., S.S., Y.H. and M.O.; validation, K.B., A.M., S.K., T.Y., R.O., Y.K. (Yui Kawamoto), H.K. and M.O.; formal analysis, K.B., A.M., S.S., T.T. and M.O.; investigation, K.B., A.M., S.K., T.Y., R.O., Y.K. (Yui Kawamoto), H.K., S.S., Y.H. and M.O.; writing—original draft preparation, K.B. and A.M.; writing—review and editing, K.B., Y.K. (Yoshiharu Kaneo), T.T. and M.O.; visualization, K.B., A.M., S.S. and M.O.; supervision, K.B. and M.O.; project administration, K.B., Y.K. (Yoshiharu Kaneo) and M.O.; funding acquisition, K.B., S.S. and M.O. All authors have read and agreed to the published version of the manuscript.

Funding: This research was funded by a research grant from the Urakami Scholarship Foundation and a donor scholarship from the Sekishinkai Medical Corporation. Both grants are for 2023, and no numbers have been assigned. This work was also supported by the Brand Research Project of Fukuyama University. This grant is for 2024, and the grant number is BRP-24-03-4.

Institutional Review Board Statement: All animal experiments described were performed following the ARRIVE guidelines, Directive 2010/63/EU of the European Parliament, European Union Council for Protection of Animals used for scientific purposes, and Fukuyama University guidelines for animal experimentation under registration number 2024-A-01 (approval date: 20 March 2024).

Informed Consent Statement: Not applicable.

Data Availability Statement: The original contributions presented in the study are included in the article/Supplementary Materials; further inquiries can be directed to the corresponding author.

Acknowledgments: We appreciate the valuable advice provided by Hiroshi Matsuoka, Yuka Nagatsuka-Handa, and Machiko Honda. We would like to thank Yuka Hokimoto, Akiho Miyamaru, Miyu Otsubo, and Hiyori Kuroki for their technical assistance.

Conflicts of Interest: The authors declare that this study received funding from Sekishinkai Medical Corporation. The funder was not involved in the study design, collection, analysis, interpretation of data, the writing of this article or the decision to submit it for publication.

References

1. Leasure, A.C.; Qureshi, A.I.; Murthy, S.B.; Kamel, H.; Goldstein, J.N.; Walsh, K.B.; Woo, D.; Shi, F.D.; Huttner, H.B.; Ziai, W.C.; et al. Intensive Blood Pressure Reduction and Perihematomal Edema Expansion in Deep Intracerebral Hemorrhage. *Stroke* **2019**, *50*, 2016–2022. [CrossRef] [PubMed]
2. Pinho, J.; Costa, A.S.; Araújo, J.M.; Amorim, J.M.; Ferreira, C. Intracerebral Hemorrhage Outcome: A Comprehensive Update. *J. Neurol. Sci.* **2019**, *398*, 54–66. [CrossRef] [PubMed]
3. Zhou, X.; Xie, L.; Altinel, Y.; Qiao, N. Assessment of Evidence Regarding Minimally Invasive Surgery vs. Conservative Treatment on Intracerebral Hemorrhage: A Trial Sequential Analysis of Randomized Controlled Trials. *Front. Neurol.* **2020**, *11*, 426. [CrossRef] [PubMed]
4. Wu, F.; Liu, Z.; Li, G.; Zhou, L.; Huang, K.; Wu, Z.; Zhan, R.; Shen, J. Inflammation and Oxidative Stress: Potential Targets for Improving Prognosis After Subarachnoid Hemorrhage. *Front. Cell Neurosci.* **2021**, *15*, 739506. [CrossRef] [PubMed]
5. Bai, Q.; Sheng, Z.; Liu, Y.; Zhang, R.; Yong, V.W.; Xue, M. Intracerebral Haemorrhage: From Clinical Settings to Animal Models. *Stroke Vasc. Neurol.* **2020**, *5*, 388–395. [CrossRef]
6. Wang, Z.; Zhou, F.; Dou, Y.; Tian, X.; Liu, C.; Li, H.; Shen, H.; Chen, G. Melatonin Alleviates Intracerebral Hemorrhage-Induced Secondary Brain Injury in Rats via Suppressing Apoptosis, Inflammation, Oxidative Stress, DNA Damage, and Mitochondria Injury. *Transl. Stroke Res.* **2018**, *9*, 74–91. [CrossRef]
7. Zheng, Y.; Li, R.; Fan, X. Targeting Oxidative Stress in Intracerebral Hemorrhage: Prospects of the Natural Products Approach. *Antioxidants* **2022**, *11*, 1811. [CrossRef]
8. Sessa, M.; Tsao, R.; Liu, R.; Ferrari, G.; Donsì, F. Evaluation of the Stability and Antioxidant Activity of Nanoencapsulated Resveratrol during in Vitro Digestion. *J. Agric. Food Chem.* **2011**, *59*, 12352–12360. [CrossRef]
9. Rauf, A.; Imran, M.; Sulera, H.A.R.; Ahmad, B.; Peters, D.G.; Mubarak, M.S. A Comprehensive Review of the Health Perspectives of Resveratrol. *Food Funct.* **2017**, *8*, 4284–4305. [CrossRef]
10. Hou, Y.; Wang, K.; Wan, W.; Cheng, Y.; Pu, X.; Ye, X. Resveratrol Provides Neuroprotection by Regulating the JAK2/STAT3/PI3K/AKT/MTOR Pathway after Stroke in Rats. *Genes Dis.* **2018**, *5*, 245–255. [CrossRef]
11. Huang, N.; Zhang, Y.; Chen, M.; Jin, H.; Nie, J.; Luo, Y.; Zhou, S.; Shi, J.; Jin, F. Resveratrol Delays 6-Hydroxydopamine-Induced Apoptosis by Activating the PI3K/Akt Signaling Pathway. *Exp. Gerontol.* **2019**, *124*, 110653. [CrossRef] [PubMed]
12. Juan, M.E.; Vinardell, M.P.; Planas, J.M. The Daily Oral Administration of High Doses of Trans-Resveratrol to Rats for 28 Days Is Not Harmful. *J. Nutr.* **2002**, *132*, 257–260. [CrossRef]
13. Singh, N.; Bansal, Y.; Bhandari, R.; Marwaha, L.; Singh, R.; Chopra, K.; Kuhad, A. Resveratrol Protects against ICV Collagenase-Induced Neurobehavioral and Biochemical Deficits. *J. Inflamm.* **2017**, *14*, 14. [CrossRef] [PubMed]
14. Walle, T.; Hsieh, F.; DeLegge, M.H.; Oatis, J.E.; Walle, U.K. High Absorption but Very Low Bioavailability of Oral Resveratrol in Humans. *Drug Metab. Dispos.* **2004**, *32*, 1377–1382. [CrossRef]
15. Amri, A.; Chaumeil, J.C.; Sfar, S.; Charrueau, C. Administration of Resveratrol: What Formulation Solutions to Bioavailability Limitations? *J. Control. Release* **2012**, *158*, 182–193. [CrossRef] [PubMed]
16. Bonsack, F.; Alleyne, C.H.; Sukumari-Ramesh, S. Resveratrol Attenuates Neurodegeneration and Improves Neurological Outcomes after Intracerebral Hemorrhage in Mice. *Front. Cell Neurosci.* **2017**, *11*, 228. [CrossRef] [PubMed]
17. Jose, S.; Anju, S.S.; Cinu, T.A.; Aleykutty, N.A.; Thomas, S.; Souto, E.B. In Vivo Pharmacokinetics and Biodistribution of Resveratrol-Loaded Solid Lipid Nanoparticles for Brain Delivery. *Int. J. Pharm.* **2014**, *474*, 6–13. [CrossRef]
18. Fan, L.; Jin, L.; Tang, T.; Zheng, Y.; Chen, Z.; Lin, H.; Ding, C.; Wang, T.; Chen, H.; Guo, Y.; et al. Neutrophil-like PH-Responsive pro-Efferocytic Nanoparticles Improve Neurological Recovery by Promoting Erythrophagocytosis after Intracerebral Hemorrhage. *Theranostics* **2024**, *14*, 283–303. [CrossRef]
19. Xu, J.; Chen, Z.; Yu, F.; Liu, H.; Ma, C.; Xie, D.; Hu, X.; Leak, R.K.; Chou, S.H.Y.; Stetler, R.A.; et al. IL-4/STAT6 Signaling Facilitates Innate Hematoma Resolution and Neurological Recovery after Hemorrhagic Stroke in Mice. *Proc. Natl. Acad. Sci. USA* **2020**, *117*, 32679–32690. [CrossRef]
20. Blanco, E.; Kessinger, C.W.; Sumer, B.D.; Gao, J. Multifunctional Micellar Nanomedicine for Cancer Therapy. *Exp. Biol. Med.* **2009**, *234*, 123–131. [CrossRef]
21. Torchilin, V.P. Micellar Nanocarriers: Pharmaceutical Perspectives. *Pharm. Res.* **2007**, *24*, 1–16. [CrossRef] [PubMed]
22. Yokoyama, M. Polymeric Micelles as a New Drug Carrier System and Their Required Considerations for Clinical Trials. *Expert. Opin. Drug Deliv.* **2010**, *7*, 145–158. [CrossRef]
23. Andreani, T.; Fangueiro, J.F.; José, S.; Santini, A.; Silva, A.M.; Souto, E.B. Hydrophilic Polymers for Modified-Release Nanoparticles: A Review of Mathematical Modelling for Pharmacokinetic Analysis. *Curr. Pharm. Des.* **2015**, *21*, 3090–3096. [CrossRef]
24. Bailon, P.; Won, C.Y. PEG-Modified Biopharmaceuticals. *Expert. Opin. Drug Deliv.* **2009**, *6*, 1–16. [CrossRef]
25. Rowe, R.C.; Sheskey, P.J.; Quinn, M.E. *Handbook of Pharmaceutical Excipients*, 6th ed.; Pharmaceutical Press: London, UK, 2009; ISBN 978 0 85369 792 3.

26. Banshoya, K.; Nakamura, T.; Tanaka, T.; Kaneo, Y. Development of α-Tocopherol Nanomicellar Formulation Using Polyethylene Glycol Monostearate for the Oxidative Stress-Related Disease. *J. Drug Deliv. Sci. Technol.* **2021**, *61*, 102310. [CrossRef]
27. Banshoya, K.; Nakamura, T.; Tanaka, T.; Kaneo, Y. Coenzyme Q10-Polyethylene Glycol Monostearate Nanoparticles: An Injectable Water-Soluble Formulation. *Antioxidants* **2020**, *9*, 86. [CrossRef]
28. Ohnishi, M.; Ohshita, M.; Tamaki, H.; Marutani, Y.; Nakayama, Y.; Akagi, M.; Miyata, M.; Maehara, S.; Hata, T.; Inoue, A. Shogaol but not gingerol has a neuroprotective effect on hemorrhagic brain injury: Contribution of the α, β-unsaturated carbonyl to heme oxygenase-1 expression. *Eur. J. Pharmacol.* **2019**, *842*, 33–39. [CrossRef] [PubMed]
29. Hijioka, M.; Matsushita, H.; Hisatsune, A.; Isohama, Y.; Katsuki, H. Therapeutic Effect of Nicotine in a Mouse Model of Intracerebral Hemorrhage. *J. Pharmacol. Exp. Ther.* **2011**, *338*, 741–749. [CrossRef]
30. Hua, Y.; Schallert, T.; Keep, R.F.; Wu, J.; Hoff, J.T.; Xi, G. Behavioral Tests after Intracerebral Hemorrhage in the Rat. *Stroke* **2002**, *33*, 2478–2484. [CrossRef]
31. Chen, W.; Xia, M.; Guo, C.; Jia, Z.; Wang, J.; Li, C.; Li, M.; Tang, X.; Hu, R.; Chen, Y.; et al. Modified Behavioural Tests to Detect White Matter Injury-Induced Motor Deficits after Intracerebral Haemorrhage in Mice. *Sci. Rep.* **2019**, *9*, 16958. [CrossRef]
32. Minchin, R.F.; Martin, D.J. Minireview: Nanoparticles for Molecular Imaging—An Overview. *Endocrinology* **2010**, *151*, 474–481. [CrossRef] [PubMed]
33. Rabanel, J.M.; Hildgen, P.; Banquy, X. Assessment of PEG on Polymeric Particles Surface, a Key Step in Drug Carrier Translation. *J. Control. Release* **2014**, *185*, 71–87. [CrossRef] [PubMed]
34. Griffin, W.C. Classification of the Surface-Active Agents by HLB. *J. Soc. Cosmet. Chem.* **1949**, *1*, 311–326.
35. Rosa, P.M.; Martins, L.A.M.; Souza, D.O.; Quincozes-Santos, A. Glioprotective Effect of Resveratrol: An Emerging Therapeutic Role for Oligodendroglial Cells. *Mol. Neurobiol.* **2018**, *55*, 2967–2978. [CrossRef] [PubMed]
36. Kuwano, M.; Ibuki, H.; Morikawa, N.; Ota, A.; Kawashima, Y. Cyclosporine A Formulation Affects Its Ocular Distribution in Rabbits. *Pharm. Res.* **2002**, *19*, 108–111. [CrossRef] [PubMed]
37. Wang, J. Preclinical and Clinical Research on Inflammation after Intracerebral Hemorrhage. *Prog. Neurobiol.* **2010**, *92*, 463–477. [CrossRef]
38. Bonnefont-Rousselot, D. Resveratrol and Cardiovascular Diseases. *Nutrients* **2016**, *8*, 250. [CrossRef]
39. Saldanha, J.F.; Leal, V.D.O.; Stenvinkel, P.; Carraro-Eduardo, J.C.; Mafra, D. Resveratrol: Why Is. It a Promising Therapy for Chronic Kidney Disease Patients? *Oxidative Med. Cell. Longev.* **2013**, *2013*, 963217. [CrossRef]
40. Almatroodi, S.A.; Alsahli, M.A.; Aljohani, A.S.M.; Alhumaydhi, F.A.; Babiker, A.Y.; Khan, A.A.; Rahmani, A.H. Potential Therapeutic Targets of Resveratrol, a Plant Polyphenol, and Its Role in the Therapy of Various Types of Cancer. *Molecules* **2022**, *27*, 2665. [CrossRef]

pharmaceutics

MDPI

Article

Combinatorial Effects of Free and Nanoencapsulated Forms of Cabazitaxel and RAS-Selective Lethal 3 in Breast Cancer Cells

Remya Valsalakumari [1,2,†], Marek Feith [1,3,†], Solveig Pettersen [1], Andreas K. O. Åslund [4], Ýrr Mørch [4,5], Tore Skotland [2,6], Kirsten Sandvig [2,6,7], Gunhild Mari Mælandsmo [1,8] and Tore-Geir Iversen [2,6,*]

1 Department of Tumor Biology, Institute for Cancer Research, Oslo University Hospital, 0379 Oslo, Norway; remyav@uio.no (R.V.); marek.feith@studmed.uio.no (M.F.); g.m.malandsmo@ous-research.no (G.M.M.)
2 Department of Molecular Cell Biology, Institute for Cancer Research, Oslo University Hospital, 0379 Oslo, Norway; tore.skotland@ous-research.no (T.S.); kirsten.sandvig@ibv.uio.no (K.S.)
3 Institute of Clinical Medicine, Faculty of Medicine, University of Oslo, 0379 Oslo, Norway
4 Department of Biotechnology and Nanomedicine, SINTEF AS, 7034 Trondheim, Norway; andreas.aaslund@sintef.no (A.K.O.Å.); yrr.morch@nadeno.com (Ý.M.)
5 NaDeNo Nanoscience AS, 7051 Trondheim, Norway
6 Centre for Cancer Cell Reprogramming, Faculty of Medicine, University of Oslo, 0379 Oslo, Norway
7 Department of Biosciences, University of Oslo, 0316 Oslo, Norway
8 Department of Medical Biology, University of Tromsø, 9019 Tromsø, Norway
* Correspondence: toregiv@medisin.uio.no
† These authors contributed equally to this work.

Abstract: Background: Combination therapies for cancer have gained considerable attention due to their potential for enhancing therapeutic efficacy and decreasing drug resistance. Introducing nanodrug delivery systems in this context may further improve the therapy due to targeted delivery, improved drug stability, sustained drug release, and prevention of rapid clearance from circulation. This study evaluates the combinatorial effects of two cytotoxic drugs, cabazitaxel (CBZ) and RSL3 (RAS-selective lethal 3), in free form as well as encapsulated within poly(2-ethyl butyl cyanoacrylate) (PEBCA) nanoparticles (NPs) in breast cancer cell lines. **Methods**: Cell proliferation was assessed using IncuCyte technology, and synergistic drug effects were determined with SynergyFinder Plus. Cell viability was measured with the MTT assay. Additionally, we investigated whether the combinatorial effects were reflected in alterations of metabolic activity or reactive oxygen species (ROS) production using Seahorse technology and the CM-H_2DCFDA assay, respectively. **Results**: The data presented reveal, for the first time, that CBZ and RSL3 exhibit synergistically or additively combinatorial effects on various breast cancer cell lines. The pattern of cytotoxic effects was consistent, whether the drugs were in free form or encapsulated in NPs. Moreover, the combinatorial effects were not observed to be associated with early changes in metabolic activity or ROS production. **Conclusion**: This study highlights the potential of CBZ and RSL3 in combinatorial nanomedicine as they may act synergistically. Further studies are warranted to better understand the mechanisms behind these combinatorial effects.

Keywords: cabazitaxel; RSL3; combinatorial effects; breast cancer; poly(2-ethyl butyl cyanoacrylate); nanoparticles

1. Introduction

Cancer is a multifactorial disease, where different cells and diverse molecules are engaged in complex regulatory networks and cross-talks between biological processes [1–3].

Hence, it is poorly manageable with a single therapeutic agent, which may lead to drug resistance [4] and tumor recurrence [5]. Combination therapy involving two or more therapeutic agents or modalities has evolved as a key strategy in cancer therapy [6]. Combining multiple agents may enhance efficacy, reduce drug resistance, and improve the overall survival of patients by targeting multiple pathways or mechanisms either synergistically or independently [6–8]. In the present scenario, chemotherapy has become an indispensable part of treatment even if the drugs used contribute to undesirable effects and adverse reactions [9]. This results in a narrow therapeutic window in addition to resistance [10]. Hence, combining chemotherapy with radiation, immunotherapy, or other targeted therapies has now been implemented in clinical practice, and novel combinations are continually being reported at the preclinical phase or undergoing clinical trials [11–14].

Cabazitaxel (CBZ) is a second-generation taxane approved for the treatment of castration-resistant prostate cancer [15] and has also undergone clinical evaluation for the treatment of breast cancer [16–18]. It is a microtubule inhibitor like other taxanes but has the advantage of having less affinity to P-glycoprotein, resulting in increased cellular retention and higher efficacy [19]. Still, low aqueous solubility and toxicity can limit clinical use, where alternate approaches are needed [20]. RSL3 is a known ferroptosis inducer [21], and it has been reported to induce improved anticancer effects when combined with other chemotherapy drugs, monoclonal antibodies, and mTOR inhibitors through different mechanisms [22–25]. It has been reported that RSL3 acts by inactivating the protein glutathione peroxidase 4 (GPX4) after binding to its active site, which in turn causes iron-dependent accumulation of ROS within the cells and eventually leads to ferroptotic cell death [21]. However, poor pharmacokinetic properties [26] and off-target side effects limit its potential for clinical applications.

In the present study, we have investigated how the combination of CBZ and RSL3 affects the viability of breast cancer cell lines. These cell lines, SKBR3, JIMT-1 and MCF7 (all human), and EO771 (murine), have different expression levels of estrogen receptors (ER), progesterone receptors (PR) and HER2 receptors, and they also have different sensitivity to CBZ and RSL3.

The effects were evaluated either when the drugs were in their free form or encapsulated within PEGylated PEBCA NPs. PEBCA NPs are synthesized by polymerization of 2-(ethyl butyl cyanoacrylate) monomers and possess excellent properties to encapsulate hydrophobic molecules such as CBZ or RSL3. We have published multiple articles evaluating the biological effects and efficacy of these NPs loaded with CBZ in different models of breast cancer, which showed promising results; see, e.g., [27–29]. Hence, it was interesting to conduct studies comparing the combinatorial effects of drugs in their free form or encapsulated within the PEBCA NPs. The present study demonstrates that CBZ and RSL3 act synergistically and are promising candidates for developing combinatorial therapy for breast cancer. Nonetheless, the mechanism of action needs further elucidation. Moreover, encapsulation of these drugs within PEBCA NPs did not hamper the combinatorial effects, indicating that these NPs may improve the pharmacokinetic properties and thus should be studied further in in vivo models.

2. Materials and Methods

2.1. Materials

RPMI 1640 medium, DMEM medium, bovine serum albumin (BSA), fetal bovine serum (FBS), dimethyl sulfoxide (DMSO), 3-(4,5-dimethyl-2-thiazolyl)-2,5-diphenyltetrazolium bromide (MTT; cat#M5655), 1S,3R-RSL3 (https://www.sigmaaldrich.com/NO/en/product/sigma/sml2234 (accessed on 24 August 2022); info of, i.e., formula and purity, >98%)

and penicillin/streptomycin (Pen/Strep P4333) were purchased from Merck (Darmstadt, Germany). DMEM F12 medium was procured from ThermoFisher Scientific (Waltham, MA, USA). CBZ was obtained from Biochempartner Co., Ltd. (Shanghai, China, https://www.biochempartner.com/183133-96-2-BCP27641 (accessed on 24 August 2022); i.e., with information on formula and purity, 98%). Milli-Q water was freshly prepared by the Millipore Milli-Q Biocell purification system (Burlington, MA, USA). All chemicals used were of analytical quality. Materials used for NP synthesis and other reagents are mentioned together with the methods.

2.2. Cell Lines

A panel of breast cancer cell lines having different molecular characteristics was used for the study. EO771 cells (murine, $ER^{\alpha-}$ PR^{+}, $HER2^{+}$) were cultured in DMEM. SKBR3 (ER^{-}, PR^{-}, $HER2^{+}$) and MCF-7 (luminal A, ER^{+}, PR^{+}, $HER2^{-}$) cells were cultured in RPMI 1640. JIMT-1 cells (ER^{-}, PR^{-}, $HER2^{+}$) were cultured using a 1:1 mixture of DMEM/F-12 medium. All media were supplemented with 10% (*v*/*v*) FBS and 100 units/mL penicillin/streptomycin. JIMT-1 cells were obtained from the German Collection of Microorganisms and Cell Cultures (DSMZ; Braunschweig, Germany). All cells were routinely subjected to mycoplasma testing.

2.3. Synthesis and Characterization of NPs

PEGylated PEBCA NPs were synthesized using a miniemulsion polymerization technique [27–30]. An oil phase (2.7 mL), containing 0.9 g of 2-ethylbutyl cyanoacrylate (EBCA) (Cuantum Medical Cosmetics, Bellaterra, Spain) and the co-stabilizer Miglyol® 812 (1.1% (*w*/*w*), Cremer Oleo GmbH & Co., KG, Hamburg, Germany), was mixed with a water phase (6.5 mL) consisting of the non-ionic PEG stabilizers Brij® L23 (5 mM, Sigma-Aldrich, St. Louis, MO, USA) and Kolliphor® HS 15 (6 mM, Sigma-Aldrich) in 0.1 M HCl. Compounds to be encapsulated were added to the oil phase (CBZ, 10% (*w*/*w* oil phase), and RSL3, 10% (*w*/*w* oil phase)) prior to mixing. The oil phase was added to the aqueous phase and immediately sonicated for 3 min on ice (6 × 30 sec intervals, 60% amplitude, Branson Ultrasonics digital sonifier 450, Brookfield, CT, USA). The solution was then rotated (15 rpm, Stuart SB3 rotator, Keison Products, Chelmsford, UK) at room temperature (RT) over two nights. The emulsions were extensively dialyzed (Spectra/Por dialysis membrane MWCO 100,000 Da, Spectrum Labs, Rancho Dominguez, CA, USA) against 1 mM HCl (pH 3.0). Loading of CBZ within PEBCA NPs was quantified by mass spectrometry analysis (Agilent, Santa Clara, CA, USA, 1290 HPLC system coupled to an Agilent 6490A triple quadrupole mass spectrometer), and encapsulation efficiency was determined as described earlier [27]. Loading of RSL3 was quantified by mass spectroscopy analysis. RSL3-PEBCA NPs were diluted in DMSO/acetonitrile/ammonium acetate (50%/45%/5%), and separation was performed using a column (Ascentis Express Phenyl Hexyl) with mobile phases A: water with 25 mM formic acid and B: acetonitrile with 25 mM formic acid. The size (diameter; z-average: intensity-based mean calculations), polydispersity index (PDI), and zeta potential (calculations based on the Helmholtz-Smoluchowski equation) of the NPs were measured in PBS, pH 7.4 (diluted 1:50, approx.NP conc. 2 µg/mL) by dynamic light scattering using a Zetasizer Nano ZS Malvern Instrument and the accompanied software package (ver. 6.01) for data acquisition and analysis (Malvern, UK). The NPs were found to be stable regarding size, drug loading and colloidal stability for a period of 2 years when stored at 4 °C in 1 mM HCl (pH 3.0), during which the experiments were performed.

2.4. Evaluation of Cell Proliferation

Cellular proliferation was evaluated by measuring cell confluence using the Incucyte technology [31]. Briefly, cells were seeded at a density of 8 × 10^3 cells/well of 96-well plates in suitable medium (200 µL). The plates were incubated for 24 h before treating them with either free drugs, PEBCA drugs or combinations in triplicate. This is followed by real-time image-based cell density measurements using the Incucyte® S3 live cell imaging and analysis system (Sartorius, Gottingen, Germany) equipped with a confluence analysis module. At least three images were captured per well. Measurements were started on day 0 of treatment and continued up to 4–6 days, depending upon the cell line. Data were plotted as % cell confluence normalized to the day of treatment initiation.

2.5. Analysis of Combinatorial Effects

The cell density measurements retrieved from the Incucyte analysis were used to assess the combinatorial effects of the treatments using the SynergyFinder Plus (version 3.10.3) web application [32]. Briefly, the normalized values of cell density measurements during the last time point of treatment were used to calculate percent inhibition with respect to untreated control. The SynergyFinder Plus application used these values to calculate the synergy scores according to different reference models. The degree of synergistic or antagonistic effect is quantified by comparing the observed combinatorial response against the expected response, calculated using a reference model that assumes no interaction between the drugs. Different reference models quantify the degree of synergy either as the excess over the maximum single drug response (highest single agent (HSA)) [33], as the multiplicative effect of single drugs as if they acted independently (Bliss) [34], as the expected response corresponding to an additive effect as if the single drugs were the same compound (Loewe) [35], or as the expected response corresponding to the effect as if the single drugs did not affect the potency of each other (zero interaction potency (ZIP)) [36]. Suitable synergy models were chosen for each cell line, according to the observed drug responses. The threshold is set in a way that synergy scores greater than 10 indicate synergy, scores between −10 and +10 indicate additivity and scores less than −10 indicate no combinatorial effects. However, synergistic effects are highly contextual; hence, this scale should be assumed as a spectrum where increasing positive values from zero span from additivity towards synergy and negative values from zero span across additivity towards no effect and then towards antagonistic effect.

2.6. Cell Viability Assay

Cell viability was assessed by the MTT assay as described elsewhere [29]. Briefly, cells were seeded and treated as described for the proliferation assay. Treated plates were incubated for 96 h at 37 °C. The medium was then aspirated, followed by the addition of MTT reagent (100 µL of 0.25 mg/mL in serum-free, phenol red-free medium) and further incubation for 3 h at 37 °C. Afterwards, the formazan crystals formed were dissolved in DMSO with 0.5% ammonia solution (200 µL/well), and the absorbance was measured at 570 nm against a 650 nm reference wavelength using a Synergy2 microplate reader (Biosys Ltd., Essex, UK). Percent cell viability was calculated with respect to the untreated control.

2.7. ROS Assay

Cells were seeded in glass-bottom white plates (Costar, Corning, NY, USA) at a density of 1.5 × 10^4/well in 200 µL media. After 24 h, the cells were gently washed with 100 µL pre-warmed medium without FBS and phenol red, followed by the addition of 10 µM ROS sensor (General Oxidative Stress Indicator, CM-H_2DCFDA, Invitrogen;

https://assets.thermofisher.com/TFS-Assets/LSG/manuals/mp36103.pdf (accessed on September 1 2022)) (50 µL/well). The plate was then incubated in the dark at 37 °C for 45 min. Subsequently, the sensor was removed, and the cells were treated with indicated concentrations of free drugs, PEBCA drugs, combinations or empty PEBCA NPs in suitable media. H_2O_2 (500 µM) was used as a positive control. Fluorescence readout was performed after 1 h and 5 h of treatment using a Victor™ X3 multiplate reader (Perkin Elmer, Waltham, MA, USA) equipped with fluorescent filters (excitation 485 nm, emission 535 nm) and a prewarmed plate holder at 37 °C. Data of treated cells are normalized to cells treated only with the sensor.

2.8. Metabolic Assay

Mitochondrial metabolism was assessed using the Seahorse XFe96 Analyzer (Agilent, Santa Clara, CA, USA) employing the Mito Stress Test according to the manufacturer's guidelines [37]. Briefly, the assay cartridge was hydrated and placed in an incubator at 37 °C without CO_2 supplementation overnight. Cells were incubated in Seahorse 96-well plates overnight and treated with either free drugs, PEBCA drugs, combinations or empty PEBCA NPs for 4 h. The cells were then rinsed with phenol red and bicarbonate-free medium (pH 7.4), supplemented with 10 mM glucose, 1 mM pyruvate, and 2 mM L-glutamine (Agilent, Santa Clara, CA, USA), and incubated for 1 h at 37 °C without CO_2. Modulating compounds used in the assay included 1.5 µM oligomycin, 0.5 µM rotenone/antimycin A (Agilent, Santa Clara, CA, USA), and 0.5 µM carbonyl cyanide 4-(trifluoromethoxy) phenylhydrazone (FCCP). Oligomycin inhibits ATP synthase, leading to a reduction in mitochondrial respiration by limiting the flow of electrons through the electron transport chain (ETC). FCCP, a protonophore, uncouples the proton gradient from respiration, enabling unrestricted electron flow through mitochondrial complexes and maximizing oxygen consumption at complex IV by disrupting the mitochondrial membrane potential. The spare respiratory capacity was determined as the difference between maximal and basal respiration rates. Additionally, the injection of rotenone and antimycin A inhibits complexes I and III, effectively halting the entire electron transport chain and allowing the measurement of non-mitochondrial respiration, driven by processes outside the mitochondria. The oxygen consumption rate (OCR) was measured using a solid-state sensor that detects changes in dissolved oxygen and free proton concentrations, which reflect oxygen consumption from mitochondrial respiration. Seahorse Wave Software v2.6.3 was utilized for data analysis. Normalization was conducted by determining total protein concentration via BCA Protein Assay (ThermoFischer Scientific, Waltham, MA, USA) at the end of the experiment, following the manufacturer's instructions.

2.9. Statistical Analysis

Statistical analysis was performed using GraphPad Prism v9 (GraphPad Software, La Jolla, CA, USA). Statistical significance in the experiments was determined either by one-way ANOVA followed by the Tukey–Kramer multiple comparison test. In some cases, a two-tailed unpaired Student's test was performed, which has been mentioned in respective figure legends. p values < 0.05 were considered statistically significant.

3. Results and Discussion

3.1. Physicochemical Characterization of NPs

The NPs used and their physicochemical properties are shown in Figure 1. When combinatory treatment with PEBCA-CBZ and PEBCA-RSL3 is performed, the terminology PEBCA drugs is used. NPs exhibit a size range of 130–175 nm with PDI less than 0.36 in

PBS, pH 7.4. They are also similar in their structure and PEGylation levels, while they differ in the percentage of drug loading, which is 6.8% and 5.1% of total dry weight for PEBCA-CBZ and PEBCA-RSL3, respectively.

A

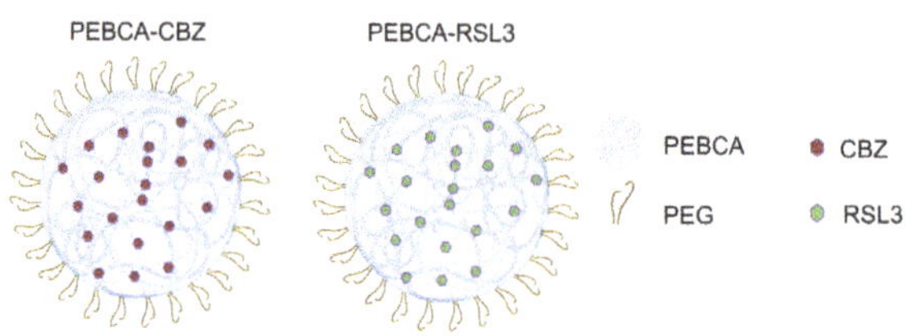

B

NPs	Size (nm)	Zeta potential (mV)	PDI	Drug loading (% w/w)
PEBCA-CBZ	149 ± 3	−5.8	0.24 ± 0.03	6.8
PEBCA-RSL3	145 ± 25	−5.1	0.36 ± 0.13	5.1
Empty PEBCA	160 ± 11	−5.4	0.33 ± 0.11	-

Figure 1. Schematic representation and physicochemical properties of NPs. (**A**) Representation of PEGylated PEBCA NPs used in different studies: empty NPs and NPs loaded either with CBZ or RSL3. (**B**) Physicochemical characteristics of NPs such as size, zeta potential, PDI and % (*w*/*w*) drug loading with respect to CBZ and RSL3. Size and PDI were measured whenever the NPs were used for experiments. Presented here are averages of 4 measurements performed from the time of preparation and for a period of 2 years. Zeta potential and drug loading were measured at one time point, within a week after the preparation of the NPs.

3.2. Combination of CBZ and RSL3 Improved Cytotoxic Effects Compared to Drugs Alone

The microtubule inhibitor, CBZ, and the ferroptosis inducer, RSL3, are well-known cytotoxic agents and have been utilized in several studies along with other therapeutic agents [15,16,21,22]. Our previous studies have shown that PEBCA NPs are suitable to encapsulate hydrophobic drugs such as CBZ, which also improved the therapeutic effects [27,29]. This prompted us to explore the combinatorial effects of two encapsulated drugs and compare them with the combination of free drugs. Thus, we decided to encapsulate RSL3 into PEBCA NPs and investigate whether a combinatory treatment with RSL3 and CBZ could improve the cytotoxic effects compared to the drugs alone. Studies were performed in four breast cancer cell lines with different origins, molecular characteristics and sensitivity to CBZ or RSL3. The treated cells were monitored for their proliferative capacity, and the combinatorial effects were analyzed using the SynergyFinder Plus application. Figures 2–5 show the effect of combinatorial treatment of CBZ and RSL3, either in their free form or encapsulated in PEBCAs in EO771, SKBR3, JIMT-1 and MCF-7 cells, respectively. The results corresponding to each cell line are discussed as below.

EO771: Figure 2A represents the synergy scores for combinations of different concentrations of CBZ and RSL3 in EO771 cells. The HSA reference model is found suitable for this cell line because one of the drugs (CBZ) is not inducing any inhibitory effect. The combination of 100 nM RSL3 and increasing concentrations of CBZ shows the highest synergy scores. Figure 2C shows the response curves for the combinations with the most pronounced synergistic effect. It is notable from the dose–response curves of CBZ and RSL3 (Figures S1 and S2) that EO771 cells are insensitive to CBZ at the concentrations studied but are sensitive to RSL3 in a time- and concentration-dependent manner. However, the sensitivity to RSL3 is transient, and the cells regained the capacity to proliferate effectively at a later time point. Hence, the evident combinatorial effects are relevant, which means that CBZ plays a significant role in maintaining the inhibition caused by RSL3, and this inhibition increases with increasing concentration of CBZ. Similar is the case of encapsulated drugs, where PEBCA-CBZ and PEBCA-RSL3 showed similar dose responses (Figures S1 and S2) and combinatorial effects as free drugs (Figure 2C,D).

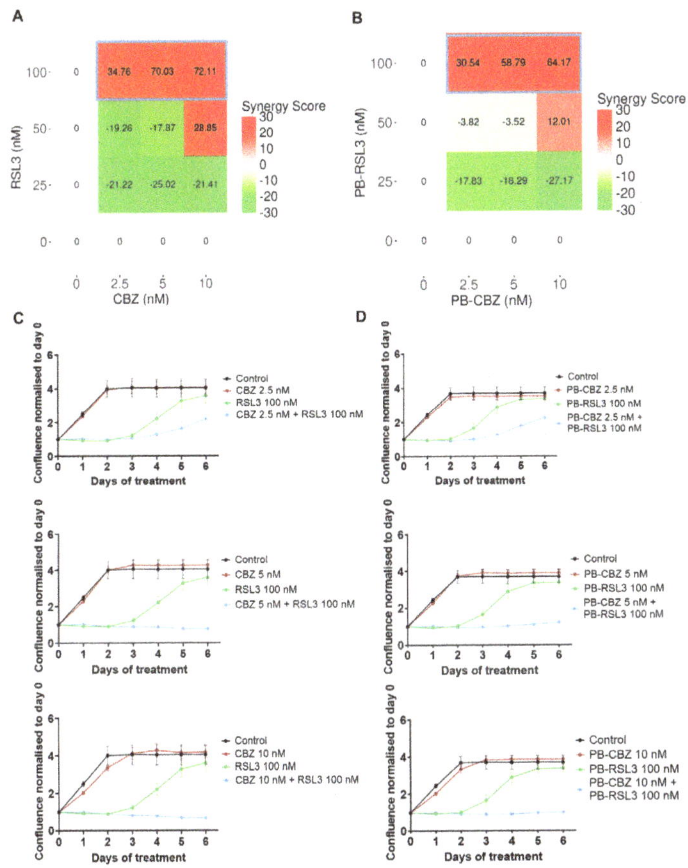

Figure 2. Combinatorial effects of CBZ and RSL3 or PEBCA-CBZ and PEBCA-RSL3 in EO771 cells. Figures shown in the left panel correspond to free drugs or their combinations, and the right panel corresponds to PEBCA drugs or their combinations. EO771 cells were treated with different concentrations of CBZ/PEBCA-CBZ and RSL3/PEBCA-RSL3 as single drugs or as combinations. Drug responses were evaluated by quantifying cell confluence through Incucyte analysis, and synergy scores were derived using SynergyFinder Plus. HSA model scores are shown here. (**A**,**B**) Synergy scores of all combinations used, respectively, for drugs alone and PEBCA drugs, confirming the synergistic effects of different concentrations of CBZ with higher concentrations of RSL3. Combinations showing high synergy scores are outlined in blue (**C**,**D**). Plots showing drug responses of combinations that showed synergistic effects and highest scores (outlined in blue in panels (**A**,**B**)), respectively, for drugs alone and PEBCA drugs. Presented here are mean values with standard deviations (error bars) of cell confluence normalized to day 0 of treatment from one out of three independent experiments, each run with triplicates. PEBCA is abbreviated to PB in graphs.

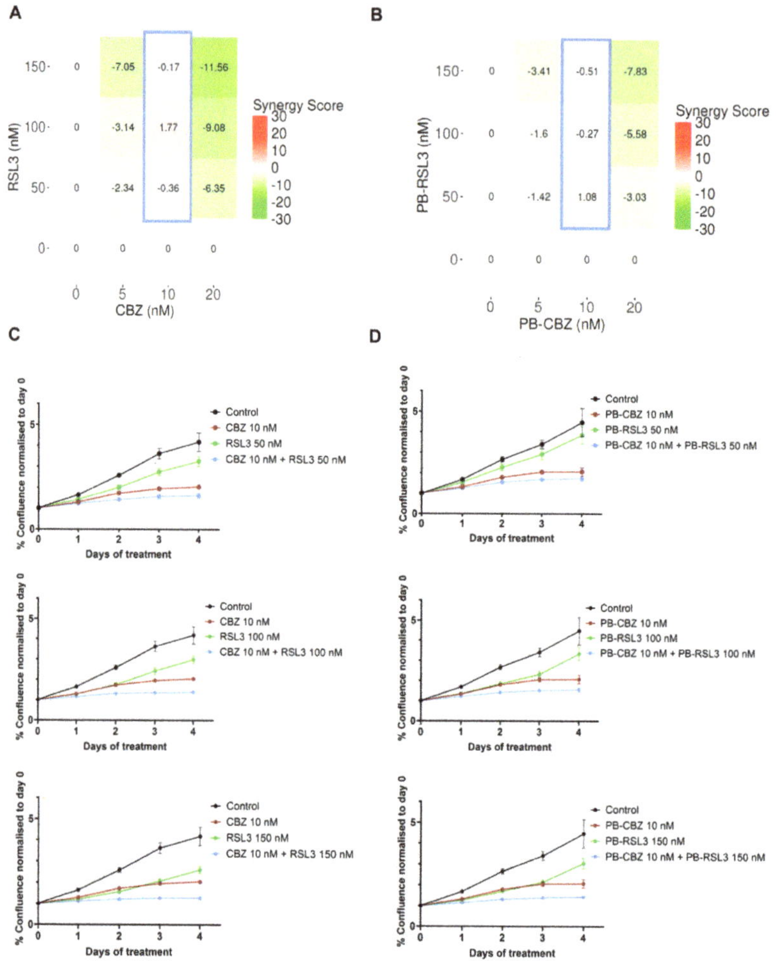

Figure 3. Combinatorial effects of CBZ and RSL3 or PEBCA-CBZ and PEBCA-RSL3 in SKBR3 cells. Figures shown in the left panel correspond to free drugs or their combinations, and the right panel corresponds to PEBCA drugs or their combinations. SKBR3 cells were treated with different concentrations of CBZ/PEBCA-CBZ and RSL3/PEBCA-RSL3 as single drugs or as combinations. Drug responses were evaluated by quantifying cell confluence through real-time Incucyte analysis for up to 4 days. Combinatorial effects were analyzed, and synergy scores were derived using SynergyFinder Plus software. Bliss model scores are shown here. (**A**) Synergy scores of all combinations used, respectively, for drugs alone and PEBCA-drugs, confirming the additive effects of certain combinations. Combinations showing the highest scores are outlined in blue. (**C**,**D**) Plots showing the drug responses of combinations that showed additive effects and highest scores (outlined in blue in panels (**A**,**B**)), respectively, for drugs alone and PEBCA drugs. Presented here are average values with standard deviations in error bars of % cell confluence over a period of 4 days from one out of three independent experiments run with triplicates. PEBCA is abbreviated to 'PB' in graphs.

SKBR3: Figure 3A,B show the synergy scores representing the effect of combinatorial treatment by free drugs and encapsulated drugs, respectively, in SKBR3 cells. Differing from EO771 cells, SKBR3 cells are sensitive to both CBZ and RSL3 with increasing concentrations (Figures S1 and S2), and the combination of drugs shows additive effects, where the synergy scores are near to zero. The Bliss reference model was used. The highest scores are shown when 10 nM CBZ was combined with different concentrations of RSL3, and this pattern is similar for both free and encapsulated drugs, as shown in Figure 3A,B. The real-time effect of the combinations showing the highest synergy scores is presented in Figure 3C,D, corresponding to free drugs and encapsulated drugs, respectively.

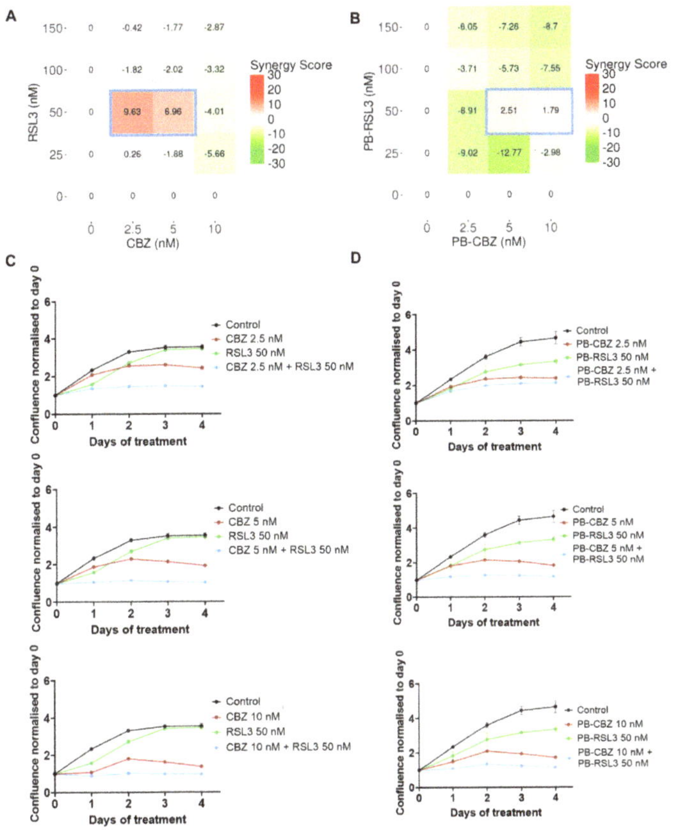

Figure 4. Combinatorial effects of CBZ and RSL3 or PEBCA-CBZ and PEBCA-RSL3 in JIMT-1 cells. Figures shown in the left panel correspond to free drugs or their combinations, and the right panel corresponds to PEBCA drugs or their combinations. JIMT-1 cells were treated with different concentrations of CBZ/PEBCA-CBZ and RSL3/PEBCA-RSL3 as single drugs or as combinations. Drug responses were evaluated by quantifying cell confluence through real-time Incucyte analysis for up to 4 days. Combinatorial effects were analyzed, and synergy scores were derived using SynergyFinder Plus software. Bliss model scores are shown here. (**A**,**B**) Synergy scores of all combinations used, respectively, for drugs alone and PEBCA drugs, confirming the additive effects of certain combinations. Combinations showing the highest scores are outlined in blue. (**C**,**D**) Plots showing drug responses of combinations that showed additive effects and highest scores (outlined in blue in panels (**A**,**B**)), respectively, for drugs alone and PEBCA drugs. Presented here are average values with standard deviation (error bars) of % cell confluence over a period of 4 days from one out of three independent experiments run with triplicates. PEBCA is abbreviated to PB in graphs.

JIMT-1: As shown in Figures S1 and S2, JIMT-1 cells are sensitive to both CBZ and RSL3 in a dose-dependent manner. Hence, the Bliss reference model was chosen. Also, the combined effect of different concentrations of the drugs either in free or encapsulated forms is shown in Figure 4A and Figure 4B, respectively, from which it is obvious that certain combinations seem to induce additive effects (scores 0–10). Plots comparing the effects of these combinations when the drugs are either in their free or encapsulated forms are shown, respectively, in Figure 4C,D. Generally, the trends are similar except for slight differences in the inhibitory effects between free RSL3 and PEBCA-RSL3, which eventually reflect in the synergy scores as well.

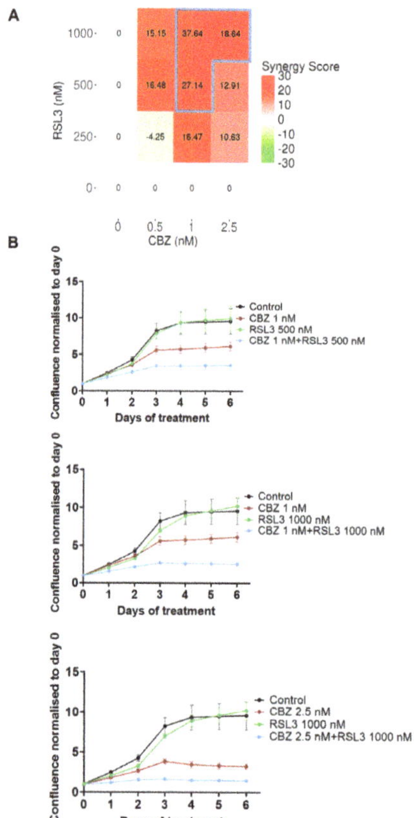

Figure 5. Combinatorial effects of CBZ and RSL3 in MCF-7 cells. MCF-7 cells were treated with different concentrations of CBZ and RSL3 as single drugs or as combinations. Drug responses were evaluated by quantifying cell confluence through real-time Incucyte analysis for up to 6 days. Combinatorial effects were analyzed, and synergy scores were derived using SynergyFinder Plus software. (**A**) Synergy scores of all combinations used, derived by the HSA model, confirm the synergistic effects of certain concentrations. Combinations showing high synergy scores are outlined in blue. (**B**) Plots showing drug responses of combinations that showed synergistic effects and highest scores (outlined in blue in panel (**A**)). Presented here are average values with standard deviations (error bars) of % cell confluence over a period of 6 days from one out of three independent experiments run with triplicates.

MCF-7: MCF-7 cells are sensitive to CBZ in a concentration-dependent manner (Figure S1) but resistant to even high concentrations of RSL3 (Figure S2). Hence, assessing combinatorial effects using the HSA model was found appropriate. The resistance to RSL3 in this cell line was correlated to the lack of endogenous acyl-CoA synthetase long-chain family member 4 (ACSL4) [38], which was reported as an essential component to induce ferroptosis [38]. Figure 5A shows the combinatorial effects of different concentrations of CBZ and RSL3, and the synergy scores show that the combination is highly synergistic. It is quite interesting that even if the cells are tolerant to RSL3 alone, this compound/drug/inhibitor sensitizes the cells to chemotherapy. We could not test encapsulated drugs in the case of MCF-7 because we knew from our previous experience [28] that empty PEBCA NPs themselves induce toxic effects in MCF-7 cells at concentrations equivalent to the higher concentrations of encapsulated RSL3. Of note, the concentrations of PEBCA used in the case of all other cell lines were below the toxic range, as is evident from Figure S3.

3.3. *In Vitro Combinatorial Effects Are Similar for Free Drugs and Encapsulated Drugs*

Combining multiple therapeutic drugs simultaneously within a drug delivery system can unify the pharmacokinetic properties of the drugs [39,40]. Moreover, it may offer protection of labile drugs along with controlled release of therapeutics for aiding combinatorial effects [41]. However, there are several challenges that make this strategy less encouraging. Vyxeos, a liposomal formulation of daunorubicin and cytarabine, is the only one that has gained approval so far [42]. Co-encapsulating drugs within proximal 3D space could promote several interactions that may affect the stability and optimal release of the drugs. Moreover, it is difficult to control all the possible interactions and their consequences to achieve optimal co-encapsulation while maintaining batch-to-batch reproducibility [41]. In the present study, the drugs are not co-encapsulated; instead, they are independently loaded within the same type of PEBCA NPs. This is mainly because it is not practical to evaluate the range of combinatorial effects in different cells when the drugs are co-loaded in a single known ratio within the NPs. This approach will also unify the drug delivery properties in the physiological system and would also allow sequential treatment of multiple therapeutics where one drug could sensitize the cells to make them more prone to the effect induced by the other. However, this may result in a higher accumulation of the NPs, and it is therefore important to investigate the safety limit of the NPs. Higher loading efficiency might be one of the criteria that need to be considered. As mentioned before, the present study utilized independently loaded CBZ and RSL3 within PEBCA NPs, and their combinatorial effects were compared with those of free drugs in three of the cell lines studied. The results shown in Figures 2–4 demonstrate that the combination of encapsulated drugs is equally effective as the combination of free drugs in all the cell lines in the in vitro conditions, also confirming that encapsulation in PEBCA does not affect the individual drug's mode of action. This study thus supports investigating the potential for improved efficacy of PEBCA drugs in the in vivo setting and perhaps also including a co-encapsulation strategy for the drug combination.

3.4. *Combinatorial Effects Are Cell Type Dependent*

The four cell lines differ in their sensitivity to single drugs. It is well known that the sensitivity of different cells to ferroptotic agents such as RSL3 depends upon various factors such as the ability of ACSL4 to modulate the composition and distribution of phospholipids [38,43], cell density [44], and the endogenous activity of GPX4 [45]. Similarly, sensitivity to CBZ is also associated with several molecular and cellular factors like activity of the tyrosine kinase receptor protein, Ror2 [46], and expression of the MDR1 gene [47]. From Figures 2–5, one could conclude that the combinatorial effects of CBZ and RSL3 are highly cell-dependent. Interestingly, the combination induced highly synergistic effects in EO771 and MCF-7 cells, where the cells have low sensitivity to one of the drugs. This may be caused by mutual sensitization as reported for RSL3, which gives a synergistic effect with many compounds, including taxanes [22–25,48]. For instance, Ye et al. reported that a low concentration of paclitaxel acts synergistically with RSL3 in hypopharyngeal squamous cell carcinoma by paclitaxel-mediated upregulation of mutant p53 expression, which in turn suppresses the transcription of SLC7A11, rendering the cells more vulnerable to RSL3-induced ferroptosis [49]. Similarly, Yuan et al. reported that RSL3 increases the chemosensitivity of triple-negative breast cancer to paclitaxel by inducing ferroptosis through NF-κB signaling [22]. Hence, single or multiple mechanisms may be involved, which may differ with the molecular characteristics of each cell line.

3.5. Cell Viability Assay Correlates with Cell Proliferation Assay

Cell proliferation assays evaluate the ability of the cells to divide, whereas cell viability assays assess single or multiple parameters of cellular function in response to treatment [50]. Evaluating drug responses utilizing different methods is valuable and will enhance our understanding of the mechanism of action. Hence, we have also evaluated the combinatorial effects by MTT assay, which assesses the metabolic activity of cells by measuring NAD(P)H-dependent oxidoreduction. Results showing effects corresponding to free drugs and PEBCA drugs are shown in Figures 6 and 7, respectively.

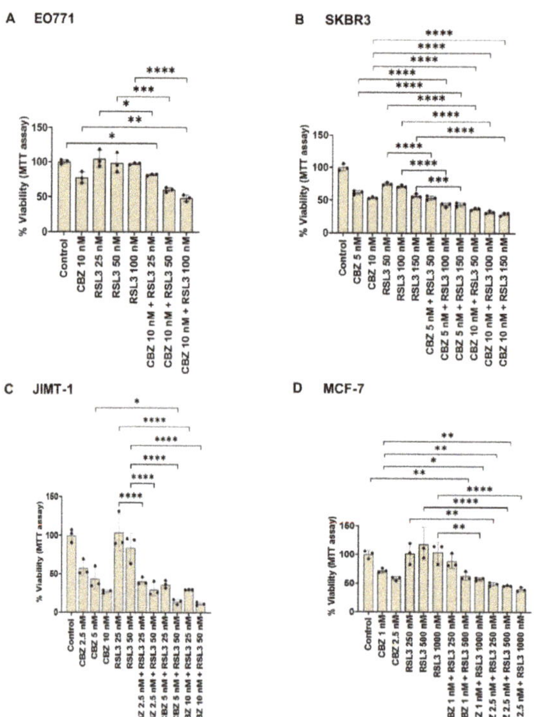

Figure 6. MTT assay showing combinatorial effects of drugs alone in breast cancer cell lines. Cells were treated with drugs alone or their combinations for 96 h, and the cytotoxicity was measured by MTT assay. Data corresponding to the range of drug concentrations that showed combinatorial effects with the highest significance are shown. Data are presented as % cell viability ± standard deviations from one out of three independent experiments, each run in triplicate. * $p < 0.05$, ** $p < 0.01$, *** $p < 0.001$, **** $p < 0.0001$.

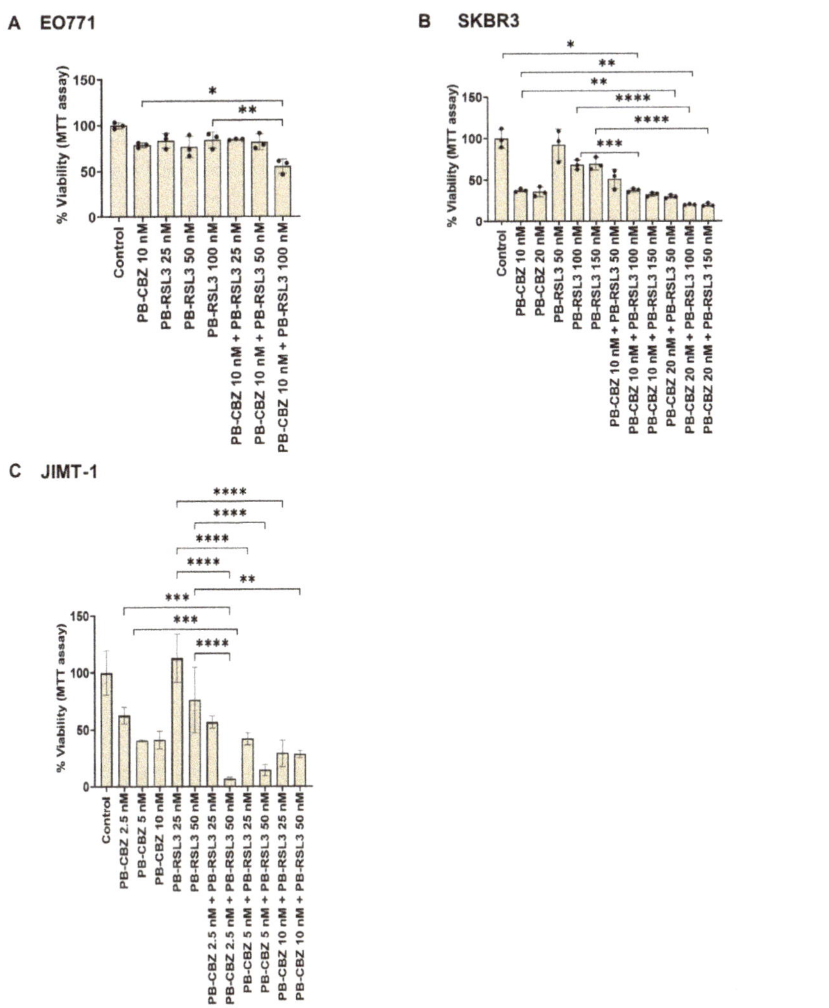

Figure 7. MTT assay showing combinatorial effects of PEBCA drugs in breast cancer cell lines. Cells were treated with PEBCA drugs or their combinations for 96 h, and the cytotoxicity was measured by MTT assay. Data corresponding to the range of PEBCA drug concentrations that showed combinatorial effects with the highest significance are shown. Data are presented as % cell viability ± standard deviations from one out of three independent experiments, each run in triplicate. * $p < 0.05$, ** $p < 0.01$, *** $p < 0.001$, **** $p < 0.0001$. PEBCA is abbreviated to PB in graphs.

As shown in Figure 6, certain combinations induced improved cytotoxic effects compared to the single drugs in all cell lines studied. It is important to note that the concentrations that showed evident combinatorial effects differ between the two assays, especially in EO771 cells. This might be because the MTT assay was assessed after 96 h and could not be continued because the cells became confluent. In contrast, the proliferation assay was continued up to 6 days, which was essential to distinguish the combinatorial effects from the transient inhibitory effect of RSL3 alone. In addition to this, the comparison between single drugs and combinations is also different in the two assays, which might be due to differences in the end points and assessed parameters. Similar patterns were observed in the case of PEBCA drugs or their combinations (Figure 7). Altogether, it can be concluded that the combinatorial effects evaluated by the two methods correlate, which confirms that the drug combinations are effective.

3.6. Combinatorial Treatment Did Not Cause Immediate Effects on ROS Levels

ROS molecules exhibit pleiotropic functions within the cell. Besides modulating cellular signaling, their excess also induces DNA damage, protein oxidation or lipid peroxidation, being responsible for different cell death mechanisms [51]. Increased ROS levels upon CBZ and RSL3 treatment have been reported in several studies. CBZ was described to have pro-oxidant effects and mediate apoptosis in cancer cells by increasing ROS and mitochondrial damage [52]. Kosaka et al. reported decreased expression of the Sestrin 3 gene leading to elevated ROS levels upon CBZ treatment in cancer cells [53]. RSL3 covalently binds to the antioxidant protein GPX4, and there are several papers reporting increased levels of ROS after RSL3 treatment [21]. GPX4 is considered a central player in regulating ferroptosis, as this protein removes lipid peroxides that are produced from polyunsaturated fatty acids by elevated ROS levels.

To investigate whether the combinatorial effects are mediated through ROS production, the ROS levels were assessed using the fluorescence probe CM-H_2DCFDA after treating the cells for up to 5 h. As shown in Figure S4, no effects on ROS production were induced by free or encapsulated CBZ and RSL3, nor their combinations, in all the cell lines studied, irrespective of their differences in H_2O_2-induced ROS production (Figure S5). However, we also observed that empty PEBCA NPs caused a slight increase in ROS levels in SKBR3 cells at the equivalent concentrations of PEBCA drugs used for the studies (Figure S6). We have earlier published that treatment with higher concentrations of empty PEBCA NPs (25 μg/mL) increased ROS production in MDA-MB-231 cells [54], whereas the concentrations used in the present studies are below 2 μg/mL. Hence, we speculate that this effect of empty PEBCA is cell dependent. To conclude, the combinatorial treatment did not cause an increase in ROS production at an early time point in any of the cell lines.

3.7. Combinatorial Effects Are Not Accompanied by Immediate Effects on Cell Metabolism

We investigated whether the combinatorial effects can be correlated with any changes in cell metabolism. Hence, we studied the effect of free drugs, PEBCA drugs and combinations on mitochondrial metabolism using a Seahorse Analyzer employing a MitoStress assay. Data are presented as oxygen consumption rate (OCR) in terms of basal respiration (Figures S7 and S8) and maximal respiration (Figures 8 and 9). As shown in Figures 8A and S7A, treatment with a higher concentration of RSL3 caused a significant decrease in OCR in EO771 cells. Otherwise, the free drugs or combinations did not induce any large effects. Nevertheless, in the case of PEBCA drugs, a gradual decrease in OCR was observed towards higher concentrations of the combinations (Figures 9A and S8A). The decrease is caused mainly by the effect of the encapsulated drugs, as the empty PEBCA NPs had no effect (Figure S9A) on the cell line. The effect of RSL3 fits with our recently published data for MDA-MB-231 cells [55]. Others have also shown that GPX4 inhibition with RSL3 increased ROS levels in mitochondria, leading to their impaired function and decrease in maximal respiration [56]. Moreover, decreased OCR for both basal metabolism and maximal respiration shows that cellular metabolism is affected by the treatment with CBZ and RSL3. This might be due to a decreased mitochondrial potential ($\Delta\Psi_M$) caused by RSL3 [55,57]. Moreover, it has been reported that CBZ also decreases $\Delta\Psi_M$, i.e., giving mitochondrial depolarization in prostate cancer cells due to increased ROS levels [52]. Hence, we speculate that drugs and NPs together contribute to the effect caused by PEBCA drugs.

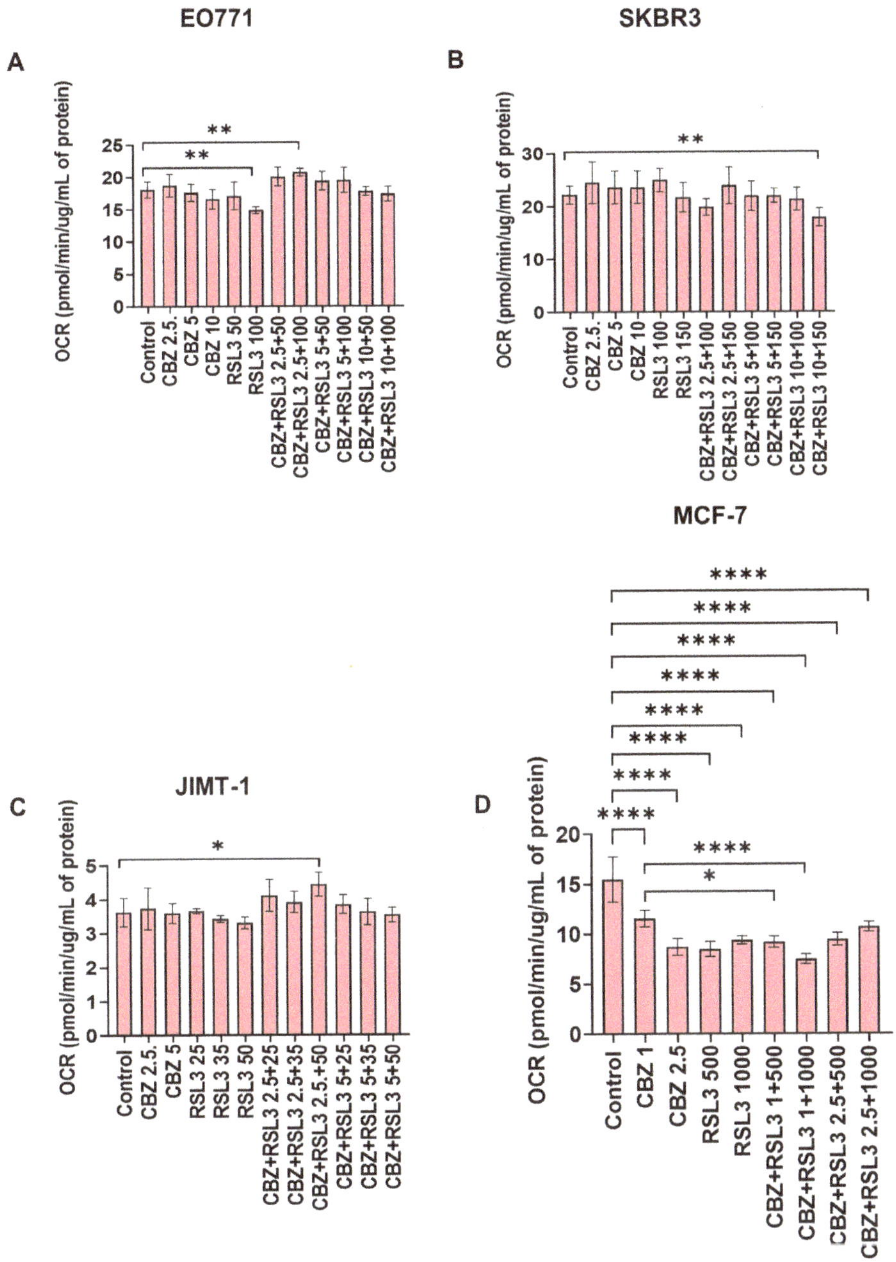

Figure 8. Effect of free drugs or combinations on mitochondrial respiration. Cells were treated for 5 h at 37 °C and were analyzed by Mito Stress Assay using a Seahorse analyzer. Shown here are data from (**A**) EO771, (**B**) SKBR3, (**C**) JIMT-1 and (**D**) MCF7 cells. The graphs show variation of oxygen consumption rate (OCR) after addition of different modulators of mitochondrial respiration in terms of maximal respiration. Values represent the mean ± SD of three technical replicates. One representative experiment is shown. * $p < 0.05$, ** $p < 0.01$, **** $p < 0.0001$.

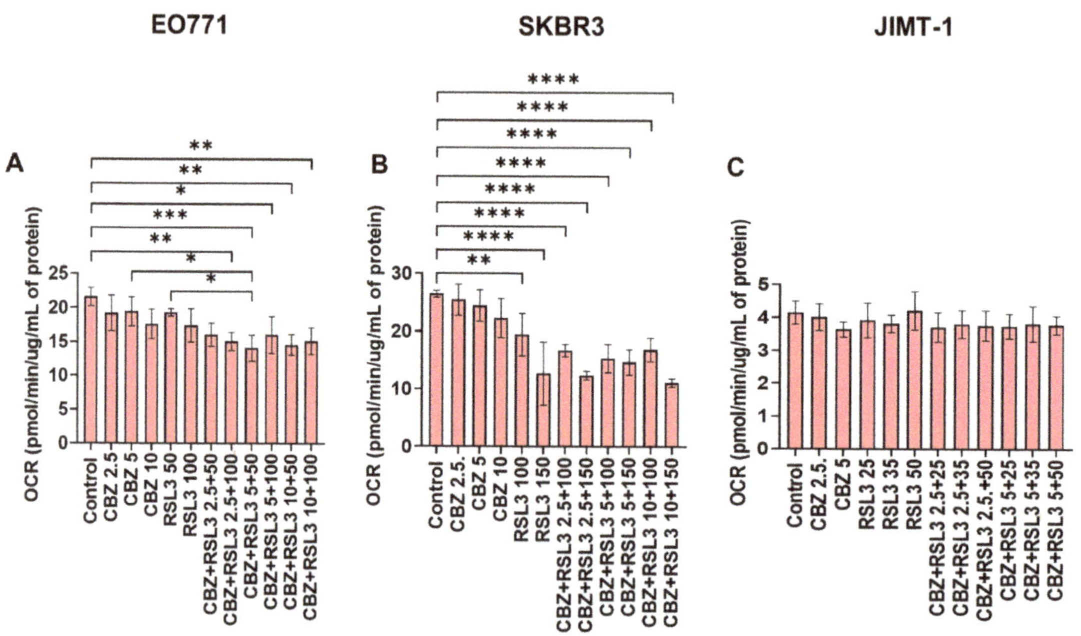

Figure 9. Effect of PEBCA drugs or their combinations on mitochondrial respiration. Cells were treated for 5 h at 37 °C and were analyzed by Mito Stress Assay using a Seahorse analyzer. Shown here are data from (**A**) EO771, (**B**) SKBR3 and (**C**) JIMT-1 cells. The graphs show variation of oxygen consumption rate (OCR) after addition of different modulators of mitochondrial respiration in terms of maximal respiration. Values represent the mean ± SD of three technical replicates. One representative experiment is shown. * $p < 0.05$, ** $p < 0.01$, *** $p < 0.001$, **** $p < 0.0001$.

JIMT-1 and SKBR3 cell lines did not show significant responses to free drugs or combinations, except that the higher concentrations of combinations significantly decreased the OCR in SKBR3 cells (Figures 9B and S7B for SKBR3 cells and Figures 9C and S7C for JIMT-1 cells, respectively). At the same time, these cell lines responded differently to the treatment with PEBCA drugs and combinations. The OCR in SKBR3 cells was significantly affected by PEBCA-RSL3 and all of the combinations (Figures 9B and S8B), while JIMT-1 cells seemed unaffected (Figures 9C and S8C). It is interesting to note that empty PEBCA also induced a decrease in OCR in SKBR3 cells and JIMT-1 cells at the highest concentrations equivalent to the combinations (Figure S9B and S9C , respectively). This did, however, not seem to have a profound impact on the effect of PEBCA drugs in JIMT-1 cells, as shown in Figures 9C and S8C. Whereas in SKBR3 cells, when comparing Figures 9B, S8B and S9B, we could conclude that both RSL3 and PEBCA NPs are responsible for the action. It should be studied whether CBZ also contributes to this effect when it is combined with RSL3 and PEBCA. Concurrently, the effects of empty PEBCA NPs on SKBR3 cells are so evident, as the results from the ROS assay (Figure S6B) correlate well with those of the Seahorse assay (Figure S9B). We already showed that empty PEBCA NPs increase ROS accumulation in MDA-MB-231 cells at a concentration of 25 µg/mL [54]. In the present study, even though we used several-fold lower concentrations of NPs, SKBR3 cells might be more susceptible to NP-induced ROS production than MDA-MB-231 cells. However, none of these effects were reflected in the proliferation and cell viability assays, where free and PEBCA drugs showed similar patterns (Figures 3, 6C and 7C). This is supported by our previous report, where

MS-based proteomics studies revealed that different breast, colon and prostate cancer cell lines have altered proteomes upon NP treatment [28]. Here, treatment with PEBCA-CBZ led to a decrease in the total amount of proteins, even though it was not reflected in the cytotoxicity pattern of free and encapsulated CBZ in all of the three cell lines studied [28]. MCF-7 cells, on the other hand, are resistant to RSL3. Still, all concentrations of RSL3, CBZ and the combinations caused a significant decrease in OCR in these cells, as shown in Figures 8D and S7D. However, there were no differences between drugs alone and combinations.

Altogether, the results from the four cell lines emphasize that the combinatorial effects are not due to immediate changes in cell metabolism as measured with the Seahorse Analyzer. However, it is possible that the treatment may induce metabolic changes after long-time exposure, which has to be explored further. Also, it is evident that PEBCA NPs alone or in combination with the drugs could influence the ROS production and cell metabolism in a cell-dependent manner, even though these changes may not be reflected in their cytotoxic effects when compared to the free drugs in the in vitro setting.

4. Conclusions

We here demonstrate that CBZ and RSL3 can give synergistic effects in vitro, indicating the possibility of developing combinatorial nanomedicine using these drugs. The combinatorial effects were similar for the free and encapsulated drugs. Remarkably, synergy was obtained even in cell lines that are non-responsive to monotreatment with one of the compounds. The drug combination did not cause immediate effects on ROS production and cell metabolism. However, we cannot exclude the possibility of an impact after a prolonged period. The fact that also encapsulated drugs give synergy could be important for the possible in vivo use, as this may provide the drug with a better pharmacokinetic profile, as documented in our previous in vivo studies with PEBCA-CBZ [27,29]. Based on the promising results shown here, future studies should address the mechanisms behind the combinatorial effects in different cell types. Importantly, drug combinations should be tested in preclinical tumor models.

Supplementary Materials: The following supporting information can be downloaded at https://www.mdpi.com/article/10.3390/pharmaceutics17050657/s1.

Author Contributions: R.V. designed the study, performed experiments and analyses, interpreted data and drafted the manuscript; M.F. performed experiments and analyses, interpreted data and drafted the manuscript. S.P. contributed to performing experiments and acquisition of data; A.K.O.Å. and Ý.M. designed, performed the manufacturing and analyzed the NPs. R.V., T.S., K.S., G.M.M. and T.-G.I. planned and supervised the studies. All authors have read and agreed to the published version of the manuscript.

Funding: The work was supported by a grant from the Norwegian Cancer Society (Project no. 208239) to KS/TGI. GMM and MF also acknowledge support from the Maria Sklodowska-Curie Actions (MSCA-ITN-2020 grant agreement ID: 956544 (DIRNANO: Directing the Immune Response through Designed Nanomaterials)). RV also acknowledges support from Astri and Birger Torsted's foundation for cancer research (Project no. 320012).

Institutional Review Board Statement: Not applicable.

Informed Consent Statement: Not applicable.

Data Availability Statement: All data supporting the findings are shown in the paper, and further queries can be directed to the corresponding author upon reasonable request.

Acknowledgments: The authors from Oslo acknowledge Marzena Szwed, Department of Medical Biophysics, University of Lodz, Poland, for the discussions regarding ROS assay.

Conflicts of Interest: Dr Ýrr Mørch is, as of August 2022, CTO and cofounder of the SINTEF spin-off NaDeNo Nanoscience AS, a company established to commercialize poly(alkyl cyanoacrylate)-encapsulated cabazitaxel NPs (currently in the pre-clinical phase). In addition, Dr Ýrr Mørch is the inventor of patent WO 2014/191502 (process of manufacturing PACA nanoparticles) and WO 2019/185685 (medical use of cabazitaxel encapsulated in PACA nanoparticles). Prof. Kirsten Sandvig, Prof. Gunhild Mælandsmo and Dr. Tore Skotland are co-inventors of the patent WO 2019/185685. The other authors declare no conflicts of interest.

References

1. Garner, H.; de Visser, K.E. Immune crosstalk in cancer progression and metastatic spread: A complex conversation. *Nat. Rev. Immunol.* **2020**, *20*, 483–497. [CrossRef]
2. Nebert, D.W. Transcription factors and cancer: An overview. *Toxicology* **2002**, *181*, 131–141. [CrossRef]
3. Slattery, M.L.; Fitzpatrick, F.A. Convergence of hormones, inflammation, and energy-related factors: A novel pathway of cancer etiology. *Cancer Prev. Res.* **2009**, *2*, 922–930. [CrossRef]
4. Nikolaou, M.; Pavlopoulou, A.; Georgakilas, A.G.; Kyrodimos, E. The challenge of drug resistance in cancer treatment: A current overview. *Clin. Exp. Metastasis* **2018**, *35*, 309–318. [CrossRef]
5. Li, S.; Kennedy, M.; Payne, S.; Kennedy, K.; Seewaldt, V.L.; Pizzo, S.V.; Bachelder, R.E. Model of tumor dormancy/recurrence after short-term chemotherapy. *PLoS ONE* **2014**, *9*, e98021. [CrossRef]
6. Mokhtari, R.B.; Homayouni, T.S.; Baluch, N.; Morgatskaya, E.; Kumar, S.; Das, B.; Yeger, H. Combination therapy in combating cancer. *Oncotarget* **2017**, *8*, 38022. [CrossRef]
7. Gotwals, P.; Cameron, S.; Cipolletta, D.; Cremasco, V.; Crystal, A.; Hewes, B.; Mueller, B.; Quaratino, S.; Sabatos-Peyton, C.; Petruzzelli, L. Prospects for combining targeted and conventional cancer therapy with immunotherapy. *Nat. Rev. Cancer* **2017**, *17*, 286–301. [CrossRef]
8. Palmer, A.C.; Sorger, P.K. Combination cancer therapy can confer benefit via patient-to-patient variability without drug additivity or synergy. *Cell* **2017**, *171*, 1678–1691.e1613. [CrossRef]
9. Behranvand, N.; Nasri, F.; Zolfaghari Emameh, R.; Khani, P.; Hosseini, A.; Garssen, J.; Falak, R. Chemotherapy: A double-edged sword in cancer treatment. *Cancer Immunol. Immunother.* **2022**, *71*, 507–526. [CrossRef]
10. Zahler, S.; Ghazi, N.G.; Singh, A.D. Principles and complications of chemotherapy. In *Clinical Ophthalmic Oncology*; Springer: Cham, Switzerland, 2019; pp. 129–142.
11. Gilad, Y.; Gellerman, G.; Lonard, D.M.; O'malley, B.W. Drug combination in cancer treatment—From cocktails to conjugated combinations. *Cancers* **2021**, *13*, 669. [CrossRef]
12. Yi, M.; Zheng, X.; Niu, M.; Zhu, S.; Ge, H.; Wu, K. Combination strategies with PD-1/PD-L1 blockade: Current advances and future directions. *Mol. Cancer* **2022**, *21*, 28. [CrossRef] [PubMed]
13. Melero, I.; Berman, D.M.; Aznar, M.A.; Korman, A.J.; Gracia, J.L.P.; Haanen, J. Evolving synergistic combinations of targeted immunotherapies to combat cancer. *Nat. Rev. Cancer* **2015**, *15*, 457–472. [CrossRef] [PubMed]
14. Yu, W.D.; Sun, G.; Li, J.; Xu, J.; Wang, X. Mechanisms and therapeutic potentials of cancer immunotherapy in combination with radiotherapy and/or chemotherapy. *Cancer Lett.* **2019**, *452*, 66–70. [CrossRef] [PubMed]
15. Paller, C.J.; Antonarakis, E.S. Cabazitaxel: A novel second-line treatment for metastatic castration-resistant prostate cancer. *Drug Des. Dev. Ther.* **2011**, *5*, 117–124.
16. Koutras, A.; Zagouri, F.; Koliou, G.-A.; Psoma, E.; Chryssogonidis, I.; Lazaridis, G.; Tryfonopoulos, D.; Kotsakis, A.; Kentepozidis, N.K.; Razis, E. Phase 2 study of Cabazitaxel as second-line treatment in patients with HER2-negative metastatic breast cancer previously treated with taxanes—A Hellenic Cooperative Oncology Group (HeCOG) Trial. *Br. J. Cancer* **2020**, *123*, 355–361. [CrossRef]
17. Yardley, D.A.; Hart, L.L.; Ward, P.J.; Wright, G.L.; Shastry, M.; Finney, L.; DeBusk, L.M.; Hainsworth, J.D. Cabazitaxel Plus Lapatinib as Therapy for HER2 Metastatic Breast Cancer With Intracranial Metastases: Results of a Dose-finding Study. *Clin. Breast Cancer* **2018**, *18*, E781–E787. [CrossRef]
18. Villanueva, C.; Awada, A.; Campone, M.; Machiels, J.P.; Besse, T.; Magherini, E.; Dubin, F.; Semiond, D.; Pivot, X. A multicentre dose-escalating study of cabazitaxel (XRP6258) in combination with capecitabine in patients with metastatic breast cancer progressing after anthracycline and taxane treatment: A phase I/II study. *Eur. J. Cancer* **2011**, *47*, 1037–1045. [CrossRef]

19. Duran, G.E.; Derdau, V.; Weitz, D.; Philippe, N.; Blankenstein, J.; Atzrodt, J.; Semiond, D.; Gianolio, D.A.; Mace, S.; Sikic, B.I. Cabazitaxel is more active than first-generation taxanes in ABCB1(+) cell lines due to its reduced affinity for P-glycoprotein. *Cancer Chemother. Pharmacol.* **2018**, *81*, 1095–1103. [CrossRef]
20. Sun, B.; Lovell, J.F.; Zhang, Y. Current development of cabazitaxel drug delivery systems. *Wiley Interdiscip. Rev. Nanomed. Nanobiotechnol.* **2023**, *15*, e1854. [CrossRef]
21. Sui, X.; Zhang, R.; Liu, S.; Duan, T.; Zhai, L.; Zhang, M.; Han, X.; Xiang, Y.; Huang, X.; Lin, H.; et al. RSL3 Drives Ferroptosis Through GPX4 Inactivation and ROS Production in Colorectal Cancer. *Front. Pharmacol.* **2018**, *9*, 1371. [CrossRef]
22. Yuan, J.; Liu, C.; Jiang, C.; Liu, N.; Yang, Z.; Xing, H. RSL3 induces ferroptosis by activating the NF-κB signalling pathway to enhance the chemosensitivity of triple-negative breast cancer cells to paclitaxel. *Sci. Rep.* **2025**, *15*, 1654. [CrossRef] [PubMed]
23. Liu, S.; Yan, S.; Zhu, J.; Lu, R.; Kang, C.; Tang, K.; Zeng, J.; Ding, M.; Guo, Z.; Lai, X.; et al. Combination RSL3 Treatment Sensitizes Ferroptosis- and EGFR-Inhibition-Resistant HNSCCs to Cetuximab. *Int. J. Mol. Sci.* **2022**, *23*, 9014. [CrossRef]
24. Li, M.; Chen, X.; Wang, X.; Wei, X.; Wang, D.; Liu, X.; Xu, L.; Batu, W.; Li, Y.; Guo, B. RSL3 enhances the antitumor effect of cisplatin on prostate cancer cells via causing glycolysis dysfunction. *Biochem. Pharmacol.* **2021**, *192*, 114741. [CrossRef] [PubMed]
25. Li, B.; Wang, Y.; Hou, F.; Du, J.; Tong, X. Rapamycin enhances inhibitory effect of RSL3 on proliferation, invasion and migration of testicular cancer I-10 cells in vitro. *Nan Fang Yi Ke Da Xue Xue Bao* **2023**, *43*, 2145–2151.
26. Peng, S.; Chen, G.; Yu, K.; Feng, Y.; Zhao, L.; Yang, M.; Cao, W.; Almahi, W.A.A.; Sun, M.; Xu, Y. Synergism of non-thermal plasma and low concentration RSL3 triggers ferroptosis via promoting xCT lysosomal degradation through ROS/AMPK/mTOR axis in lung cancer cells. *Cell Commun. Signal.* **2024**, *22*, 112. [CrossRef] [PubMed]
27. Fusser, M.; Overbye, A.; Pandya, A.D.; Morch, Y.; Borgos, S.E.; Kildal, W.; Snipstad, S.; Sulheim, E.; Fleten, K.G.; Askautrud, H.A.; et al. Cabazitaxel-loaded Poly(2-ethylbutyl cyanoacrylate) nanoparticles improve treatment efficacy in a patient derived breast cancer xenograft. *J. Control. Release* **2019**, *293*, 183–192. [CrossRef]
28. Øverbye, A.; Torgersen, M.L.; Sønstevold, T.; Iversen, T.G.; Mørch, Ý.; Skotland, T.; Sandvig, K. Cabazitaxel-loaded poly (alkyl cyanoacrylate) nanoparticles: Toxicity and changes in the proteome of breast, colon and prostate cancer cells. *Nanotoxicology* **2021**, *15*, 865–884. [CrossRef]
29. Valsalakumari, R.; Pandya, A.D.; Prasmickaite, L.; Kvalvaag, A.; Myrann, A.G.; Aslund, A.K.O.; Kjos, M.S.; Fontecha-Cuenca, C.; Haroon, H.B.; Ribeiro, A.R.S.; et al. Preclinical Efficacy of Cabazitaxel Loaded Poly(2-alkyl cyanoacrylate) Nanoparticle Variants. *Int. J. Nanomed.* **2024**, *19*, 3009–3029. [CrossRef]
30. Sulheim, E.; Iversen, T.G.; To Nakstad, V.; Klinkenberg, G.; Sletta, H.; Schmid, R.; Hatletveit, A.R.; Wagbo, A.M.; Sundan, A.; Skotland, T.; et al. Cytotoxicity of Poly(Alkyl Cyanoacrylate) Nanoparticles. *Int. J. Mol. Sci.* **2017**, *18*, 2454. [CrossRef]
31. Lyles, R.D.Z.; Martinez, M.J.; Sherman, B.; Schürer, S.; Burnstein, K.L. Automation, live-cell imaging, and endpoint cell viability for prostate cancer drug screens. *PLoS ONE* **2023**, *18*, e0287126. [CrossRef]
32. Zheng, S.; Wang, W.; Aldahdooh, J.; Malyutina, A.; Shadbahr, T.; Tanoli, Z.; Pessia, A.; Tang, J. SynergyFinder plus: Toward better interpretation and annotation of drug combination screening datasets. *Genom. Proteom. Bioinform.* **2022**, *20*, 587–596. [CrossRef] [PubMed]
33. Berenbaum, M.C. What is synergy? *Pharmacol. Rev.* **1989**, *41*, 93–141. [CrossRef]
34. Bliss, C.I. The toxicity of poisons applied jointly 1. *Ann. Appl. Biol.* **1939**, *26*, 585–615. [CrossRef]
35. Loewe, S. The problem of synergism and antagonism of combined drugs. *Arzneimittel-Forschung* **1953**, *3*, 285–290.
36. Yadav, B.; Wennerberg, K.; Aittokallio, T.; Tang, J. Searching for drug synergy in complex dose–response landscapes using an interaction potency model. *Comput. Struct. Biotechnol. J.* **2015**, *13*, 504–513. [CrossRef]
37. Zhang, J.; Zhang, Q. Using seahorse machine to measure OCR and ECAR in cancer cells. *Cancer Metab. Methods Protoc.* **2019**, *1928*, 353–363.
38. Doll, S.; Proneth, B.; Tyurina, Y.Y.; Panzilius, E.; Kobayashi, S.; Ingold, I.; Irmler, M.; Beckers, J.; Aichler, M.; Walch, A.; et al. ACSL4 dictates ferroptosis sensitivity by shaping cellular lipid composition. *Nat. Chem. Biol.* **2017**, *13*, 91–98. [CrossRef]
39. Hu, Q.; Sun, W.; Wang, C.; Gu, Z. Recent advances of cocktail chemotherapy by combination drug delivery systems. *Adv. Drug Deliv. Rev.* **2016**, *98*, 19–34. [CrossRef]
40. Kemp, J.A.; Shim, M.S.; Heo, C.Y.; Kwon, Y.J. "Combo" nanomedicine: Co-delivery of multi-modal therapeutics for efficient, targeted, and safe cancer therapy. *Adv. Drug Deliv. Rev.* **2016**, *98*, 3–18. [CrossRef]
41. Bhattacharjee, S. Craft of co-encapsulation in nanomedicine: A struggle to achieve synergy through reciprocity. *ACS Pharmacol. Transl. Sci.* **2022**, *5*, 278–298. [CrossRef]
42. Tzogani, K.; Penttila, K.; Lapvetelainen, T.; Hemmings, R.; Koenig, J.; Freire, J.; Marcia, S.; Cole, S.; Coppola, P.; Flores, B.; et al. EMA Review of Daunorubicin and Cytarabine Encapsulated in Liposomes (Vyxeos, CPX-351) for the Treatment of Adults with Newly Diagnosed, Therapy-Related Acute Myeloid Leukemia or Acute Myeloid Leukemia with Myelodysplasia-Related Changes. *Oncologist* **2020**, *25*, e1414–e1420. [CrossRef] [PubMed]

43. Wiernicki, B.; Dubois, H.; Tyurina, Y.Y.; Hassannia, B.; Bayir, H.; Kagan, V.E.; Vandenabeele, P.; Wullaert, A.; Vanden Berghe, T. Excessive phospholipid peroxidation distinguishes ferroptosis from other cell death modes including pyroptosis. *Cell Death Dis.* **2020**, *11*, 922. [CrossRef] [PubMed]
44. Panzilius, E.; Holstein, F.; Dehairs, J.; Planque, M.; von Toerne, C.; Koenig, A.-C.; Doll, S.; Bannier-Hélaouët, M.; Ganz, H.M.; Hauck, S.M. Cell density-dependent ferroptosis in breast cancer is induced by accumulation of polyunsaturated fatty acid-enriched triacylglycerides. *BioRxiv* **2018**, preprint. [CrossRef]
45. Ursini, F.; Maiorino, M. Lipid peroxidation and ferroptosis: The role of GSH and GPx4. *Free Radic. Biol. Med.* **2020**, *152*, 175–185. [CrossRef]
46. Sennoune, S.R.; Nelius, T.; Jarvis, C.; Pruitt, K.; Kottapalli, K.R.; Filleur, S. The Wnt non-canonical signaling modulates cabazitaxel sensitivity in prostate cancer cells. *PLoS ONE* **2020**, *15*, e0234078. [CrossRef]
47. Machioka, K.; Izumi, K.; Kadono, Y.; Iwamoto, H.; Naito, R.; Makino, T.; Kadomoto, S.; Natsagdorj, A.; Keller, E.T.; Zhang, J. Establishment and characterization of two cabazitaxel-resistant prostate cancer cell lines. *Oncotarget* **2018**, *9*, 16185. [CrossRef]
48. Huang, W.; Guo, Y.; Qian, Y.; Liu, X.; Li, G.; Wang, J.; Yang, X.; Wu, M.; Fan, Y.; Luo, H.; et al. Ferroptosis-inducing compounds synergize with docetaxel to overcome chemoresistance in docetaxel-resistant non-small cell lung cancer cells. *Eur. J. Med. Chem.* **2024**, *276*, 116670. [CrossRef]
49. Ye, J.; Jiang, X.; Dong, Z.; Hu, S.; Xiao, M. Low-concentration PTX and RSL3 inhibits tumor cell growth synergistically by inducing ferroptosis in mutant p53 hypopharyngeal squamous carcinoma. *Cancer Manag. Res.* **2019**, *11*, 9783–9792. [CrossRef]
50. Nozhat, Z.; Khalaji, M.S.; Hedayati, M.; Kia, S.K. Different methods for cell viability and proliferation assay: Essential tools in pharmaceutical studies. *Anti-Cancer Agents Med. Chem. (Former. Curr. Med. Chem.-Anti-Cancer Agents)* **2022**, *22*, 703–712. [CrossRef]
51. Nakamura, H.; Takada, K. Reactive oxygen species in cancer: Current findings and future directions. *Cancer Sci.* **2021**, *112*, 3945–3952. [CrossRef]
52. Eryilmaz, I.E.; Egeli, U.; Cecener, G. Association between the apoptotic effect of Cabazitaxel and its pro-oxidant efficacy on the redox adaptation mechanisms in prostate cancer cells with different resistance phenotypes. *Cancer Biol. Ther.* **2024**, *25*, 2329368. [CrossRef] [PubMed]
53. Kosaka, T.; Hongo, H.; Miyazaki, Y.; Nishimoto, K.; Miyajima, A.; Oya, M. Reactive oxygen species induction by cabazitaxel through inhibiting Sestrin-3 in castration resistant prostate cancer. *Oncotarget* **2017**, *8*, 87675–87683. [CrossRef] [PubMed]
54. Szwed, M.; Sønstevold, T.; Øverbye, A.; Engedal, N.; Grallert, B.; Mørch, Ý.; Sulheim, E.; Iversen, T.-G.; Skotland, T.; Sandvig, K. Small variations in nanoparticle structure dictate differential cellular stress responses and mode of cell death. *Nanotoxicology* **2019**, *13*, 761–782. [CrossRef] [PubMed]
55. Feith, M.; Das Sajib, S.; Myrann, A.G.; Høgset, A.; Garrido, P.; Martinez, A.; Knutsen, E.; Sandvig, K.; Skotland, T.; Maelandsmo, G.M.; et al. Induction of Cell Death by Combined Treatment with Photosensitizer-Chitosan Nanoparticles and the Ferroptosis Inducer RSL3 in Breast Cancer Cell Lines. *Adv. NanoBiomed Res.* **2025**, *5*, 2400208. [CrossRef]
56. Merkel, M.; Goebel, B.; Boll, M.; Adhikari, A.; Maurer, V.; Steinhilber, D.; Culmsee, C. Mitochondrial Reactive Oxygen Species Formation Determines ACSL4/LPCAT2-Mediated Ferroptosis. *Antioxidants* **2023**, *12*, 1590. [CrossRef]
57. Liang, H.; Remmen, H.V.; Frohlich, V.; Lechleiter, J.; Richardson, A.; Ran, Q. Gpx4 protects mitochondrial ATP generation against oxidative damage. *Biochem. Biophys. Res. Commun.* **2007**, *356*, 893–898. [CrossRef]

pharmaceutics

MDPI

Review

Emerging Trends in Snake Venom-Loaded Nanobiosystems for Advanced Medical Applications: A Comprehensive Overview

Álisson E. F. Alves [1,2], Anne B. C. Barros [1], Lindomara C. F. Silva [1], Lucas M. M. Carvalho [1], Graziela M. A. Pereira [1], Ana F. C. Uchôa [1], José M. Barbosa-Filho [2], Marcelo S. Silva [2], Karla P. O. Luna [3], Karla S. R. Soares [1] and Francisco H. Xavier-Júnior [1,2,*]

1 Laboratory of Pharmaceutical Biotechnology (BioTecFarm), Department of Pharmacy, Federal University of Paraíba (UFPB), Campus I-Castelo Branco III., Joao Pessoa 58051-900, PB, Brazil; aefa@ltf.ufpb.br (Á.E.F.A.); annebcbarross@gmail.com (A.B.C.B.); lindomaracristina16@gmail.com (L.C.F.S.); lucas.medeiros2@academico.ufpb.br (L.M.M.C.); graziela.maria.dearaujo@gmail.com (G.M.A.P.); anaflaviauchoauf@gmail.com (A.F.C.U.); karllasamara@yahoo.com.br (K.S.R.S.)

2 Post-Graduated Program in Natural and Synthetic Bioactive Products, Federal University of Paraíba (UFPB), Campus I-Castelo Branco III., Joao Pessoa 58051-900, PB, Brazil; jbarbosa@ltf.ufpb.br (J.M.B.-F.); marcelosobral.ufpb@gmail.com (M.S.S.)

3 Venomics Laboratory (LabVenom), Center for Biological and Health Sciences, State University of Paraíba (UEPB), Campus I, Bodocongó, Campina Grande 58429-600, PB, Brazil; karlaceatox@yahoo.com.br

* Correspondence: fhxj@academico.ufpb.br

Abstract: Advances in medical nanobiotechnology have notably enhanced the application of snake venom toxins, facilitating the development of new therapies with animal-derived toxins. The vast diversity of snake species and their venom complexities underline the need for ongoing research. This review is dedicated to exploring the integration of snake venom with nanoparticles to enable their use in human therapies aiming to develop treatments. The complex mixture of snake venom not only inflicts significant pathological effects but also offers valuable insights for the creation of innovative therapies, particularly in the realm of nanobiotechnology. Nanoscale encapsulation not only mitigates the inherent toxicity of snake venom but also amplifies their antitumoral, antimicrobial, and immunomodulatory properties. The synergy between venom-derived macromolecules and nanotechnology offers a novel pathway for augmenting the efficacy and safety of conventional antivenom therapies, extending their applicability beyond treating bites to potentially addressing a myriad of health issues. In conclusion, nanotechnology presents a compelling therapeutic frontier that promises to improve current treatment modalities and ameliorate the adverse effects associated with venomous snakebites.

Keywords: drug delivery; drug discovery; nanomedicine; nanoformulation; venom peptides

1. Introduction

Snakebite accidents represent a major public health problem worldwide [1,2]. However, it was only recently recognized as a neglected tropical disease by the World Health Organization (WHO) in 2017 [3]. Due to their physiology, snakes are abundant in warm climates, particularly in tropical regions with higher rates of snakebite accidents [4]. Many victims do not opt for healthcare services and instead resort to traditional treatments [5] and different needs are required in different communities [6]. Nevertheless, available data show that, annually, 4.5 to 5.4 million people are affected by snakebites. Of these, 1.8 to 2.7 million develop clinical illnesses, and 81,000 to 138,000 die due to the disease's implications [3]. The WHO has subsequently developed a comprehensive plan known as the snakebite roadmap, aiming to achieve the ambitious objective of reducing the global burden of snakebites by 50% by the year 2030 [7].

Brazil is recognized for its vast biodiversity, of which Viperidae and Elapidae are snake families considered venomous and of medical importance [8–10]. Venomous and poisonous animals provide a wide variety of compounds, among which proteins, peptides, and neurotransmitters are the ones that draw the most attention for scientific research [11]. Nanotechnology has been applied for various purposes within the healthcare field with the production of nanomaterials. In the context of snakebite, nanoparticles have shown promising results in the development of more efficient therapies [12]. The encapsulation of snake venom in polymeric nanoparticles, as an example, is well studied due to its biodisponibility and selectivity for treatment, as well as the decrease in systemic toxicity [13].

Within this context, the present review holds significant promise in elucidating the therapeutic potential of nanosystems in conjunction with snake venom, as documented in the existing literature (Figure 1).

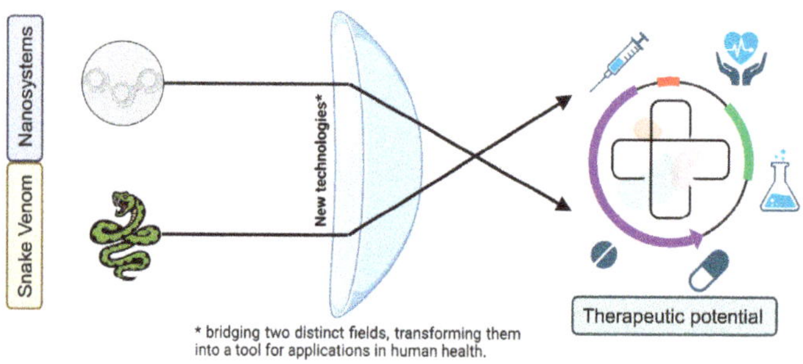

Figure 1. Uniting nature and technology for health purposes.

2. Snake Venom Composition

Venoms and/or poisons are used for predatory and defensive purposes and have evolved independently in a wide phylogenetic range of organisms, including snakes, spiders, scorpions, and jellyfish, among others [14–16]. Ophidian venoms are innovative sources found scattered throughout the animal evolutionary tree [17]. The bioactive properties of these toxins are key points for the discovery of new drugs for possible human therapies [18]. Numerous molecules with different therapeutic approaches have been discovered due to the great biological diversity that is capable of providing us with various solutions to public health problems [19].

It has been scientifically determined that there is a direct correlation between the level of biodiversity within a biome and the complexity of chemical–ecological interactions among organisms occupying the same niche. Furthermore, this biodiversity directly contributes to an increased abundance of diverse molecules and pharmacological activities within these cohabiting organisms. Consequently, it is widely acknowledged that the likelihood of successful drug discovery from natural sources is significantly enhanced when maximal levels of both biodiversity and chemical diversity are present [20].

A specific group of molecules derived from animal venom, which has consistently garnered attention, consists of bioactive peptides. These molecules exhibit high selectivity for specific cell receptors and target specificity, alongside structural stability in bodily fluids and suitability for genetic and synthetic engineering. Due to these characteristics, natural venom peptides serve as promising frameworks for the development of biopharmaceuticals or biotherapeutics. Examples of such peptides include cysteine-stabilized and linear helical peptides [21–23].

In venom, the primary enzymatic classes are oxidases and hydrolases [24]. Snake venoms contain various compounds, mostly protein mixtures including phospholipase A2 (PLA2) that can cause myotoxicity and inflammation; snake venom metalloproteinases (SVMPs) with pathological effects like hemorrhage and coagulation inhibition; snake venom serine proteases (SVSPs) that can cause coagulation inhibition; L-amino acid oxidases (LAAOs) causing cytotoxicity and oxidative stress; disintegrins (DISs) causing inhibition of platelet aggregation; and C-type lectins (CTLs) causing hemostasis modulation and blood coagulation [25,26]. Three-finger toxins (3FTxs) (that cause neurotoxicity and paralysis), PLA2s, SVMPs, and SVSPs are the most abundant components of these venom protein families, which exhibit multifunctionality and are encoded by multilocus gene families [27,28]. Figure 2 summarizes the composition and pathological effects of these components.

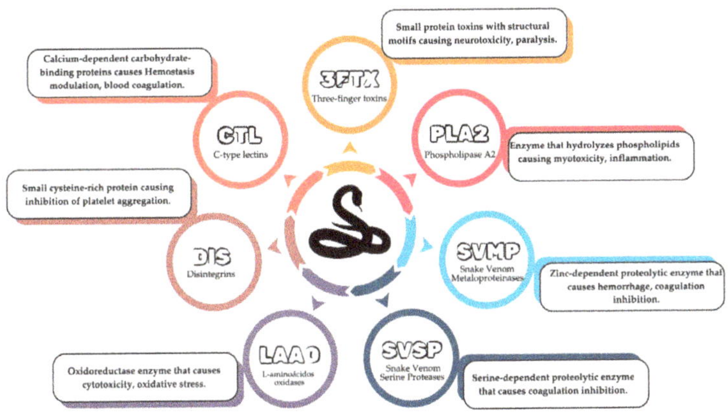

Figure 2. Components, compositions, and the pathological effects of snake venom.

These snake venom components play a crucial role in snakebite accidents and have been adopted in various therapies, including anticancer agents, antihemorrhagic agents, and antihypertensive agents [24,29–31]. As an example of an antihypertensive agent, there is the contemporary pharmaceutical known as captopril which was synthesized using the venom obtained from a type of pit viper that is endemic to Brazil, scientifically referred to as *Bothrops jararaca*, commonly known as the Brazilian lancehead snake [32].

The pathological effects of snakebites are diverse and can include neuromuscular paralysis (neurotoxicity), hemorrhage and coagulopathy (hemotoxicity), and/or local swelling, blisters, and tissue necrosis (cytotoxicity) around the bite site [4]. Most venoms of the Viperidae family and some Elapidae induce local tissue damage. The PLA2 present in these venoms is mainly responsible for the myotoxic action leading to myonecrosis by binding to the plasma membrane of muscle fibers, causing its integrity to rupture [33,34].

The development of Small Molecule Therapeutics (SMT) for the initial and adjunctive treatment of snakebites has been shown by Bulfone et al. [35] and these types of review studies are related to enabling successful strategies and potential uses such as drug targets on venom molecules.

3. Snake Venom as Treatment

As venoms are a cocktail mixed with toxins with plenty of physiological activities, some products from snake venoms are available for a range of disease treatments. As cited before, captopril is an example of an antihypertensive agent, acting as an angiotensin-converting inhibitor, produced from research on bradykinin potentiator peptides (BPPs). Snake *Bothrops jararaca* venom's BPPs aid in capturing the target by lowering its blood pressure. In pharmaceuticals, captopril is used in a similar manner for humans: it relaxes blood vessels, which is indicated for hypertensive patients [16,32,36,37].

Eptifibatide, a synthetic cyclic heptapeptide disintegrin, used to prevent the formation of blood clots, is a peptide derived from the *Sistrurus miliarius barbourin* snake, a rattlesnake species [38,39]. Tirofiban, inspired by the echistatin structure, also a short monomeric disintegrin from the venom of the snake species *Echis carinatus*, is also a blood thinner; similar to Eptifibatide, this medicine can also prevent the formation of blood clots in the medical condition that leads to acute blood blockage of blood flow to the heart and heart attack [38,40].

Another important medicine produced from snake venom is Batroxobin, isolated from *Bothrops moojeni* snake venom. Its pathway includes the dissolution of fibrinogen to fibrin and D-dimer and the mobilization of endothelial cells to promote thrombolysis [41]. In nature, it enables the snake to capture the prey by activating blood clotting. Clinically, this capability was explored to treat different types of bleeding and thrombosis [16].

A PLA_2 from *Bothrops erythromelas*, an endemic snake in northeast Brazil, was isolated, and its biological activity was tested confirming antibiofilm activity against *Acinetobacter baumannii* and antibacterial activity against *Staphylococcus aureus* [42,43].

Still under research are compounds from snake venoms such as Collinein-1, crotamine, and Crotalphine, which have shown potential for some clinical disorders. First, Collinein-1 is a serine protease from *Crotalus durissus colllineatus*, a species that occurs all over Brazil as well as in other countries in South America. This toxin presents the ability to block a specific potassium channel expressed in some cancer lines, reducing the tumor cells' viability [11,37,44]. Second, crotamine, a compound from the venom of the *Crotalus durissus terrificus* sp., widely distributed in Brazil [44], is a peptide with the ability of membrane translocation with cytoplasmic, vesicular, and nuclear distributions. In melanoma and lymphoma cells, it can undergo an intake process during cell proliferation, constituting a potential anticancer drug [37,45–47]. Crotalphine, also isolated from the snake *Cortalus durissus terrificus* venom, is a peptide that has shown activity against peripheral opioid receptors and has been investigated as a potent painkiller for cancer pain [47–49].

In antivenom serum, IgG is fractionated to produce F(ab) and F(ab′)2, which reduces adverse reactions and increases efficacy. Commercial antivenoms constitute polyclonal mixtures of antibodies or their fractions raised against all toxin antigens in a specific venom. Improvements in antivenoms aiming for safety and more effectiveness have taken place in recent years. Proteomics and transcriptomics are two of these improvements that have been applied to venom toxin composition (venomics), guiding the understanding of medically important toxins. Also, it is important to point out that these methodologies can identify toxins that contain epitopes recognized by antivenom molecules (antivenomics). Furthermore, humanized and fully human monoclonal antibodies and their fractions as well as enzyme inhibitors have been experimentally developed against venom toxins [50].

4. Challenges in Venom Therapy: Addressing Formulation Needs and the Role of Nanotechnology

Venom composition has been studied and it has been demonstrated that it is composed of 100 to 500 pharmacologically active compounds. As shown, there are millions of natural products that can be used for drug discovery. Nevertheless, less than 0.01%

have been identified and characterized and a large proportion of toxins act on unknown receptors. Well-known toxins, usually, are described incompletely; the reasons for this are the difficulties in obtaining reliable sources of venoms, inadequate use of screening tests, difficulties in purifying and characterizing the given toxin in detail, and the limited number of academic or industrial research groups working, which is increasing significantly [51,52].

Although there is a larger number of molecules in the venom pool to be analyzed, multiple sophisticated and complementary bioanalytical methodologies and approaches are employed for the isolation, characterization, and assessment of the pharmacological attributes of proteins and peptides found in snake venom. These efforts are directed towards advancing drug discovery investigations. Purification of the peptide requires a combination of chromatographic methods, and, recently, a new online microfluidic high-resolution screening method was employed to identify neurotoxic components [53]. Search-based sequencing uses bioinformatics tools, such as Mascot (Matrix Science) or Protein Pilot (ABSciex), modern mass spectrometers (MSs), and ESI or MALDI TOF mass spectrometers. MS/MS, $(MS)_n$, or cleaving of the peptide backbone is performed using a variety of methods, including post-source decay (PSD), collision-induced dissociation (CID), and high collision energy dissociation (HCD). The radical fragmentation technique is used to analyze peptides with labile posttranslational modifications, such as electron capture dissociation (ECD) and electron transfer dissociation (ETD) [54]. As shown briefly, a considerable number of techniques are required to guarantee an optimal characterization of animal toxins.

Other approaches can be explored in the field of mass spectrometry with direct infusion nanoelectrospray ionization (nano-ESI-MS) to evaluate peptide distributions in *Bothrops* snake venoms, as shown by Souza et al. [55], or even the use of proline present in *B. jararaca* venom, as carriers of therapeutic agents, which cannot cross the cell membrane, increasing their applicability in drug delivery [56].

Furthermore, identification, isolation, and purification are not the only concerns in this field nowadays, but the therapeutic potential of these venoms has also been considered, e.g., as shown in Biswas et al. and Munawar et al. [57,58]; this method is illustrated in Figure 3.

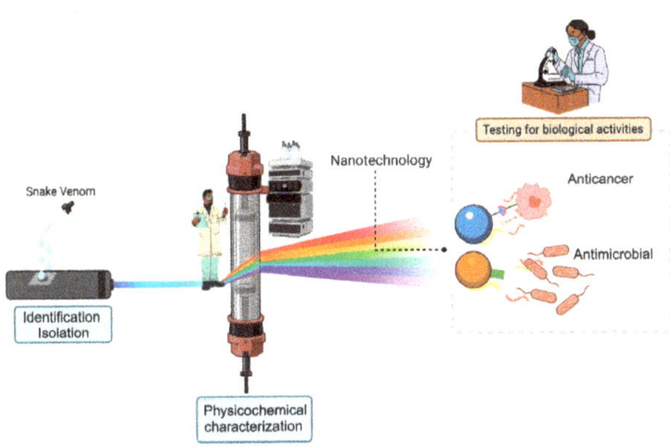

Figure 3. Representation of snake venom characterization and evaluation for biological tests.

According to Biswas et al. [57], nanosized particles exhibit unique properties that can be harnessed to enhance the pharmacological and therapeutic efficacy of medications. The process of nanoencapsulation for these potent therapeutic compounds serves not only as a platform for improved drug delivery but also contributes to heightened stability, increased bioavailability, and targeted application of the drugs. While larger molecules may be subject to elimination from the body, nanoparticles, due to their size, are effectively taken up by cells. A variety of hydrophilic nanoparticles, including chitosan, nano-gold, nano-silver, magnetic and superparamagnetic nanoparticles, and dendrimers, among others, are currently under extensive investigation for their role as drug delivery vehicles. This involves the conjugation of these nanoparticles with various therapeutically potent substances, including venoms and toxins, particularly peptides, proteins, and antigens.

5. Commonly Used Nanosystems to Encapsulate Snake Venom

Nanomedicine emerges as a pioneering field that holds great promise for the twenty-first century, employing nanotechnology to enhance healthcare and pharmaceutical products [59,60]. Consequently, it has been determined that diverse nanocarriers present numerous benefits, including (i) ensuring consistent drug dosage and targeted delivery while avoiding concentration fluctuations, (ii) maximizing therapeutic efficacy while minimizing potential side effects and risks of toxicity, and (iii) providing drug protection against enzymatic degradation [61].

In the field of drug delivery systems, nanomaterials have emerged as key contributors to enhancing drug stability, solubility, release control, minimizing toxicity, and achieving enhanced therapeutic effects [62]. Various nanocarrier systems (Figure 4), including polymeric nanoparticles, inorganic nanoparticles, and nanohydrogels, have been developed to address these challenges [63].

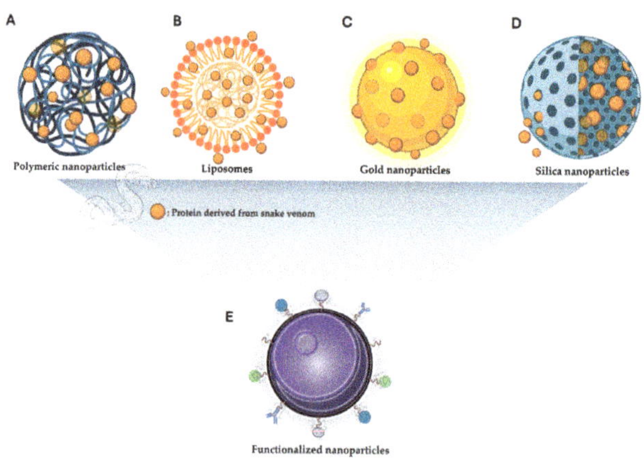

Figure 4. The use of nanotechnology is based on the incorporation of proteins derived from snakes. (**A**) Polymeric nanoparticles; (**B**) liposomes nanoparticles; (**C**) gold nanoparticles (GNPs); and (**D**) mesoporous silica nanoparticles. (**E**) Schematic representation showing that nanoparticles, regardless of their nature, can be modified, thus improving their physical, chemical, and biological properties.

The progress made in the field of nanotechnology, specifically in nanovaccinology, has mitigated the limitations associated with conventional vaccines and paved the way for the creation of an innovative nanoparticle-based delivery system for vaccines. Nanotechnology allows designing nanoparticles with varying characteristics such as composition, size, shape, and surface properties to meet specific medical needs [64].

Currently, there is a significant research trend using nanotechnology in venom and toxin studies [65]. The research on nanotechnology and snake venom/toxin has two distinct dimensions: (i) the use of nanotechnology in developing antivenom for snake venom and (ii) the application of nanotechnology in developing drug indicators against physiopathologies such as arthritis, cancer, and HIV. So far, antivenom snake serum (ASVS) has been the only clinically used antidote against snakebite, which was developed in 1894 at the Pasteur Institute in Paris by Dr. Calmette. The limitations (cost factor, shelf life, no protection against local effects, hemorrhage, and organ toxicity) and side effects (anaphylactic shock, pyrogenic reaction, and serum sickness) on patients are well known [66].

Snakebites can be treated by the administration of antivenom serum and can be associated with adjuvants such as immunoadjuvants that increase the immune response to antigens which are unable to stimulate the immune system with their properties [67]. In this sense, the use of nanotechnology has been investigated aiming at inducing effective adaptive immune responses and reducing tissue damage caused by the toxins present in the venom [68]. In addition, these toxins can be evaluated for their therapeutic efficacy, against emerging diseases [69]. Approaches using nanotechnology emerge as trends and strategies in current medicine.

With a wide range of applications in nanotechnology in biomedical sciences, it is essential to understand how these nanoparticles function within the biological system. It is a huge commercial market today, with billions of investments in global prospects, and it is expected to be a giant market in the future. Various nanoproducts are already commercially available, and numerous products are under research and development. Therefore, nanotechnology can be considered the next technological revolution that has a direct impact on human health and the environment [66].

Over the years, snake venoms have been used as medical research tools or therapeutic/diagnostic agents [57,70]. Calmette et al. [71] proposed that the physiologically active components of these venoms may have therapeutic potential. They demonstrated that these venoms could kill cancer cells in animal models. Subsequently, many reports established the anticancer potential of different species of Elapidae, Viperidae, and Crotalidae snake venoms [72–74]. It has also been observed that these compounds are highly toxic, producing adverse neurological, cardiac, and respiratory effects [75].

Many attempts have been made to increase the therapeutic potential and decrease the toxicity of venom toxins in biological systems. One such approach in modern research has been nanotechnology, which can provide higher efficacy and lower toxicity [66]. A common approach is to attach the protein present in the venom to a nanoparticle through the process of non-covalent electrostatic adsorption [76]. However, this non-covalent interaction is not stable at high salt concentrations and pH values above the protein's isoelectric point. A covalent bond between the toxin protein of the venom and the nanoparticle is more stable, and the binding site to the nanoparticle can be controlled. Conjugation of nanoparticles with snake venom can cause protein folding/unfolding [77]. It is advised to verify the function of the peptide/protein structure after conjugation with nanoparticles. Co-functionalization of nanoparticles with polyethylene glycol can aid in protein unfolding and can be used in the formation of nanoparticles loaded with snake venom [66].

In the pharmaceutical field, there are several approaches involving polymers, and one alternative is poly(ε-caprolactone) (PCL), which has been used as a material for controlled release to produce micro- and nanospheres, reducing toxicity and increasing hydrophobic solubility and passivation, segmentation, or action [78]. PCL's high permeability to various drugs and molecules allows them to be released from the matrix by diffusion [79,80].

The biodegradability of PCL has garnered considerable attention in academic and research circles [81,82]. Numerous research investigations have documented the effective utilization of multi-arm PCL-based nanoparticles in conveying a spectrum of medications, encompassing anticancer agents, antibiotics, and anti-inflammatory substances [83,84]. Colloidal systems utilizing PCL have sparked significant interest owing to their application as drug delivery vehicles [85]. Self-assembled structures such as micelles have been extensively investigated for their efficacy in encapsulating a diverse array of drugs and proteins [86].

The parameters of each technique must be adjusted to obtain stable spheres with sufficient electric charge and solubility and with the smallest diameter. These factors are important because they determine the drug target in the body, the type of cellular interaction, and its release [87].

6. Encapsulation of Venom Fractions with Bioactivity and Its Applications

6.1. Nanosystems Containing Bothrops Snake Venom

In the context of improving antivenom serums, nanoparticles are colloidal dispersions that can be targeted through a modified release of protein biomolecules [88]. In a study by Santos-Silva et al. [68], a new immunoadjuvant was prepared based on cationic PLGA nanoparticles (CNps), which are biocompatible with *Bothrops jararaca* venom, thus comparing antibody production with a conventional immunoadjuvant such as aluminum hydroxide. The particles showed positive charges by the addition of polyethyleneimine (PEI), where an increase in zeta potential was observed with an increasing PEI/polymer ratio. The concentrations of 0.5 and 1.0% were used for *B. jararaca* (BJ) venom in the CNps, obtaining an encapsulation efficiency higher than 98%. In the presence of venom, the nanoparticles showed little variation in size, given in nanometers, for CNps without the venom (140.0 $\pm$ 1.8), CNps + 0.5%BJ venom (167.3 $\pm$ 6.3), and CNps + 1.0%BJ venom (168.7 $\pm$ 3.8), and zeta potential, given in millivolts, as 47.38 $\pm$ 6.24, 42.83 $\pm$ 14.6, and 35.6 $\pm$ 13.4, respectively, with more than 98% encapsulation efficiency for both concentrations. From the results in the in vivo studies, it was observed that, in the immunization protocol, animals treated with only CNps did not show an immune response, whereas when treated with aluminum hydroxide or nanoparticles containing the venom, they produced detectable antibodies at the lowest dilution tested (1 to 51,200); no statistical difference was shown between aluminum hydroxide and CNps, making cationic nanoparticles promising in the study of nanoformulations containing bioactive molecules as immunoadjuvants.

Among the nanosystems related to the *Bothrops* genus, nanoparticles are found in different studies, such as the one using chitosan, which is a polysaccharide obtained by partial N-deacetylation of chitin and has also been investigated as an immunoadjuvant for reducing side effects and improving the efficiency of serums and vaccines [89,90]. Thus, CNPs loaded with venoms of *B. jararaca* and *B. erythromelas* were evaluated by Soares et al. [67] for the production of serum against these venoms. According to these authors, the particles were obtained by the ionic gelation technique, through titrations of a tripolyphosphate (TPP) in a chitosan solution, with an average size of 167.5 nm, zeta potential (ZP) of +24.5 mV, and polydispersity index (PdI) of $\leq$0.3. To incorporate the proteins in the formulation, they were carried in the particles through the incorporation method using a TPP solution, causing no harm to the formation of CNPs. The encapsulation efficiency showed values of 87% for all formulations tested containing *B. erythromelas* venom. For *B. jararaca*, the protein encapsulation efficiency reached levels higher than 67%. The results showed that chitosan is a less inflammatory biopolymer that requires a lower dose of antigen, with equivalent or better performance than aluminum hydroxide as an immunoadjuvant.

Another applicability for bioactive components presents in snakes of the genus *Bothrops* is in the development of drugs with bacterial resistance. In this sense, *Bothrops jararacussu* venom-loaded CNPs were studied by Ribeiro et al. [91] with potential antibacterial activity.

The nanoparticles were obtained by ionic gelation, with a size of approximately 160 nm (PdI < 0.5) and a protein loading efficiency higher than 70%. The antibacterial activity was performed by broth microdilution on *Escherichia coli* and *Staphylococcus aureus* strains. *Bothrops jararacussu* venom showed antimicrobial activity and the nanoparticle containing 0.5 and 1.0% of venom showed great potential as an antibacterial agent against Gram-positive strains.

Another approach to the utilization of *Bothrops* venoms is related to the induction of oral tolerance through the previous oral administration of an antigen, as seen in the work of Tsuruta et al. [92], who evaluated the induction of tolerance to snake venoms adsorbed to/encapsulated in nanostructured SBA-15 silica in BALB/c mice, observing specific tolerance and cross-reactivity with other toxins of *Bothrops* species. The authors developed a protocol in BALB/c mice for oral tolerance, by administration of *B. jararaca* venom, measuring antibody titers in antisera after immunization with snake venoms, to understand the mechanisms involved. With this, it was observed that the induction of oral tolerance by the venom will depend on the dose administered and is specific for each snake species, showing that tolerance was induced more effectively when the same snake venom was used in oral administration and immunization, even when the antiserum showed cross-reaction with venom from other snakes. The study of Wambre and Jeong [93] demonstrates the application of oral tolerance as a non-invasive form, allowing studies in the prophylaxis and treatment of diseases.

Nanotechnology also allows for a reduction in the toxicity of venoms and for use in the immunization of horses for the development of antisera to treat snakebites, reducing the stress and suffering of the horse, as reported by Costa et al. [94], who proposed a modification of the immunization protocol, with inactivated venoms and encapsulation in stabilized liposomes. The particles showed an encapsulation efficiency of 59% to liposomes with venom of *Bothrops* (LB) and 99% to liposomes with crotalic venom (LC), as followed by filtration on Sephacryl S1000. Light scattering measurements led them to conclude that both LB (119 $\pm$ 47 nm) and LC (147 $\pm$ 56 nm) were stable for 22 days at 4 °C, even after lyophilization. The venoms used were from five different *Bothrops* species (*B. alternatus*, *B. jararaca*, *B. jararacussu*, *B. moojeni*, and *B. neuwiedi*) and *Crotalus durissus terrificus*, which were solubilized in an MPB-pBB/EDTA buffer (20 mM phosphate, 295 mM mannitol, 5 mM EDTA, 3 mM pBB, pH 7.2). Therefore, the aim of this study was to inactivate the venom in order to reduce the toxicity and inhibition of PLA_2 resulting in the stabilization of liposomes, using the chelating agent $CaNa_2EDTA$ and EDTA to increase the inhibition of PLA_2. The authors also pointed to the importance of formulation stability after freeze-drying, which was demonstrated in another study by Oliver et al. [95]: freeze-dried liposomes containing PLA_2 and partially rehydrated at controlled humidities did not show PLA_2 activated during freeze-drying but were more active during rehydration. Another problem found was the stability of the membrane after freeze-drying, which was overcome by using mannitol as a cryoprotectant. Bothropic venom-loaded liposomes showed an efficiency of 99%, while crotalic venoms showed 59%. After the procedures, no signs of venom toxicity were observed in the horses immunized with the liposomal formulations containing snake venom, due to venom alkylation and delivery within stable nanosystems, showing no differences in their reactivity. In addition, the lipolytic activity associated with venoms is directly related to the PLA_2, which can induce membrane rupture through fusogenic agents such as lysophosphatidylcholine [96–98].

According to Carvalho et al. [99], another possibility for the use of liposomes is the use of sphingomyelin in its composition for the encapsulation of *B. jararaca*, resistant to phospholipase, with a high cholesterol content to increase membrane stability. The authors used a mouse model to evaluate the toxicity of *B. jararaca* venom encapsulated in liposomes, in order to protect animals against the lethal effects of poisoning. The encapsulation method was dehydration–rehydration vesicles according to Kirby and Gregoriadis [100]. Two immunostimulants were used: a hydrophilic lipopolysaccharide (LPS), which was solubilized together with the venom, and hydrophobic monophosphoryl lipid A (MPLA), solubilized with the lipids. There was a reduction in the clinical toxicity of encapsulated *B. jararaca* (BJ) venom, where no mice died after injection of a dose equivalent to 6×LD 50 of liposome-encapsulated BJ venom. The formulations were stable at 37 °C, with a retention of 65% of the originally encapsulated *B. jararaca* venom, allowing future studies to examine the use of liposomes as immunoadjuvant systems for commercial antivenoms.

The PLA_2 isolated from *Bothrops jararacussu* venom can also be evaluated for its anti-*Leishmania amazonensis* activity in vitro according to de Barros et al. [101] who added 2 mg/mL of the PLA_2 to a lipid mixture, which was solubilized in chloroform and dried under nitrogen flow, with the formation of the vesicles by the extrusion method. The liposomes showed monodisperse populations with 205.2 nm of the encapsulated Asp49 liposomes. The zeta potential was −25.4 mV and there was an encapsulation rate of 69.6%, meaning 1.4 mg of the toxin was detectable in 1 mL of the liposomes produced. The formed nanosystems reduced 78% of promastigote forms and preserved 82% of J774 macrophage viability. In addition, there was a considerable reduction (55%; $p < 0.05$) in amastigote forms, compared to the control group, conferring the possibility of biotechnological studies using nanotechnology against neglected diseases, such as leishmaniasis. The same liposome nanosystem was later used by the same research group of de Barros et al. [102] in another study to evaluate the in vivo anti-*Leishmania amazonensis* activity. In comparison to the control group, a 57.1% reduction was observed in the paw tissue. The infected groups treated with Asp49-PLA2 liposomes exhibited a notable increase in TNF-α production in the lymph nodes and paws (43.73 pg/mL ± 2.25 and 81.03 pg/mL ± 5.52, respectively), along with elevated nitrite levels (31.28 μM ± 0.58 and 35.64 μM ± 5.08 in lymph nodes and paw tissues, respectively) when compared to the untreated groups.

6.2. Nanosystems Containing Crotalus Snake Venom

Chitosan nanoparticles were developed by Glaucia-Silva et al. [103] through an ionic gelatinization method containing *Crotalus durissus cascavella* crude snake venom capable of inducing murine immunization, incorporating auxiliary components to enhance the immunogenicity of the serum or vaccine. Chitosan nanoparticles (CNPs) exhibit characteristics such as biocompatibility, biodegradability, mucoadhesive properties, and controlled drug release. Physicochemical properties of these CNPs were determined without venom (size: 155.5 ± 7.6 nm; PdI: 0.296 ± 0.007; zeta potential: +30.8 ± 12.9 mV), with 0.5% of *C. d. cascavella* venom (size: 147.4 ± 9.0 nm; PdI: 0.316 ± 0.004; zeta potential: +36.0 ± 18.6 mV), with 1.0% of *C. d. cascavella* venom (size: 147.7 ± 12.0 nm; PdI: 0.323 ± 0.009; zeta potential: +41.80 ± 14.1 mV), and with an encapsulation efficacy over 95% for both concentrations. Immunization with CNPs incorporated with *C. d. cascavella* in BALB mice exhibited high effectiveness when compared with the control, producing high IgG titters until the dilution of 1 to 6400, positioning them as a promising pharmaceutical agent.

Crotoxin, a major component of the venom of *Crotalus durissus terrificus*, demonstrated in vivo immunomodulatory and anti-inflammatory characteristics [104]. Thus, in order to nanoencapsulate crotoxin, Sant'Anna et al. [105] employed nanostructured mesoporous silica (SBA-15) and evaluated its responsiveness in the context of experimental autoimmune encephalomyelitis (EAE) in a murine model of multiple sclerosis; SBa-15 was utilized, with the incorporation of crotoxin diluted in PBS at a ratio of 1:10 (toxin to silica) maintained for 24 h at 2 °C to 8 °C with sporadic agitation. It was observed that the use of crotoxin+SBA-15 in the initial phase of development could ameliorate EAE-induced neuroinflammation by preventing the recruitment of peripheral Th17 cells, reducing cellular infiltration into the central nervous system, decreasing IL-17, and inhibiting glial cell activation. This, in turn, led to a reduction in the intensity and incidence of clinical signs while preserving muscular function. Furthermore, it demonstrated potential as a nanocarrier for use with biological compounds, exhibiting significant efficacy in reducing toxicity, allowing for increased dosage without a corresponding rise in toxic effects, and enhancing the delivery and effectiveness of crotoxin compared to its non-encapsulated form [105–107].

In a study conducted by Karpel et al. [108], gold nanoparticles (GNPs) were employed to be optimized with crotamine. Crotamine was adsorbed onto GNPs through the interaction involving the covalent bonding of the peptide and a polyethylene glycol (PEG) ligand, resulting in the formation of the PEG–crotamine–GNP transport complex. The particles exhibited a zeta potential close to zero (−0.6 mV) and a crotamine-free surface charge (−25 mV), with diameters for GNPs, GNP–PEG, and GNP–PEG–crotamine measured at 10, 34, and 43 nm, respectively. GNPs associated with crotamine demonstrated high efficiency in being absorbed by HeLa cells, indicating the sustained cell-penetrating nature of crotamine even after complex association. In this context, GNPs are emerging as increasingly significant players in multimodal and multifunctional molecular imaging for the early-stage detection of diseases such as cancer. Moreover, the research highlighted that GNP–crotamine complexes have evolved into potential platforms for therapeutic agents.

In a study using chitosan nanoparticles (CNPs), Jimenez-Canale et al. [109] studied the venom of *Crotalus molossus molossus* for cytotoxic activity against T-47D breast carcinoma cells. The particles, derived from the venom (primarily composed of VAP2A, Ruberlysin, Apoxin I, and PLA2), exhibited dimensions of 415.9 ± 21.67 nm, a polydispersity index (PdI) of 0.44 ± 0.03, and a zeta potential of +28.3 ± 1.17 mV. They displayed distinct features of a smooth and semi-spherical surface. Moreover, these particles demonstrated non-hemolytic properties in human red blood cells and reduced cell viability in T-47D breast carcinoma cells.

Utilizing liposomes as a nanoencapsulated system, Magalhães et al. and Freitas and Frézard [110,111] generated nanoliposomes encapsulating crotoxin to observe the effects of these nanoparticles as immunostimulants and their immunomodulatory impacts for potential antivenom therapy or vaccination purposes, respectively. In both studies, liposomes associated with crotoxin elicited immunization effects in rats against the respective venom, concurrently reducing toxicity and allowing for increased doses with diminished toxic effects. Magalhães et al. [110] showed that the toxicity of liposomes containing crotoxin was influenced by the composition of the membrane. Thus, encapsulating crotoxin in liposomes composed of sphingomyelin significantly diminished its toxicity; even when mice were inoculated with more than 50 times the lethal dose (LD_{50}) of crotoxin, none succumbed. When crotoxin was encapsulated into dehydration–rehydration vesicles, a protection level exceeding 50% was achieved for both Magalhães et al. and Freitas and Frézard [110,111]. Immunization with 100% protection was achieved by Magalhães et al. when the immunization with liposomes encapsulating crotoxin was prepared using a non-phospholipid mixture and without immunostimulants, resulting in this intriguing outcome. This method proved to be cost-effective, with a preparation cost of approximately USD 80 per liter of liposome preparation.

By employing the Stöber method, Baudou et al. [112] investigated the adsorption of *Crotalus durissus terrificus* snake venom onto silica nanoparticles (SiNPs), resulting in the formation of nanovenoms. Through experimentation with various enzymatic activities, the researchers demonstrated that SiNPs effectively bound snake venom proteins while preserving their biological activity. This study demonstrated that venom nanoparticles conserve the cytotoxic effect that was tested in THP-1 cells (human monocytic cell line derived from acute monocytic leukemia), and a significant decrease in proliferation ($\sim$84%) was constated in different concentrations tested for nanoparticles venom (5 and 10 μg/mL).

Improved biodistribution is observed when the venom is encapsulated. Gomes et al. [113] conducted a comparative study on the biodistribution of free and encapsulated *Crotalus durissus terrificus* venom in mice. Upon subcutaneous inoculation of the encapsulated venom at a higher dose of 25 μg/mouse (2 LD_{50}), no *C. d. terrificus* was detected in the kidney, spleen, brain, or other tissues. On the other hand, free *C. d. terrificus* venom following subcutaneous injection was detected in the serum 15 min after administration, reaching its peak concentration at 30 min (45 ± 5 ng/mL), and was cleared from circulation within 6 h. Two hours post-inoculation, the venom was found in the kidney (57 ± 9 ng/g of tissue), spleen (18 ± 4 ng/g of tissue), and brain (14 ± 6 ng/g of tissue). However, after intravenous injection of the encapsulated venom at a dose of 5 μg/mouse (LD_{50}), venom was observed in liver and spleen tissues. Whether administered subcutaneously or intravenously, free *C. d. terrificus* venom was initially detected in the kidney. The paper focuses on the implications of encapsulated venom biodistribution in relation to the reduction in toxicity and enhancement of adjuvanticity effects. All the information about the nanosystems used to encapsulate the different snake venoms is summarized and highlighted in Table 1. Some nanosystems of Elapidic venom can also be found in the literature.

6.3. *Nanosystems Containing Lachesis Snake Venom*

We conducted a comprehensive search across various available databases aiming to identify studies that explored the utilization of *Lachesis* snake venom in nanosystems. However, no studies reporting this specific association have been found thus far. The absence of identified studies highlights a current gap in scientific knowledge regarding the utilization of *Lachesis* venom in nanotechnology. This finding underscores the need for future research in this area to explore the potential of this association and its possible applications in nanomedicine and nanotherapy.

To the best of our knowledge so far, it is noteworthy to mention that although no studies investigating the combination of *Lachesis* venom with nanotechnology were found, in our search, there exist numerous studies reporting on the biological activities of this venom. These studies have elucidated various pharmacological and therapeutic properties associated with *Lachesis* venom. Future research endeavors could draw upon these biological activities as a foundation for bridging these two scientific domains, thus exploring the potential synergistic effects and novel applications that may arise from integrating nanotechnology with the known biological activities of *Lachesis* venom.

In the realm of *Lachesis* venom's biological activities, several studies have shed light on its remarkable properties. For instance, research by Stransky et al. [114] demonstrated the potent cytotoxic effect of different cell lines, the toxicity observed in this context encompasses the integration of diverse mechanisms of cell death. *Lachesis muta muta* venom decreased, in a concentration-dependent manner, the cell viability of the tumor (MGSO-3 and HeLa), as well as immortalized (VERO, EA.hy 926) and normal cells (keratinocytes). Different levels of toxicity were observed for each cell type. Vero cells (EC_{50} = 0.83 μg/mL) and MGSO-3 (EC_{50} = 2.26 μg/mL) were the most sensitive to venom and HeLa cells (EC_{50} = 7.14 μg/mL) were more resistant.

Table 1. Overview of snake venom and nanobiotechnology-derived nanobiosystems being used for biological application and its nanosystem types and sources.

Molecule/Toxin	Classification/Source	Nanosystems	Characterization Size (nm); PDI; PZ (mV)	Encapsulation Efficiency (%)	Biological Activities	References
Whole venom	*Bothrops jararaca* (BJ)	Cationic PLGA nanoparticles (CNps) +1.0% BJ venom	168.7 ± 3.8 nm, 0.09 ± 0.01 +35.6 ± 13.4 mV	-	Immunoadjuvant	[68]
Whole venom	*Bothrops jararaca* and *Bothrops erythromelas (Bery)*	Chitosan nanoparticles (CNPs)	BJ 10%: 174.7 ± 5.0 nm, 0.203 ± 0.07 +24.91 ± 2.91 mV Bery 10%: 160.0 ± 2.3 nm, 0.302 ± 0.01 +19.00 ± 2.76 mV	BJ 67.7 Bery 87.6	Immunoadjuvant	[67]
Whole venom	*Bothrops jararacussu*	CNPs	160 nm Inferior to 0.5	More than 70%	Antimicrobial activity against Gram-positive bacteria	[91]
Whole venom	*Bothrops jararaca*	Nanostructured SBA-15 silica	-	1 μg of venom adsorbed /encapsulated	Oral tolerance induction	[92]
Whole venom	*Pool of venom containing Bothrops alternatus, Bothrops jararaca, Bothrops jararacussu, Bothrops Bothrops moojeni, Bothrops neuwiedi*, and *Crotalus durissus terrificus*	Liposomes with *Bothrops* venom (LB), and with crotalic venom (LC)	LB (119 ± 47 nm) and LC (147 ± 56 nm)	The liposomal efficiencies of protein incorporation were 99% for the bothropic and 59% for the crotalic venoms	Development of antisera to treat snakebites, reducing the stress of the horse	[94]
Whole venom	*Bothrops jararaca*	Liposomes	-	-	Immunostimulants	[99]
PLA_2 toxin	*Bothrops jararacussu*	Liposomes	Size: 241.9 and 205.2 nm Zeta: −18 mV and −25.4 mV	-	Anti-*Leishmania amazonensis* activity in vitro and in vivo	[101,102]
Whole venom	*Crotalus molossus molossus*	CNPs	Size: 415.9 ± 21.67 nm PdI: 0.44 ± 0.03 Zeta: +28.3 ± 1.17 mV	48.29 ± 3.84	Cytotoxic activity against T-47D breast carcinoma cells	[109]
Whole venom	*Crotalus durissus cascavella*	CNPs + *C. d. cascavella* venom 1.0%	Size: 147.7 nm PdI: 0.323 Zeta: +41.80 mV	98.9	Immunostimulants	[103]
Crotoxin	*Crotalus durissus Terrificus*	Nanostructured mesoporous silica SBA-15	-	-	Immunomodulatory and anti-inflammatory activity in multiple sclerosis	[105]
Crotamine	*Crotalus durissus Terrificus*	Gold nanoparticle GNPs	Size: 43 nm Zeta: −0.6 ± 3.9 mV	-	Cell-penetrating ability	[108]
Crotoxin	*Crotalus durissus Terrificus*	Liposomes	-	-	Immunostimulants	[110,111]
Whole venom	*Crotalus durissus terrificus*	Liposomes	-	-	Biodistribution of free and encapsulated liposomes	[113]

Additionally, studies by Fuly et al. [115] highlighted the procoagulant, proteolytic, and phospholipase A_2 activities. Furthermore, according to Almeida et al. (2015) [116], investigations on patent application fields showed anti-aging antioxidant preparations [application number: RU20010122763], prevention and treatment of viral and bacterial infections in avian species [application number: RU20020111635], antimicrobial agents [application number: IE20070000737, WO2008EP08602], and analgesic composition [application number: US2015030691]. These notable findings not only underscore the diverse biological activities associated with *Lachesis* venom but also serve as a foundation for future studies that can harness these activities in tandem with nanotechnology, opening new avenues for therapeutic interventions and biomedical applications.

7. Theragnostic Applications of Polymeric Nanosystems

Crotamine exhibits distinct selectivity towards actively proliferating (AP) cells [117–119] and is primarily internalized via endocytosis, even when loaded with a cargo [120]. It selectively facilitates the delivery of plasmid DNA into AP mammalian cells [118,120]. The capacity to transport cargo without requiring covalent bond formation [120], coupled with its preferential cytotoxicity towards tumor cells [121], indicates the potential utility of crotamine in cancer therapy and/or diagnostics.

Short peptides derived from crotamine, such as CyLoPs (Cytosol Localizing Peptides) and Nucleolar Targeting Peptides (NrTPs), have been extensively studied [122]. Notably, NrTP-1, a prototype of the NrTP series, was originally designed [123] and subsequently modified by replacing the cysteine residue at position 4 with a serine, resulting in an analog called NrTP-6 (YKQSHKKGGKKGSG) [124,125]. NrTP-6 has demonstrated efficient penetration and accumulation in specific cancer cell lines and populations, making it a promising theragnostic probe for imaging cancer heterogeneity and targeted delivery of chemotherapeutics to specific subpopulations of cancer cells [126].

Crotalicidin, a vipericidin [127], and an Asian elapid CRAMP (cathelicidin-related antimicrobial peptide) have exhibited notable biological activities, particularly in relation to their antipathogenic and antitumoral properties, as extensively reviewed elsewhere [128–130]. The interaction of crotalicidin and elapid CRAMPs with biological membranes and biomimetics has been extensively demonstrated, encompassing not only the native, full sequences but also short derived peptide fragments. Their ability to induce perturbations in the plasma membrane is closely associated with the mechanisms underlying their antimicrobial activity [131–135], as well as their pronounced antitumor effects, which have been experimentally validated in vitro using various cancer cell lines [136,137].

Cardiotoxin VII4, obtained from the Mozambique spitting cobra *Naja mossambica mossambica*, has been identified as a cell-penetrating peptide with the capability to enter mitochondria. Its application led to mitochondrial fragmentation in both mouse primary cortical neurons and human neuroblastoma (SH-SY5Y) cells [138]. A short cationic [13-mer] peptide derived from the C-terminal region of the Lys49 PLA2 from the venom of the broad-banded copperhead snake *Agkistrodon contortrix laticinctus* was prepared and this imparted engineered peptide characteristics of CPPs and AMPs and revealed an effective leishmanicidal activity [139].

According to Sciani et al. [56], Bradykinin-Potentiating Peptides (BPPs) derived from the venom of *Bothrops jararaca* have exhibited notable cell-penetrating capabilities. Fluorescence microscopy experiments revealed the translocation of these peptides into the cytoplasm of human melanoma tumor cells (SK-MEL-28 and A2058), without any detectable cytotoxic effects even when 1 μM peptide was used. Thus, no change in metabolism, cell cycle, or morphology was observed with either labeled or unlabeled CPPs, clearly demonstrating their role as cargo carriers rather than drugs themselves. A label-free method for identifying CPP peptides, regardless of biological source, has been developed and validated, potentially opening the way for the discovery of new and more effective drug delivery systems.

8. Clinical Trial, Patents, and Market

Increasingly sustainable economies are being represented by the bioeconomy. The focus is on bio-innovations that explore renewable natural resources for bio-inputs crucial to the development of new bioprocesses and bioproducts [140,141]. The enhancement of bio-entrepreneurship [142] with biotechnology plays a pivotal role in shaping this new knowledge-based bioeconomy [143–145]. Innovative products emerge from the integration of cellular and molecular biotechnology techniques with bio-resources [146]. In this context, both known and yet-to-be-discovered organisms that make up our natural biodiversity could serve as primary sources for novel bio-compounds with boundless applications in what is nowadays called One Health.

An example of bioactive compounds that have been used in the design of new therapeutic agents and products with diverse pharmacological applications, cosmeceuticals, and diagnostic tools can be found in a review carried out by [11].

A considerable number of patents applied to the discovery of medicines based on bioactive compounds from venomous animals have been developed [116]. New technologies, such as nanotechnology, have been used for the targeted delivery of medicines based on the delivery of materials from synthetic and natural sources to their clinical applications, aiming to improve pharmacokinetics and pharmacodynamics, which is a very prosperous area for the development and study of new products [62].

Snake venom is a rich source of biological compounds. Patents referring to toxins present in the genus *Bothrops* are related to hypotensive and vasodilatory formulations, medicines for the treatment of systemic–degenerative diseases, modulators of acetylcholine receptors, anticoagulant and thrombolytic formulations, bactericides, and others [116]. The captopril was developed based on the bradykinin potentiating factor obtained from *Bothrops* jararaca snake venom and it was the first animal toxin-based drug approved for human use [147,148].

Almeida and collaborators [116] report that *Crotalus* genus snakes are the most investigated in Brazil for drug discovery, and patent registrations for bioactive compounds originating from the venom of this type of snake are used to obtain various pharmaceutical products with different applications, such as antimicrobial drugs, drugs for cancer treatment, analgesic drugs to treat AIDS, anti-inflammatory agents, and others. A study developed by Boldrine-Franca et al. [149] demonstrated that a toxin isolated from the venom of the snake *Crotalus durissus collilineatus* demonstrated important activity against several strains of cancer, being a potential candidate for the development of a pharmacological product.

The United States Patent and Trademark Office [150] stated, by a cooperative patent classification (CPC), that patents including snake venom are one of the most frequently reported patents, with 7242 patents reported until the year 2020 in the field of agriculture [151].

Clinical trials using snake venom are challenging due to their complex protein and peptide toxins, making them clinically challenging and scientifically fascinating [152]. High variability in snake venom composition is responsible for the various clinical manifestations in envenoming, making it difficult to find clinical trials using snake venom [4]. Clinical trials using snake venom are difficult to find due to the lack of relevant studies in the literature [153,154]. However, a recent review by Matkar et al. highlights the significance of research on natural products, emphasizing the role of serine proteases and their applications in human therapies, including amyloid formation [155]. This underscores the critical need to explore the diverse therapeutic potential of natural product-derived compounds, particularly those sourced from animals, when starting new clinical trials.

9. Conclusions

The application of snake venom in medical therapies has garnered global attention due to its potential to advance our understanding of biological activity and enhance immunization strategies against toxic effects. Among the most promising developments is the utilization of polymeric, lipidic, or metal nanoparticles to encapsulate and/or incorporate venom toxins from various snake species. This innovative approach offers significant therapeutic potential by improving treatment efficacy and reducing the damage caused by envenomation.

Despite these advancements, the vast diversity of snake species and the intricate composition of their venoms highlight the substantial knowledge gaps that remain. Continued exploration, particularly focusing on less studied species, is essential to deepen our understanding of venom components and their therapeutic applications. This research not only aims to refine treatments for snakebite envenomation but also opens new avenues for the development of novel therapies. Ultimately, these efforts contribute to expanding scientific knowledge and advancing human health in multiple domains.

Author Contributions: Á.E.F.A.: investigation, writing—original draft preparation. A.B.C.B.: investigation, writing—original draft preparation. L.C.F.S.: investigation, writing—original draft preparation. L.M.M.C.: investigation, writing—original draft preparation. G.M.A.P.: investigation, writing—original draft preparation. A.F.C.U.: investigation, writing—original draft preparation. J.M.B.-F.: conceptualization, methodology, formal analysis, writing—review, editing, supervision. M.S.S.: conceptualization, methodology, formal analysis, writing—review, editing, supervision. K.P.O.L.: conceptualization, methodology, formal analysis, writing—review, editing, supervision. K.S.R.S.: review, editing, data curation. F.H.X.-J.: conceptualization, methodology, formal analysis, writing—review, editing, supervision. All authors have read and agreed to the published version of the manuscript.

Funding: This work was supported by Paraíba State Research Foundation [FAPESQ] granter number #1866/2022, and Fundação Coordenação de Aperfeiçoamento de Pessoal de Nível Superior [CAPES], Brazil.

Institutional Review Board Statement: Not applicable.

Data Availability Statement: Not applicable.

Conflicts of Interest: The authors declare that this study received funding from FAPESQ. The funder was not involved in the study design; the collection, analysis, or interpretation of the data; the writing of this article; or the decision to submit it for publication. The authors declare no conflicts of interest.

References

1. Mirtschin, P.J.; Dunstan, N.; Hough, B.; Hamilton, E.; Klein, S.; Lucas, J.; Millar, D.; Madaras, F.; Nias, T. Venom yields from australian and some other species of snakes. *Ecotoxicology* **2006**, *15*, 531–538. [CrossRef] [PubMed]
2. Pucca, M.B.; Knudsen, C.; Oliveira, I.S.; Rimbault, C.; Cerni, F.A.; Wen, F.H.; Sachett, J.; Sartim, M.A.; Laustsen, A.H.; Monteiro, W.M. Current knowledge on snake dry bites. *Toxins* **2020**, *12*, 668. [CrossRef] [PubMed]
3. Williams, D.; World Health Organization. Snakebite Envenoming. Available online: https://www.who.int/health-topics/snakebite (accessed on 6 September 2023).
4. Gutiérrez, J.M.; Calvete, J.J.; Habib, A.G.; Harrison, R.A.; Williams, D.J.; Warrell, D.A. Snakebite envenoming. *Nat. Rev. Dis. Primers* **2017**, *3*, 17063. [CrossRef]

5. Aron, M.B.; Mulwafu, M.; Mailosi, B.; Kreuels, B.; Dullie, L.; Kachimanga, C.; Blessmann, J.; Ndarama, E.; Sambani, C.; Munyaneza, F.; et al. Experiences and practices of traditional healers on snakebite treatment and prevention in rural Malawi. *PLoS Negl. Trop. Dis.* **2023**, *17*, e0011653. [CrossRef] [PubMed]
6. Strand, E.; Murta, F.; Tupetz, A.; Barcenas, L.; Phillips, A.J.; Farias, A.S.; Santos, A.C.; Rocha, G.S.; Staton, C.A.; Ramos, F.R.; et al. Perspectives on snakebite envenoming care needs across different sociocultural contexts and health systems: A comparative qualitative analysis among us and brazilian health providers. *Toxicon X* **2023**, *17*, 100143. [CrossRef] [PubMed]
7. Williams, A.; Paterson, C. What future for policing? Some reflections on the concept and purpose of policing and their implications for police reform in England and Wales. *Int. J. Law Public Adm.* **2019**, *2*, 12–22. [CrossRef]
8. Schneider, M.C.; Min, K.; Hamrick, P.N.; Montebello, L.R.; Ranieri, T.M.; Mardini, L.; Camara, V.M.; Raggio Luiz, R.; Liese, B.; Vuckovic, M.; et al. Overview of snakebite in Brazil: Possible drivers and a tool for risk mapping. *PLoS Negl. Trop. Dis.* **2021**, *15*, e0009044. [CrossRef]
9. Araújo, S.C.M.; Câmara, J.T.; Guedes, T.B. Snakebites in northeastern Brazil: Accessing clinical-epidemiological profile as a strategy to deal with neglected tropical diseases. *Rev. Soc. Bras. Med. Trop.* **2023**, *56*, e02242023. [CrossRef] [PubMed]
10. Duque, B.R.; Bruno, S.F.; Ferreira, V.; Guedes, T.B.; Machado, C.; Hamdan, B. Venomous snakes of medical importance in the brazilian state of Rio de Janeiro: Habitat and taxonomy against ophidism. *Braz. J. Biol.* **2023**, *83*, e272811. [CrossRef]
11. Bordon, K.D.C.F.; Cologna, C.T.; Fornari-Baldo, E.C.; Pinheiro-Júnior, E.L.; Cerni, F.A.; Amorim, F.G.; Anjolette, F.A.P.; Cordeiro, F.A.; Wiezel, G.A.; Cardoso, I.A.; et al. From animal poisons and venoms to medicines: Achievements, challenges and perspectives in drug discovery. *Front. Pharmacol.* **2020**, *11*, 1132. [CrossRef]
12. Werner, R.M.; Soffa, A.N. Considerations for the development of a field-based medical device for the administration of adjunctive therapies for snakebite envenoming. *Toxicon X* **2023**, *20*, 100169. [CrossRef] [PubMed]
13. Hamzaoui, A.; Laraba-Djebari, F. Development and evaluation of polymeric nanoparticles as a delivery system for snake envenoming prevention. *Biologicals* **2021**, *70*, 44–52. [CrossRef] [PubMed]
14. Fry, B.G.; Roelants, K.; Champagne, D.E.; Scheib, H.; Tyndall, J.D.A.; King, G.F.; Nevalainen, T.J.; Norman, J.A.; Lewis, R.J.; Norton, R.S.; et al. The toxicogenomic multiverse: Convergent recruitment of proteins into animal venoms. *Annu. Rev. Genom. Hum. Genet.* **2009**, *10*, 483–511. [CrossRef] [PubMed]
15. Fry, B.G.; Vidal, N.; van-der Weerd, L.; Kochva, E.; Renjifo, C. Evolution and diversification of the toxicofera reptile venom system. *J. Proteom.* **2009**, *72*, 127–136. [CrossRef]
16. Jenner, R.; Undheim, E. *Venom: The Secrets of Nature's Deadliest Weapon*; Smithsonian Books: Washington, DC, USA, 2017.
17. Schendel, R.; Jenner, R.; Undheim, E. The diversity of venom: The importance of behavior and venom system morphology in understanding its ecology and evolution. *Toxins* **2019**, *11*, 666. [CrossRef] [PubMed]
18. Clark, G.C.; Casewell, N.R.; Elliott, C.T.; Harvey, A.L.; Jamieson, A.G.; Strong, P.N.; Turner, A.D. Friends or foes? Emerging impacts of biological toxins. *Trends Biochem. Sci.* **2019**, *44*, 365–379. [CrossRef]
19. Chen, N.; Xu, S.; Zhang, Y.; Wang, F. Animal protein toxins: Origins and therapeutic applications. *Biophys. Rep.* **2018**, *4*, 233–242. [CrossRef]
20. Singh, S.B.; Pelaez, F. Biodiversity, chemical diversity and drug discovery. In *Natural Compounds as Drugs, Volume I*; Birkhäuser: Basel, Switzerland, 2008; pp. 141–174.
21. Kessler, P.; Marchot, P.; Silva, M.; Servent, D. The three-finger toxin fold: A multifunctional structural scaffold able to modulate cholinergic functions. *J. Neurochem.* **2017**, *142* (Suppl. S2), 7–18. [CrossRef] [PubMed]
22. Shafee, T.M.A.; Lay, F.T.; Phan, T.K.; Anderson, M.A.; Hulett, M.D. Convergent evolution of defensin sequence, structure and function. *Cell. Mol. Life Sci.* **2017**, *74*, 663–682. [CrossRef]
23. Zhang, S.-K.; Song, J.; Gong, F.; Li, S.-B.; Chang, H.-Y.; Xie, H.-M.; Gao, H.-W.; Tan, Y.-X.; Ji, S.-P. Design of an α-helical antimicrobial peptide with improved cell-selective and potent anti-biofilm activity. *Sci. Rep.* **2016**, *6*, 27394. [CrossRef] [PubMed]
24. McCleary, R.J.R.; Kini, R.M. Non-enzymatic proteins from snake venoms: A gold mine of pharmacological tools and drug leads. *Toxicon* **2013**, *62*, 56–74. [CrossRef] [PubMed]
25. Xiao, H.; Pan, H.; Liao, K.; Yang, M.; Huang, C. Snake Venom PLA_2, a promising target for broad-spectrum antivenom drug development. *Biomed. Res. Int.* **2017**, *2017*, 6592820. [CrossRef]
26. Tasoulis, T.; Isbister, G. A review and database of snake venom proteomes. *Toxins* **2017**, *9*, 290. [CrossRef] [PubMed]
27. Casewell, N.R.; Jackson, T.N.W.; Laustsen, A.H.; Sunagar, K. Causes and consequences of snake venom variation. *Trends Pharmacol. Sci.* **2020**, *41*, 570–581. [CrossRef] [PubMed]
28. Tasoulis, T.; Isbister, G.K. A current perspective on snake venom composition and constituent protein families. *Arch. Toxicol.* **2023**, *97*, 133–153. [CrossRef] [PubMed]
29. Oliveira, A.L.; Viegas, M.F.; Silva, S.L.; Soares, A.M.; Ramos, M.J.; Fernandes, P.A. The chemistry of snake venom and its medicinal potential. *Nat. Rev. Chem.* **2022**, *6*, 451–469. [CrossRef]

30. Slagboom, J.; Mladić, M.; Xie, C.; Kazandjian, T.D.; Vonk, F.; Somsen, G.W.; Casewell, N.R.; Kool, J. high throughput screening and identification of coagulopathic snake venom proteins and peptides using nanofractionation and proteomics approaches. *PLoS Negl. Trop. Dis.* **2020**, *14*, e0007802. [CrossRef]
31. Koh, C.Y.; Kini, R.M. from snake venom toxins to therapeutics—Cardiovascular examples. *Toxicon* **2012**, *59*, 497–506. [CrossRef] [PubMed]
32. Wolf, G.; Wenzel, U.; Ziyadeh, F.N.; Stahl, R.A.K. Angiotensin converting-enzyme inhibitor treatment reduces glomerular P16 INK4 and P27 Kip1 expression in diabetic BBdp rats. *Diabetologia* **1999**, *42*, 1425–1432. [CrossRef] [PubMed]
33. Dixon, R.W.; Harris, J.B. Myotoxic activity of the toxic phospholipase, notexin, from the venom of the australian tiger snake. *J. Neuropathol. Exp. Neurol.* **1996**, *55*, 1230–1237. [CrossRef]
34. Gutiérrez, J.M.; Ponce-Soto, L.A.; Marangoni, S.; Lomonte, B. Systemic and local myotoxicity induced by snake venom group II Phospholipases A2: Comparison between crotoxin, crotoxin B and a Lys49 PLA2 homologue. *Toxicon* **2008**, *51*, 80–92. [CrossRef]
35. Bulfone, T.C.; Samuel, S.P.; Bickler, P.E.; Lewin, M.R. Developing small molecule therapeutics for the initial and adjunctive treatment of snakebite. *J. Trop. Med.* **2018**, *2018*, 4320175. [CrossRef] [PubMed]
36. Greener, M. The next generation of venom-based drugs. *Prescriber* **2020**, *31*, 28–32. [CrossRef]
37. Oliveira, A.N.; Soares, A.M.; Silva, S.L. Why to study peptides from venomous and poisonous animals? *Int. J. Pept. Res. Ther.* **2023**, *29*, 76. [CrossRef]
38. Trim, C.M.; Byrne, L.J.; Trim, S.A. Utilisation of compounds from venoms in drug discovery. *Prog. Med. Chem.* **2021**, *60*, 1–66. [PubMed]
39. Lazarovici, P.; Marcinkiewicz, C.; Lelkes, P.I. From snake venom's disintegrins and C-type lectins to anti-platelet drugs. *Toxins* **2019**, *11*, 303. [CrossRef] [PubMed]
40. Coulter-Parkhill, A.; McClean, S.; Gault, V.A.; Irwin, N. Therapeutic potential of peptides derived from animal venoms: Current views and emerging drugs for diabetes. *Clin. Med. Insights Endocrinol. Diabetes* **2021**, *14*, 117955142110060. [CrossRef] [PubMed]
41. Lan, D.; Song, S.; Liu, Y.; Jiao, B.; Meng, R. Use of Batroxobin in central and peripheral ischemic vascular diseases: A systematic review. *Front. Neurol.* **2021**, *12*, 716778. [CrossRef]
42. Faioli, C.N.; Domingos, T.F.S.; Oliveira, E.C.; Sanchez, E.F.; Ribeiro, S.; Muricy, G.; Fuly, A.L. appraisal of antiophidic potential of marine sponges against *Bothrops jararaca* and *Lachesis muta* venom. *Toxins* **2013**, *5*, 1799–1813. [CrossRef] [PubMed]
43. Nunes, E.; Frihling, B.; Barros, E.; de Oliveira, C.; Verbisck, N.; Flores, T.; de Freitas Júnior, A.; Franco, O.; de Macedo, M.; Migliolo, L.; et al. Antibiofilm activity of acidic phospholipase isoform isolated from *Bothrops erythromelas* snake venom. *Toxins* **2020**, *12*, 606. [CrossRef]
44. Uetz, P. The Reptile Database. Available online: http://www.reptile-database.org/ (accessed on 2 February 2024).
45. Pompeia, C.; Frare, E.O.; Peigneur, S.; Tytgat, J.; Silva, Á.P.; Oliveira, E.B.; Pereira, A.; Kerkis, I.; Kolonin, M.G. Synthetic polypeptide crotamine: Characterization as a myotoxin and as a target of combinatorial peptides. *J. Mol. Med.* **2022**, *100*, 65–76. [CrossRef] [PubMed]
46. Si, H.; Yin, C.; Wang, W.; Davies, P.; Sanchez, E.; Suntravat, M.; Zawieja, D.; Cromer, W. Effect of the snake venom component crotamine on lymphatic endothelial cell responses and lymph transport. *Microcirculation* **2023**, *30*, e12775. [CrossRef] [PubMed]
47. Hayashi, M.A.F.; Campeiro, J.D.; Yonamine, C.M. Toxicon revisiting the potential of south american rattlesnake *Crotalus durissus terrificus* toxins as therapeutic, theranostic and/or biotechnological agents. *Toxicon* **2022**, *206*, 1–13. [CrossRef]
48. Giardini, A.C.; Evangelista, B.G.; Sant'Anna, M.B.; Martins, B.B.; Lancellotti, C.L.P.; Ciena, A.P.; Chacur, M.; Pagano, R.L.; Ribeiro, O.G.; Zambelli, V.O.; et al. Crotalphine attenuates pain and neuroinflammation induced by experimental autoimmune encephalomyelitis in mice. *Toxins* **2021**, *13*, 827. [CrossRef] [PubMed]
49. Konno, K.; Picolo, G.; Gutierrez, V.P.; Brigatte, P.; Zambelli, V.O.; Camargo, A.C.M.; Cury, Y. Crotalphine, a novel potent analgesic peptide from the venom of the south american rattlesnake *Crotalus durissus terrificus*. *Peptides* **2008**, *29*, 1293–1304. [CrossRef]
50. Silva, A.; Isbister, G.K. Current research into snake antivenoms, their mechanisms of action and applications. *Biochem. Soc. Trans.* **2020**, *48*, 537–546. [CrossRef] [PubMed]
51. Silva, S.L.; Rowan, E.G.; Albericio, F.; Stábeli, R.G.; Calderon, L.A.; Soares, A.M. Animal toxins and their advantages in biotechnology and pharmacology. *Biomed. Res. Int.* **2014**, *2014*, 951561. [CrossRef]
52. Peigneur, S.; Tytgat, J. Toxins in drug discovery and pharmacology. *Toxins* **2018**, *10*, 126. [CrossRef]
53. Slagboom, J.; Otvos, R.A.; Cardoso, F.C.; Iyer, J.; Visser, J.C.; van Doodewaerd, B.R.; McCleary, R.J.R.; Niessen, W.M.A.; Somsen, G.W.; Lewis, R.J.; et al. Neurotoxicity fingerprinting of venoms using on-line microfluidic AChBP profiling. *Toxicon* **2018**, *148*, 213–222. [CrossRef] [PubMed]
54. Dutertre, S.; Undheim, E.A.B.; Pineda, S.S.; Jin, A.-H.; Lavergne, V.; Fry, B.G.; Lewis, R.J.; Alewood, P.F.; King, G.F. Venoms-based drug discovery: Proteomic and transcriptomic approaches. In *Venom to Drugs*; King, G.F., Ed.; Royal Society of Chemistry: London, UK, 2015; pp. 80–96.

55. Souza, G.H.M.F.; Catharino, R.R.; Ifa, D.R.; Eberlin, M.N.; Hyslop, S. Peptide fingerprinting of snake venoms by direct infusion nanoelectrospray ionization mass spectrometry: Potential use in venom identification and taxonomy. *J. Mass Spec.* **2008**, *43*, 594–599. [CrossRef] [PubMed]
56. Sciani, J.M.; Vigerelli, H.; Costa, A.S.; Câmara, D.A.D.; Junior, P.L.-S.; Pimenta, D.C. An unexpected cell-penetrating peptide from *Bothrops jararaca* venom identified through a novel size exclusion chromatography screening. *J. Pept. Sci.* **2017**, *23*, 68–76. [CrossRef] [PubMed]
57. Biswas, A.; Gomes, A.; Sengupta, J.; Datta, P.; Singha, S.; Dasgupta, A.K.; Gomes, A. Nanoparticle-conjugated animal venom-toxins and their possible therapeutic potential. *J. Venom. Res.* **2012**, *3*, 15–21.
58. Munawar, A.; Ali, S.; Akrem, A.; Betzel, C. Snake venom peptides: Tools of biodiscovery. *Toxins* **2018**, *10*, 474. [CrossRef]
59. Hamimed, S.; Abdeljelil, N.; Landoulsi, A.; Chatti, A.; Aljabali, A.A.A.; Barhoum, A. Bacterial cellulose nanofibers. In *Handbook of Nanocelluloses*; Springer: Cham, Switzerland, 2022; pp. 1–38.
60. Langer, R.; Weissleder, R. Nanotechnology. *JAMA* **2015**, *313*, 135. [CrossRef] [PubMed]
61. Jong, W.H.; Borm, P.J.A. Drug delivery and nanoparticles: Applications and hazards. *Int. J. Nanomed.* **2008**, *3*, 133–149. [CrossRef] [PubMed]
62. Patra, J.K.; Das, G.; Fraceto, L.F.; Campos, E.V.R.; Rodriguez-Torres, M.D.P.; Acosta-Torres, L.S.; Diaz-Torres, L.A.; Grillo, R.; Swamy, M.K.; Sharma, S.; et al. Nano based drug delivery systems: Recent developments and future prospects. *J. Nanobiotech.* **2018**, *16*, 71. [CrossRef]
63. Jacob, J.; Haponiuk, J.T.; Thomas, S.; Gopi, S. Biopolymer based nanomaterials in drug delivery systems: A review. *Mater. Today Chem.* **2018**, *9*, 43–55. [CrossRef]
64. Moghimi, S.M.; Hunter, A.C.; Murray, J.C. Nanomedicine: Current status and future prospects. *FASEB J.* **2005**, *19*, 311–330. [CrossRef] [PubMed]
65. Utkin, Y.N. Modern trends in animal venom research—Omics and nanomaterials. *World J. Biol. Chem.* **2017**, *8*, 4–12. [CrossRef]
66. Gomes, A.; Ghosh, S.; Sengupta, J.; Saha, K.; Gomes, A. Nanotechnology in snake venom research—An overview. *Indian J. Exp. Biol.* **2018**, *56*, 707–715.
67. Soares, K.; Gláucia-Silva, F.; Silva, A.D.; Torres-Rêgo, M.; Araújo, N.; Menezes, Y.; Damasceno, I.; Tambourgi, D.; Silva-Júnior, A.; Fernandes-Pedrosa, M. Antivenom production against *Bothrops jararaca* and *Bothrops erythromelas* snake venoms using cross-linked chitosan nanoparticles as an immunoadjuvant. *Toxins* **2018**, *10*, 158. [CrossRef]
68. Santos-Silva, E.d.; Torres-Rêgo, M.; Gláucia-Silva, F.; Feitosa, R.C.; Lacerda, A.F.; Rocha, H.A.d.O.; Fernandes-Pedrosa, M.d.F.; Silva-Júnior, A.A. Cationic PLGA nanoparticle formulations as biocompatible immunoadjuvant for serum production and immune response against *Bothrops jararaca* venom. *Toxins* **2022**, *14*, 888. [CrossRef] [PubMed]
69. Oliveira, I.C.F.; Paula, M.O.; Lastra, H.C.B.; Alves, B.B.; Moreno, D.A.N.; Yoshida, E.H.; Amaral Filho, J.; Cogo, J.C.; Varanda, E.A.; Rai, M.; et al. Activity of silver nanoparticles on prokaryotic cells and *Bothrops jararacussu* snake venom. *Drug Chem. Toxicol.* **2019**, *42*, 60–64. [CrossRef] [PubMed]
70. Pal, S.K.; Gomes, A.; Dasgupta, S.C.; Gomes, A. Snake venom as therapeutic agents: From toxin to drug development. *Indian J. Exp. Biol.* **2002**, *40*, 1353–1358.
71. Calmette, A.; Saenz, A.; Costil, L. Effects du venin de cobra sur les greffes cancereuses et sur le cancer spontane adenocarcinome de la souris. *Comptes Rendus Acad. Sci.* **1933**, *197*, 205–210.
72. Tu, A.T.; Giltner, J.B. Cytotoxic effects of snake venoms on kb and yoshida sarcoma cells. *Res. Commun. Chem. Pathol. Pharmacol.* **1974**, *9*, 783–786. [PubMed]
73. Iwaguchi, T.; Takechi, M.; Hayashi, K. Cytolytic activity of cytotoxin isolated from indian cobra venom against experimental tumor cells. *Biochem. Int.* **1985**, *10*, 343–349. [PubMed]
74. Debnath, A.; Saha, A.; Gomes, A.; Biswas, S.; Chakrabarti, P.; Giri, B.; Biswas, A.K.; Gupta, S.D.; Gomes, A. A lethal cardiotoxic–cytotoxic protein from the indian monocellate cobra *Naja kaouthia* venom. *Toxicon* **2010**, *56*, 569–579. [CrossRef] [PubMed]
75. Saha, P.P.; Bhowmik, T.; Dasgupta, A.K.; Gomes, A. In vivo and in vitro toxicity of nanogold conjugated snake venom protein toxin GNP-NKCT1. *Toxicol. Rep.* **2014**, *1*, 74–84. [CrossRef]
76. Aubin-Tam, M.-E. Conjugation of nanoparticles to proteins. In *Nanomaterial Interfaces in Biology*; Methods in Molecular Biology; Humana Press: Totowa, NJ, USA, 2013; pp. 19–27.
77. Aubin-Tam, M.-E.; Hamad-Schifferli, K. Structure and function of nanoparticle–protein conjugates. *Biomed. Mater.* **2008**, *33*, 034001. [CrossRef]
78. Pohlmann, A.R.; Fonseca, F.N.; Paese, K.; Detoni, C.B.; Coradini, K.; Beck, R.C.; Guterres, S.S. Poly-ε-caprolactone microcapsules and nanocapsules in drug delivery. *Expert Opin. Drug Deliv.* **2013**, *10*, 623–638. [CrossRef]
79. Rosenberg, R.T.; Siegel, S.J.; Dan, N. Release of highly hydrophilic drugs from Poly-ε-caprolactone matrices. *J. Appl. Polym. Sci.* **2008**, *107*, 3149–3156. [CrossRef]

80. Lin, M.; Meng, S.; Zhong, W.; Li, Z.; Du, Q.; Tomasik, P. Novel biodegradable blend matrices for controlled drug release. *J. Pharm. Sci.* **2008**, *97*, 4240–4248. [CrossRef] [PubMed]
81. Nait Mohamed, F.A.; Laraba-Djebari, F. Development and characterization of a new carrier for vaccine delivery based on calcium-alginate nanoparticles: Safe immunoprotective approach against scorpion envenoming. *Vaccine* **2016**, *34*, 2692–2699. [CrossRef]
82. Archer, E.; Torretti, M.; Madbouly, S. Biodegradable polycaprolactone PCL based polymer and composites. *Phys. Sci. Rev.* **2023**, *8*, 4391–4414. [CrossRef]
83. Kakkar, A.; Traverso, G.; Farokhzad, O.C.; Weissleder, R.; Langer, R. Evolution of macromolecular complexity in drug delivery systems. *Nat. Rev. Chem.* **2017**, *1*, 0063. [CrossRef] [PubMed]
84. Lotocki, V.; Kakkar, A. Miktoarm star polymers: Branched architectures in drug delivery. *Pharmaceutics* **2020**, *12*, 827. [CrossRef] [PubMed]
85. Jeong, J.-C.; Lee, J.; Cho, K. Effects of crystalline microstructure on drug release behavior of polyε-caprolactone microspheres. *J. Control. Release* **2003**, *92*, 249–258. [CrossRef] [PubMed]
86. Ahmed, F.; Discher, D.E. Self-porating polymersomes of PEG–PLA and PEG–PCL: Hydrolysis-triggered controlled release vesicles. *J. Control. Release* **2004**, *96*, 37–53. [CrossRef]
87. Dash, T.K.; Konkimalla, V.B. Poly-ε-caprolactone based formulations for drug delivery and tissue engineering: A review. *J. Control. Release* **2012**, *158*, 15–33. [CrossRef]
88. Najafi-Hajivar, S.; Zakeri-Milani, P.; Mohammadi, H.; Niazi, M.; Soleymani-Goloujeh, M.; Baradaran, B.; Valizadeh, H. Overview on experimental models of interactions between nanoparticles and the immune system. *Biomed. Pharmacother.* **2016**, *83*, 1365–1378. [CrossRef] [PubMed]
89. Zhao, K.; Han, J.; Zhang, Y.; Wei, L.; Yu, S.; Wang, X.; Jin, Z.; Wang, Y. Enhancing mucosal immune response of newcastle disease virus DNA vaccine using n-2-hydroxypropyl trimethylammonium chloride chitosan and n,o-carboxymethyl chitosan nanoparticles as delivery carrier. *Mol. Pharm.* **2018**, *151*, 226–237. [CrossRef] [PubMed]
90. Zhao, H.; Wang, H.; Hu, Y.; Xu, D.; Yin, C.; Han, Q.; Zhang, J. chitosan nanovaccines as efficient carrier adjuvant system for il-12 with enhanced protection against hbv. *Int. J. Nanomed.* **2021**, *16*, 4913–4928. [CrossRef] [PubMed]
91. De França, A.C.R.B.; Soares, K.S.R.; Silva, F.G.; Silva, A.D.; Rêgo, M.T.; Silva, D.P.; Ferreira, S.S.; Furtado, A.A.; Silva-Júnior, A.A.; Pedrosa, M.F.F. Evaluation of self-assembled *Bothrops jararacussu* snake venom proteins cross-linked chitosan nanoparticles activity for use as a potential antibacterial. *Toxicon* **2019**, *168*, S32–S33. [CrossRef]
92. Tsuruta, L.R.; Moro, A.M.; Tambourgi, D.V.; Sant'Anna, O.A. Oral tolerance induction by *Bothrops jararaca* venom in a murine model and cross-reactivity with toxins of other snake venoms. *Toxins* **2021**, *13*, 865. [CrossRef] [PubMed]
93. Wambre, E.; Jeong, D. Oral tolerance development and maintenance. *Immunol. Allergy Clin. N. Am.* **2018**, *38*, 27–37. [CrossRef] [PubMed]
94. Costa, M.H.B.; Sant'Anna, O.A.; Quintilio, W.; Schwendener, R.A.; Araujo, P.S. A rational design for the nanoencapsulation of poisonous animal venoms in liposomes prepared with natural phospholipids. *Curr. Drug Deliv.* **2012**, *9*, 637–644. [CrossRef] [PubMed]
95. Oliver, A.E.; Fisk, E.; Crowe, L.M.; Araujo, P.S.; Crowe, J.H. Evidence of phospholipase activity in phospholipid bilayers under conditions of low hydration. *J. Plant Physiol.* **1997**, *150*, 661–667. [CrossRef]
96. Ferreira, T.L.; Ward, R.J. The interaction of bothropstoxin-I Lys49-PLA_2 with liposome membranes. *Toxicon* **2009**, *54*, 525–530. [CrossRef] [PubMed]
97. Leite, L.C.C.; Fatima, M.; Furtado, D.; Correa, T.C.; Raw, I. Characterization of the snake venoms from seven brazilian species of *Bothrops* by FPLC anion-exchange chromatography. *Comp. Biochem.* **1992**, *102*, 515–520. [CrossRef] [PubMed]
98. Oliver, A.E.; Crowe, L.M.; Araujo, P.S.; Fisk, E.; Crowe, J.H. Arbutin inhibits PLA_2 in partially hydrated model systems. *Biochim. Biophys. Acta* **1996**, *1302*, 69–78. [CrossRef] [PubMed]
99. Carvalho, V.T.; Gomes, R.T.; Viotti, A.P.; Freitas, T.V. Immunization with liposome-encapsulated *Bothrops jararaca* venom. *Toxicon* **2000**, *38*, 881–886. [CrossRef]
100. Kirby, C.; Gregoriadis, G. Dehydration-rehydration vesicles: A simple method for high yield drug entrapment in liposomes. *Nat. Biotechnol.* **1984**, *2*, 979–984. [CrossRef]
101. Barros, N.B.; Macedo, S.R.A.; Ferreira, A.S.; Tagliari, M.P.; Zanchi, F.B.; Kayano, A.M.; Soares, A.M.; Nicolete, R. Liposomes containing an ASP49-Phospholipase A_2 from *Bothrops jararacussu* snake venom as experimental therapy against cutaneous leishmaniasis. *Int. Immunopharmacol.* **2016**, *36*, 225–231. [CrossRef]
102. Barros, N.B.; Macedo, S.R.A.; Ferreira, A.S.; Tagliari, M.P.; Kayano, A.M.; Nicolete, L.D.F.; Soares, A.M.; Nicolete, R. ASP49-Phospholipase A_2-loaded liposomes as experimental therapy in cutaneous leishmaniasis model. *Int. Immunopharmacol.* **2018**, *55*, 128–132. [CrossRef] [PubMed]

103. Gláucia-Silva, F.; Torres-Rêgo, M.; Rocha Soares, K.S.; Damasceno, I.Z.; Tambourgi, D.V.; Silva-Júnior, A.A.; Fernandes-Pedrosa, M.F. A biotechnological approach to immunotherapy: Antivenom against *Crotalus durissus cascavella* snake venom produced from biodegradable nanoparticles. *Int. J. Biol. Macromol.* **2018**, *120*, 1917–1924. [CrossRef] [PubMed]
104. Sartim, M.A.; Menaldo, D.L.; Sampaio, S.V. immunotherapeutic potential of crotoxin: Anti-inflammatory and immunosuppressive properties. *J. Venom. Anim. Toxins Incl. Trop. Dis.* **2018**, *24*, 39. [CrossRef] [PubMed]
105. Sant'Anna, M.B.; Giardini, A.C.; Ribeiro, M.A.C.; Lopes, F.S.R.; Teixeira, N.B.; Kimura, L.F.; Bufalo, M.C.; Ribeiro, O.G.; Borrego, A.; Cabrera, W.H.K.; et al. The Crotoxin:SBA-15 complex down-regulates the incidence and intensity of experimental autoimmune encephalomyelitis through peripheral and central actions. *Front. Immunol.* **2020**, *11*, 591563. [CrossRef] [PubMed]
106. Sant'Anna, M.B.; Lopes, F.S.R.; Kimura, L.F.; Giardini, A.C.; Sant'Anna, O.A.; Picolo, G. Crotoxin conjugated to SBA-15 nanostructured mesoporous silica induces long-last analgesic effect in the neuropathic pain model in mice. *Toxins* **2019**, *11*, 679. [CrossRef]
107. Lu, J.; Liong, M.; Li, Z.; Zink, J.I.; Tamanoi, F. Biocompatibility, biodistribution, and drug-delivery efficiency of mesoporous silica nanoparticles for cancer therapy in animals. *Small* **2010**, *6*, 1794–1805. [CrossRef]
108. Karpel, R.L.; Silva Liberato, M.; Campeiro, J.D.; Bergeon, L.; Szychowski, B.; Butler, A.; Marino, G.; Cusic, J.F.; Oliveira, L.C.G.; Oliveira, E.B.; et al. Design and characterization of crotamine-functionalized gold nanoparticles. *Colloids Surf. B Biointerfaces* **2018**, *163*, 1–8. [CrossRef]
109. Jimenez-Canale, J.; Fernandez-Quiroz, D.; Teran-Saavedra, N.G.; Diaz-Galvez, K.R.; Gallegos-Tabanico, A.; Burgara-Estrella, A.J.; Sarabia-Sainz, H.M.; Guzman-Partida, A.M.; Robles-Burgueño, M.d.R.; Vazquez-Moreno, L.; et al. Cytotoxic activity of a northern black-tailed rattlesnake *Crotalus molossus molossus* venom-loaded in chitosan nanoparticles as a potential antitumoral system. *Acta Biochim. Pol.* **2022**, *69*, 233–243. [CrossRef]
110. Magalhães, T.; Viotti, A.P.; Gomes, R.T.; Freitas, T.V. Effect of membrane composition and of coencapsulation of immunostimulants in a liposome-entrapped crotoxin. *Biotechnol. Appl. Biochem.* **2001**, *33*, 61–64. [CrossRef] [PubMed]
111. Freitas, T.V.; Frézard, F. encapsulation of native crotoxin in liposomes: A safe approach for the production of antivenom and vaccination against *Crotalus durissus terrificus* venom. *Toxicon* **1997**, *35*, 91–100. [CrossRef] [PubMed]
112. Baudou, F.G.; Fusco, L.; Giorgi, E.; Diaz, E.; Municoy, S.; Desimone, M.F.; Leiva, L.; De Marzi, M.C. Physicochemical and biological characterization of nanovenoms, a new tool formed by silica nanoparticles and *Crotalus durissus terrificus* venom. *Colloids Surf. B Biointerfaces* **2020**, *193*, 111128. [CrossRef]
113. Gomes, R.T.; Camargos, R.P.F.; Viotti, A.P.; Tavares, A.P.; Revelo, M.P.; Freitas, T.V. Comparison of the biodistribution of free or liposome-entrapped *Crotalus durissus terrificus* south american rattlesnake venom in mice. *Comp. Biochem. Physiol. Part C Toxicol. Pharmacol.* **2002**, *131*, 295–301. [CrossRef]
114. Stransky, S.; Costal-Oliveira, F.; -Souza, L.L.; Duarte, C.G.; Chávez-Olórtegui, C.; Braga, V.M.M. In vitro assessment of cytotoxic activities of *Lachesis muta muta* snake venom. *PLoS Negl. Trop. Dis.* **2018**, *12*, e0006427. [CrossRef] [PubMed]
115. Fuly, A.L.; Machado, O.L.; Alves, E.W.; Carlini, C.R. Mechanism of inhibitory action on platelet activation of a phospholipase A_2 isolated from *Lachesis muta* bushmaster snake venom. *Thromb. Haemost.* **1997**, *78*, 1372–1380. [CrossRef] [PubMed]
116. Almeida, F.M.; Pimenta, A.M.C.; Oliveira, M.C.; Lima, M.E. Venoms, toxins and derivatives from the brazilian fauna: Valuable sources for drug discovery. *Sheng Li Xue Bao* **2015**, *67*, 261–270.
117. Kerkis, A.; Kerkis, I.; Rádis-Baptista, G.; Oliveira, E.B.; Vianna-Morgante, A.M.; Pereira, L.V.; Yamane, T. Crotamine is a novel cellpenetrating protein from the venom of rattlesnake *Crotalus durissus terrificus*. *FASEB J.* **2004**, *18*, 1407–1409. [CrossRef] [PubMed]
118. Kerkis, A.; Hayashi, M.A.F.; Yamane, T.; Kerkis, I. Properties of cell penetrating peptides CPPs. *IUBMB Life* **2006**, *58*, 7–13. [CrossRef] [PubMed]
119. Porta, L.C.; Campeiro, J.D.; Hayashi, M.A.F. A native CPP from rattlesnake with therapeutic and theranostic properties. In *Cell Penetrating Peptides*; Methods in Molecular Biology; Humana: New York, NY, USA, 2022; pp. 91–104.
120. Nascimento, F.D.; Hayashi, M.A.F.; Kerkis, A.; Oliveira, V.; Oliveira, E.B.; Rádis-Baptista, G.; Nader, H.B.; Yamane, T.; Tersariol, I.L.S.; Kerkis, I. Crotamine mediates gene delivery into cells through the binding to heparan sulfate proteoglycans. *J. Bio. Chem.* **2007**, *282*, 21349–21360. [CrossRef]
121. Hayashi, M.A.; Nascimento, F.D.; Kerkis, A.; Oliveira, V.; Oliveira, E.B.; Pereira, A.; Rádis-Baptista, G.; Nader, H.B.; Yamane, T.; Kerkis, I.; et al. Cytotoxic effects of crotamine are mediated through lysosomal membrane permeabilization. *Toxicon* **2008**, *52*, 508–517. [CrossRef] [PubMed]
122. Ponnappan, N.; Budagavi, D.P.; Chugh, A. CyLoP-1: Membrane-active peptide with cell-penetrating and antimicrobial properties. *Biochim. Biophys. Acta* **2017**, *1859*, 167–176. [CrossRef]
123. Rádis-Baptista, G.; LaTorre, B.G.; Andreu, D. A Novel Cell-penetrating peptide sequence derived by structural minimization of a snake toxin exhibits preferential nucleolar localization. *J. Med. Chem.* **2008**, *51*, 7041–7044. [CrossRef]

124. Rodrigues, M.; LaTorre, B.G.; Rádis-Baptista, G.; Santos, N.C.; Andreu, D. efficient cellular delivery of β-galactosidase mediated by NrTPs, a new family of cell-penetrating peptides. *Bioconjug. Chem.* **2011**, *22*, 2339–2344. [CrossRef] [PubMed]
125. Rodrigues, M.; Santos, A.; LaTorre, B.G.; Rádis-Baptista, G.; Andreu, D.; Santos, N.C. molecular characterization of the interaction of crotamine-derived nucleolar targeting peptides with lipid membranes. *Biochim. Biophys. Acta* **2012**, *1818*, 2707–2717. [CrossRef] [PubMed]
126. Tansi, F.L.; Filatova, M.P.; Koroev, D.O.; Volpina, O.M.; Lange, S.; Schumann, C.; Teichgräber, U.K.; Reissmann, S.; Hilger, I. New generation cpps show distinct selectivity for cancer and noncancer cells. *J. Cell. Biochem.* **2019**, *120*, 6528–6541. [CrossRef] [PubMed]
127. Falcao, C.B.; LaTorre, B.G.; Pérez-Peinado, C.; Barron, A.E.; Andreu, D.; Rádis-Baptista, G. Vipericidins: A novel family of cathelicidin-related peptides from the venom gland of south american pit vipers. *Amino Acids* **2014**, *46*, 2561–2571. [CrossRef]
128. Falcao, C.B.; Radis-Baptista, G. Crotamine and Crotalicidin, membrane active peptides from *Crotalus durissus terrificus* rattlesnake venom, and their structurally-minimized fragments for applications in medicine and biotechnology. *Peptides* **2020**, *126*, 170234. [CrossRef] [PubMed]
129. Peinado, C.P.; Defaus, S.; Andreu, D. Hitchhiking with nature: Snake venom peptides to fight cancer and superbugs. *Toxins* **2020**, *12*, 255. [CrossRef] [PubMed]
130. VanHoek, M.L. Antimicrobial peptides in reptiles. *Pharmaceuticals* **2014**, *7*, 723–753. [CrossRef]
131. Zhou, H.; Dou, J.; Wang, J.; Chen, L.; Wang, H.; Zhou, W.; Li, Y.; Zhou, C. The antibacterial activity of bf-30 in vitro and in infected burned rats is through interference with cytoplasmic membrane integrity. *Peptides* **2011**, *32*, 1131–1138. [CrossRef] [PubMed]
132. Juba, M.L.; Porter, D.K.; Williams, E.H.; Rodriguez, C.A.; Barksdale, S.M.; Bishop, B.M. Helical cationic antimicrobial peptide length and its impact on membrane disruption. *Biochim. Biophys. Acta* **2015**, *1848*, 1081–1091. [CrossRef]
133. Cavalcante, C.S.P.; Aguiar, F.L.L.; Fontenelle, R.O.S.; Menezes, R.R.P.P.B.; Martins, A.M.C.; Falcão, C.B.; Andreu, D.; Rádis-Baptista, G. Insights into the candidacidal mechanism of ctn[15–34]—A carboxyl-terminal, crotalicidin-derived peptide related to cathelicidins. *J. Med. Microbiol.* **2018**, *67*, 129–138. [CrossRef]
134. Pérez-Peinado, C.; Dias, S.A.; Domingues, M.M.; Benfield, A.H.; Freire, J.M.; Rádis-Baptista, G.; Gaspar, D.; Castanho, M.A.R.B.; Craik, D.J.; Henriques, S.T.; et al. Mechanisms of bacterial membrane permeabilization by crotalicidin ctn and its fragment ctn15–34, antimicrobial peptides from rattlesnake venom. *J. Bio. Chem.* **2018**, *293*, 1536–1549. [CrossRef]
135. Aguiar, F.L.L.D.; Santos, N.C.; de Paula Cavalcante, C.S.; Andreu, D.; Baptista, G.R.; Gonçalves, S. Antibiofilm activity on candida albicans and mechanism of action on biomembrane models of the antimicrobial peptide Ctn[15–34]. *Int. J. Mol. Sci.* **2020**, *21*, 8339. [CrossRef] [PubMed]
136. Falcao, C.B.; Peinado, C.P.; La Torre, B.G.; Mayol, X.; Carreras, H.Z.; Jiménez, M.Á.; Rádis-Baptista, G.; Andreu, D. Structural dissection of crotalicidin, a rattlesnake venom cathelicidin, retrieves a fragment with antimicrobial and antitumor activity. *J. Med. Chem.* **2015**, *58*, 8553–8563. [CrossRef] [PubMed]
137. Wang, L.; Chan, J.Y.W.; Rêgo, J.V.; Chong, C.-M.; Ai, N.; Falcão, C.B.; Rádis-Baptista, G.; Lee, S.M.Y. rhodamine b-conjugated encrypted vipericidin nonapeptide is a potent toxin to zebrafish and associated with in vitro cytotoxicity. *Biochim. Biophys. Acta* **2015**, *1850*, 1253–1260. [CrossRef]
138. Zhang, E.; Xing, R.; Liu, S.; Qin, Y.; Li, K.; Li, P. advances in chitosan-based nanoparticles for oncotherapy. *Carbohydr. Polym.* **2019**, *222*, 115004. [CrossRef]
139. Mendes, B.; Almeida, J.R.; Vale, N.; Gomes, P.; Gadelha, F.R.; Silva, S.L.; Miguel, D.C. Potential use of 13-mer peptides based on phospholipase and oligoarginine as leishmanicidal agents. *Comp. Biochem. Physiol.* **2019**, *226*, 108612. [CrossRef] [PubMed]
140. Aguilar, A.; Twardowski, T.; Wohlgemuth, R. Bioeconomy for sustainable development. *Biotechnol. J.* **2019**, *14*, e1800638. [CrossRef]
141. Woźniak, E.; Tyczewska, A.; Twardowski, T. Bioeconomy development factors in the european union and poland. *New Biotechnol.* **2021**, *60*, 2–8. [CrossRef]
142. Kuckertz, A.; Berger, E.S.C.; Brändle, L. Entrepreneurship and the sustainable bioeconomy transformation. *Environ. Innov. Soc. Transit.* **2020**, *37*, 332–344. [CrossRef]
143. Bugge, M.; Hansen, T.; Klitkou, A. What is the bioeconomy? a review of the literature. *Sustainability* **2016**, *8*, 691. [CrossRef]
144. Patermann, C.; Aguilar, A. The origins of the bioeconomy in the european union. *New Biotechnol.* **2018**, *40*, 20–24. [CrossRef]
145. Vivien, F.D.; Nieddu, M.; Befort, N.; Debref, R.; Giampietro, M. The hijacking of the bioeconomy. *Ecol. Econ.* **2019**, *159*, 189–197. [CrossRef]
146. Alamgir, A.N.M. Biotechnology, in vitro production of natural bioactive compounds, herbal preparation, and disease management (treatment and prevention). In *Therapeutic Use of Medicinal Plants and their Extracts: Volume 2*; Progress in Drug Research; Springer: Cham, Switzerland, 2018; pp. 585–664.

147. Ferreira, S.H. A bradykinin-potentiating factor BPF present in the venom of *Bothrops jararaca*. *Br. J. Pharmacol. Chemother.* **1965**, *24*, 163–169. [CrossRef] [PubMed]
148. Camargo, A.C.M.; Ianzer, D.; Guerreiro, J.R.; Serrano, S.M.T. Bradykinin-potentiating peptides: Beyond captopril. *Toxicon* **2012**, *59*, 516–523. [CrossRef] [PubMed]
149. Boldrini-França, J.; Pinheiro-Junior, E.L.; Peigneur, S.; Pucca, M.B.; Cerni, F.A.; Borges, R.J.; Costa, T.R.; Carone, S.E.I.; Fontes, M.R.D.M.; Sampaio, S.V.; et al. Beyond hemostasis: A snake venom serine protease with potassium channel blocking and potential antitumor activities. *Sci. Rep.* **2020**, *10*, 4476. [CrossRef]
150. USPTO. Cooperative Patent Classification—CPC. Available online: https://www.uspto.gov/web/patents/classification/cpc/html/cpc.html (accessed on 6 March 2024).
151. Oliveira, Á.S.; Fantinel, A.L.; Artuzo, F.D.; Oliveira, L.; Singer, R.B.; da Frota Júnior, M.L.C.; Dewes, H.; Talamini, E. Applications of venom biodiversity in agriculture. *EFB Bioecon. J.* **2021**, *1*, 100010. [CrossRef]
152. Warrell, D.A. Snake Bite. *Lancet* **2010**, *375*, 77–88. [CrossRef] [PubMed]
153. Bolton, F.; Casewell, N.; Al-Abdulla, I. Snake antivenom trial. *Vet. Rec.* **2014**, *174*, 126. [CrossRef]
154. Williams, D.J.; Habib, A.G.; Warrell, D.A. Clinical studies of the effectiveness and safety of antivenoms. *Toxicon* **2018**, *150*, 1–10. [CrossRef]
155. Metkar, S.; Udayakumar, S.; Girigoswami, A.; Girigoswami, K. Natural serine proteases and their applications in combating amyloid formation: Review article. *ADMET DMPK* **2024**, *12*, 797–820.

pharmaceutics

MDPI

Review

Non-Invasive Delivery of CRISPR/Cas9 Ribonucleoproteins (Cas9 RNPs) into Cells via Nanoparticles for Membrane Transport

Toshihiko Tashima

Tashima Laboratories of Arts and Sciences, 1239-5 Toriyama-cho, Kohoku-ku, Yokohama 222-0035, Japan; tashima_lab@yahoo.co.jp

Abstract: The clustered regularly interspaced short palindromic repeats/CRISPR-associated protein 9 (CRISPR/Cas9) system is a promising biotechnology tool for genome editing. However, in living organisms, several pharmacokinetic challenges arise, including off-target side effects due to incorrect distribution, low bioavailability caused by membrane impermeability, and instability resulting from enzymatic degradation. Therefore, innovative delivery strategies must be developed to address these issues. Modified nanoparticles offer a potential solution for the non-invasive delivery of CRISPR/Cas9 ribonucleoproteins (Cas9 RNPs). Cas9 RNPs encapsulated in nanoparticles are protected from enzymatic degradation, similar to how microRNAs are shielded within exosomes. It is well-established that certain materials, including proteins, are expressed selectively in specific cell types. For example, the α-7 nicotinic receptor is expressed in endothelial and neuronal cells, while the αvβ3 integrin is expressed in cancer cells. These endogenous materials can facilitate receptor-mediated endocytosis or transcytosis. Nanoparticles encapsulating Cas9 RNPs and coated with ligands targeting such receptors may be internalized through receptor-mediated mechanisms. Once internalized, Cas9 RNPs could perform the desired gene editing in the nucleus after escaping the endosome through mechanisms such as the proton sponge effect or membrane fusion. In this review, I discuss the potential and advantages of delivering Cas9 RNP-encapsulated nanoparticles coated with ligands through receptor-mediated endocytosis or transcytosis.

Keywords: CRISPR/Cas9; drug delivery system; nanodelivery; transmembrane substance delivery; receptor-mediated endocytosis/transcytosis

1. Introduction

Biotechnology has significantly contributed to the advancement of medicine, but the creation of genetically modified animals required for in vivo testing often presents challenges, such as the lack of desired embryonic stem cells. Therefore, efficient methods for obtaining knockout animals are needed. Currently, the clustered regularly interspaced short palindromic repeats (CRISPR)/CRISPR-associated (Cas) system is a revolutionary and simple tool for genome editing [1,2], alongside zinc finger nucleases (ZFNs) [3] and transcription activator-like effector nucleases (TALENs) [4]. In recognition of their pioneering work on this genome editing system, the 2020 Nobel Prize in Chemistry was awarded to Dr. Emmanuelle Charpentier and Dr. Jennifer Doudna. Although various Cas protein families have been discovered, Cas9 is a CRISPR-associated endonuclease that facilitates

RNA-guided, site-specific DNA cleavage. In its natural form from *Streptococcus pyogenes*, Cas9 functions in conjunction with a dual-guide RNA, which consists of CRISPR RNA (crRNA) and trans-activating crRNA (tracrRNA). The mechanism of the type II CRISPR-Cas system can be summarized in three steps [5]: (i) Adaptation Stage: during the first infection, bacteria incorporate a new genetic spacer, derived from invading bacteriophages (phages), into the CRISPR array. (ii) crRNA Biogenesis: Upon reinfection, precursor CRISPR RNAs (pre-crRNAs) are transcribed from the CRISPR array. These pre-crRNAs are then processed into mature crRNAs with the help of tracrRNA and Cas9, facilitated by RNase III. (iii) Interference Stage (Figure 1): A complex of Cas9, tracrRNA, and crRNA binds to a short protospacer adjacent motif (PAM) sequence (3–6 nucleotides) located next to the target DNA protospacer sequence, enabling site-specific DNA targeting. Double-strand DNA is unwound around the PAM sequence. The single-stranded complementary DNA, which forms base pairs with the spacer of crRNA, is cleaved by the HNH domain (an endonuclease domain named for its characteristic histidine and asparagine residues), while the single-stranded non-complementary DNA is cleaved by the RuvC domain (an endonuclease domain named after an *Escherichia coli* protein involved in DNA repair). Ultimately, the targeted double-strand DNA break occurs in this manner [6]. Genome editing is achieved through non-homologous end joining (NHEJ) and homology-directed repair (HDR) following double-strand DNA cleavage. NHEJ is a repair mechanism that directly joins the ends of cleaved double-strand DNA. Occasionally, this process results in the insertion of unnecessary sequences, leading to repair errors. In bacteria, an excess of Cas9 can cause repeated double-strand DNA breaks, ultimately leading to repair errors [7]. On the other hand, HDR is a repair mechanism that uses a complementary sequence to guide precise DNA repair, allowing for accurate genome editing [8]. To improve efficiency, single-guide RNA (sgRNA) is artificially formed by connecting crRNA and tracrRNA (Figure 2). It is believed that CRISPR/Cas9 ribonucleoprotein (Cas9 RNP) is essential for the execution of gene editing in advance, providing an advantage over drug delivery and functional expression. However, the delivery of Cas9 into cells across the plasma membrane poses a challenge due to its large protein size. Nevertheless, in vitro or ex vivo gene editing can be performed through methods such as injection into fertilized eggs, chimeric antigen receptor (CAR)-T cells, or mesenchymal stem cells (MSCs) using techniques like lipofection (cationic lipid-mediated transfection), electroporation (electric pulse method), viral vector-based gene delivery, or microinjection [9]. However, in vivo gene editing in a non-invasive manner, without off-target side effects, presents several challenges that differ from in vitro gene editing using methods such as lipofection, electroporation, or microinjection. RNAs are degraded by RNases, while proteins are degraded by proteases. Interestingly, microRNAs (miRNAs) in exosomes released from donor cells into the systemic circulation are protected from such enzymatic degradation. A nanodelivery system using nanoparticles could potentially protect Cas9 RNP from enzymatic degradation. Furthermore, Cas9 RNP alone lacks the ability for cell internalization and cell selectivity. Currently, selective drug delivery systems, particularly those utilizing biological machinery such as receptor-mediated endocytosis/transcytosis, have been developed [10,11]. Therefore, in this perspective review, I explore the potential and implementation of selective, effective Cas9 RNP delivery into cells using nanoparticles to achieve non-invasive in vivo gene editing.

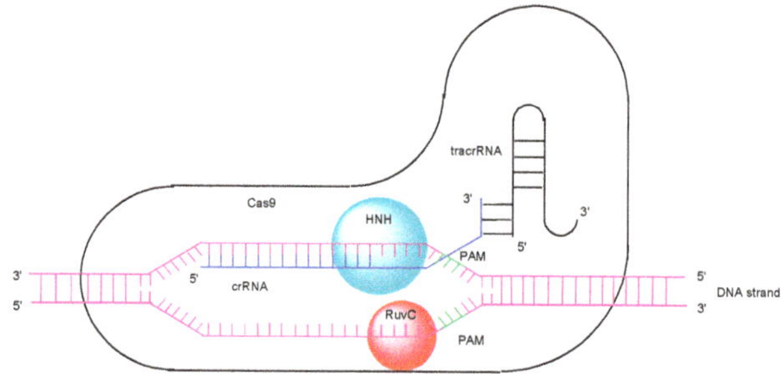

Figure 1. The interference mechanism of the type II CRISPR-Cas system involves the cleavage of target double-stranded DNA, originating from bacteria, at two endonuclease domains on the Cas9 protein: the HNH domain (named for its characteristic histidine and asparagine residues) and the RuvC domain (named after an *Escherichia coli* protein involved in DNA repair). CRISPR RNA (crRNA) and trans-activating CRISPR RNA (tracrRNA) recognize the target gene based on a specific protospacer adjacent motif (PAM).

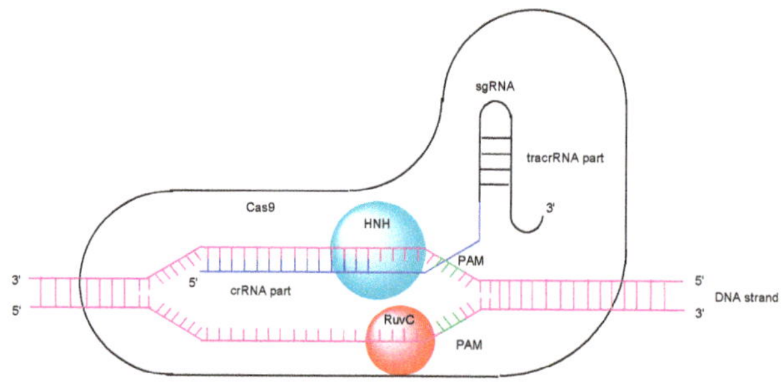

Figure 2. Single-guide RNA (sgRNA) is an engineered RNA molecule formed by combining CRISPR RNA (crRNA) and trans-activating CRISPR RNA (tracrRNA). PAM refers to the protospacer adjacent motif. The HNH domain is an endonuclease domain named for its characteristic histidine and asparagine residues, while the RuvC domain is an endonuclease domain named after a protein in *Escherichia coli* that plays a role in DNA repair.

2. Discussion

2.1. Membrane Impermeability/Permeability of Substances

In drug discovery and development, membrane impermeability poses a significant pharmacokinetic challenge in living organisms. This is particularly evident for hydrophobic low-molecular-weight compounds and hydrophilic high-molecular-weight compounds, which cannot cross the blood–brain barrier (BBB) to enter the brain [12]. The BBB's function is primarily attributed to physical barriers, including (i) the hydrophobic lipid bilayers of the capillary endothelial cell plasma membranes, (ii) the tight junctions between capillary endothelial cells, and (iii) periendothelial accessory structures supported by pericytes and astrocytes. Additionally, biological barriers play a role, such as (iv) excretion by efflux transporters like multidrug resistance protein 1 (MDR1, P-glycoprotein), which capture hydrophobic low-molecular-weight compounds that pass through the apical membrane of capillary endothelial cells. Consequently, the BBB significantly hinders the development of drugs for central nervous system (CNS) diseases. Moreover, cancer drug resistance is often attributed to the overexpression of MDR1 proteins at the apical membrane [13], posing a significant challenge to the development of effective cancer therapies. Therefore, it is crucial to establish selective and efficient transmembrane drug delivery systems by leveraging existing biological machinery, as guided by the principles of structuralism proposed by Dr. Claude Lévi-Strauss [14,15]. For example, nutrients such as glucose and amino acids are transported from systemic circulation into the brain across the BBB through solute carrier (SLC) transporters, including glucose transporter 1 (GLUT1) and the large neutral amino acid transporter (LAT). The mechanisms by which substances enter tissues, such as the brain, from the bloodstream across the endothelium or epithelium can be categorized into the following pathways: the paracellular pathway, including (i) passive diffusion through tight junctions, and the transcellular pathway, including (ii) passive diffusion through the lipid bilayers, (iii) carrier-mediated transport via solute carrier (SLC) transporters, (iv) receptor-mediated endocytosis/transcytosis involving specific receptors, (v) bystander macropinocytosis, and (vi) other less understood or unconventional transport methods [10]. The specific internalization mechanisms that occur depend on the properties of the substances, such as their molecular size, hydrophobicity or hydrophilicity, the structural environment in which they are situated, and their administration site. In some cases, multiple mechanisms occur simultaneously. Generally, substances are categorized based on molecular weight into low-molecular-weight compounds (molecular weight (MW) < approximately 500), high-molecular-weight compounds (MW > approximately 3000), and middle-molecular-weight compounds (MW approximately 500–3000). High-molecular-weight monoclonal antibodies, such as IgG molecules (approximately 14.2 nm in diameter and 150 kDa) [16], cannot pass through the pores of SLC transporters due to their size. Additionally, their hydrophilic nature prevents them from penetrating the membrane via passive diffusion. Nanoparticles (10–100 nm in diameter) face greater size limitations than monoclonal antibodies. It is well-known that nanoparticles coated with vectors such as anti-receptor monoclonal antibodies, cell-penetrating peptides (CPPs), or tumor-homing peptides (THPs) can cross cell membranes via receptor-mediated endocytosis. Among the various types of endocytosis, (i) clathrin-dependent endocytosis (with endosomal diameters of 85–150 nm) is the most prevalent. Other types include (ii) caveolae-dependent endocytosis (endosomal diameters of 50–100 nm), (iii) clathrin- and caveolae-independent endocytosis (endosomal diameters of approximately 90 nm), (iv) macropinocytosis (endosomal diameters of 0.2–5 μm), and (v) other less understood or mechanically complex forms of endocytosis [10]. Thus, Cas9 RNP-encapsulated nanoparticles can effectively cross membranes through carefully designed strategies that leverage structuralism to control biological machinery systems. The use of nanoparticles not only protects Cas9 RNP from degradative enzymes but also allows for easy surface modifications, enabling additional functionalities such as enhanced membrane internalization, cell selectivity, and immune tolerance. Indeed, nanoparticles encapsulating Cas9 RNP have been developed following these principles, as detailed below.

2.2. *Nanoparticles Encapsulating Cas9 RNPs*

2.2.1. Distribution to Tissues, Such as the Liver, Involved in the Reticuloendothelial System

The establishment of in vivo genome editing has the potential to significantly expand therapeutic options for patients. Notably, (i) NTLA-2001, an intravenously administered lipid nanoparticle formulation encapsulating Cas9 messenger RNA (mRNA) and sgRNA targeting transthyretin (*TTR*), has been studied in a phase 1 clinical trial since 2020 for hereditary TTR amyloidosis with polyneuropathy (ATTRv-PN) and TTR amyloidosis-related cardiomyopathy (ATTR-CM) (NCT04601051). Preclinical studies demonstrated that, by day 28, the mean reduction in serum TTR protein concentration was 52% in the group receiving a dose of 0.1 mg per kilogram and 87% in the group receiving a dose of 0.3 mg per kilogram in humans [17]. TTR is a plasma transport protein responsible for carrying retinol (vitamin A) and thyroxine, primarily synthesized in the liver and the choroid plexus [18]. TTR amyloidosis is a gradually progressive condition resulting from the accumulation of misfolded TTR protein in tissues. Preclinical studies also revealed that NTLA-2001 complexes with apolipoprotein E (APOE) entered hepatocytes via receptor-mediated endocytosis utilizing low-density lipoprotein receptor (LDLR), leading to a reduction in serum TTR protein expression due to TTR gene knockout (Figure 3). It is believed that captured APOE acts as a vector facilitating internalization into hepatocytes. Endosomal escape is thought to occur through the breakdown of the lipid nanoparticle (LNP) and the disruption of the endosomal membrane. NTLA-2001 is formulated with proprietary ionizable lipids, combined with 2,3-dioleoyl-glycero-1-phosphocholine (DOPC), 1,2-dimyristoyl-rac-glycero-3-methoxypolyethylene glycol-2000 (DMG-PEG), and cholesterol (Figure 4) [19]. DOPC is a unique zwitterionic lipid that adopts a disordered superstructure characteristic of a fluid membrane [20]. Fundamentally, the delivery methods for Cas9 mRNA/sgRNA are the same as those for Cas9 RNPs, as both are encapsulated in nanoparticles. In the case of lipid nanoparticles, endosomal escape occurs through membrane fusion, membrane disruption, or the proton sponge effect. However, after being released from endosomes, Cas9 RNPs can immediately perform genome editing. In contrast, Cas9 mRNA must first undergo translation to produce the Cas9 protein. Table 1 presents studies that utilize either RNP delivery or mRNA/sgRNA delivery.

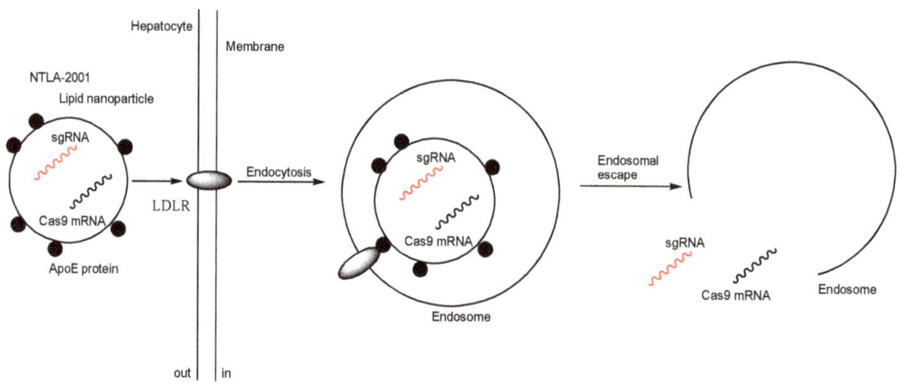

Figure 3. The pathway of NTLA-2001 involves lipid nanoparticles encapsulating Cas9 messenger RNA (mRNA) and single-guide RNA (sgRNA) targeting transthyretin (TTR), coated with apolipoprotein E (APOE) protein. APOE binds to low-density lipoprotein receptor (LDLR) to facilitate endocytosis. Endosomal escape is likely achieved through the breakdown of the lipid nanoparticles and the disruption of the endosomal membrane.

Additionally, (ii) a lipid nanoparticle (LNP)-based Cas RNP delivery system incorporating optimally designed single-stranded oligonucleotides (ssODNs) was developed (Figure 5). The LNP formulation consisted of 1,10-[[2-[4-[2-[2-[bis(2-hydroxydodecyl)amino] ethylamino]ethyl]piperazin-1-yl]ethyl]azanediyl]bis(dodecan-2-ol) (C12-200), 1,2-dioleoyl-sn-glycero-3-phosphoethanolamine (DOPE), cholesterol, DMG-PEG, and 1,2-dioleoyl-3-trimethylammonium-propane (DOTAP) (Figure 4), with molar ratios of 35:16:46.5:2.5:0.25, respectively. Cas9 RNP-ssODN complexes remain stable at temperatures below 25 °C but release ssODNs in the cytosol at 37 °C under physiological conditions. Regarding gene editing efficiency, RNP-loaded LNPs containing anti-TTR sgRNA (sgTTR-G269) and ssODNs designed with a 50% complementation rate (calculated Tm: 30 °C) achieved the highest suppression of TTR protein levels in serum, reducing them by 34% one week after intravenous administration in mice. Similarly, RNP-loaded LNPs containing anti-TTR sgRNA (sgTTR-G211) and ssODNs designed with a 40% complementation rate (calculated Tm: 26 °C) achieved the highest suppression of TTR protein levels, reducing them by 56% [21]. It has been suggested that the affinity between ssODNs and RNPs plays a role in suppression activity. An LNP-based Cas RNP delivery system utilizing ssODNs may help avoid interactions with off-target sequences within cells. Typically, ssODNs, composed of no more than 200 nt, contain a ≥30-nt homology arm at both the 5′ and 3′ ends, which can pair with both the DNA donor of sgRNA and the target site [22]. The mechanism of endosomal escape in this case is unknown; however, it is likely due to membrane fusion, although this is not certain. It is known that multicomponent lipoplexes, incorporating three to six lipid species simultaneously, were fused to the endosomal membrane [23]. It was not believed that PEGs played a role in endosomal escape because DMG-PEG was present with a relatively low content (0.25%). Alternatively, ionic lipids might induce local membrane disruption [24]. From the perspective of absorption, distribution, metabolism, and excretion (ADME), both RNP-loaded LNPs and free RNP predominantly migrated to the liver after intravenous administration, although they were also observed in the spleen, kidney, and lungs to a lesser extent. RNP-loaded LNPs accumulated in the liver, as it is a key organ for metabolism within the reticuloendothelial system. It is also suggested that phagocytic cells, such as macrophages, engulf LNPs. PEGs on nanoparticles help evade macrophage uptake by reducing protein attachment. It has been reported that polyglycerol on nanoparticles prevents macrophage uptake more effectively than PEG [25]. In other cases of nanodelivery, it is well-known that nanoparticles are spontaneously accumulated in solid cancer tissues due to passive targeting, which is based on the enhanced permeability and retention (EPR) effect [26,27]. Nanoparticles pass through the leaky vessels of neovasculature into solid tumor parenchyma, where the lymphatic system is underdeveloped. Eventually, in this case, with respect to distribution, intravenously injected LNPs encapsulating Cas9-RNP (termed pLNP) and HDR template (pLNP-HDR) (245 ± 4 nm in diameter) via tail vein injections delivered Cas9-RNPs primarily to the liver, with additional presence in the spleen and lungs of Ai9 female mice. On the other hand, intravenously injected LNPs encapsulating Cas9 mRNA and sgRNA (termed mLNP) and HDR template (mLNP-HDR) (150 ± 2 nm in diameter) via tail vein injections delivered Cas9 mRNA mainly to the liver. mLNP resulted in 60% gene knockout in hepatocytes in mice. Both LNPs were composed of C12-200, DOPE, cholesterol, DMG-PEG, and DOTAP. It is likely that Cas9 mRNA was translated into Cas9 protein in situ, while Cas9-RNP was short-lived. In fact, mLNP-HDR surpassed pLNP-HDR in gene editing efficiency in in vitro tests using HEK293T or HEPA1-6 cells [21]. It is believed that the spontaneous delivery of nanoparticles to the liver via the reticuloendothelial system is a form of passive targeting in a broad sense. Nonetheless, nanoparticles were also observed in the spleen, kidney, and lung, in addition to the liver. Therefore, distribution based on more active targeting is expected to help avoid serious off-target side effects.

(iii) Ionizable cationic lipids with a pKa of approximately 6.4 bind to negatively charged RNAs at the pH used during mixing, thereby protecting the RNA. Tumor suppressor genes such as *P53*, *PTEN*, and *RB1* were targeted by RNPs. The 5A2-DOT type of LNP was composed of 5A2-SC8/DOPE/cholesterol/DMG-PEG/DOTAP in a molar ratio of 15:15:30:3:7 (Figure 6). As a component, 5A2-SC8 is a degradable dendrimer containing multiple ester groups, while DOTAP is a permanently cationic lipid. The notation 5A2-DOT-n refers to LNPs with n mole% DOTAP incorporation. Intravenous administration of 5A2-DOT-5 LNPs encapsulating Cas9/sgP53/sgPTEN/sgRB1 RNPs resulted in visible tumor formation on the liver in adult C57BL/6 mice. Similarly, intravenous administration of 5A2-DOT-50 LNPs encapsulating Cas9/sgEml/sgAlk RNPs induced Eml4-Alk rearrangements, leading to lung tumor formation in adult C57BL/6 mice (Figure 7) [28]. The formation of tumors in the liver and lungs following administration of LNPs encapsulating RNPs targeting tumor suppressor genes suggests a lack of tissue selectivity in these LNPs. The internalization mechanism was identified as lipid raft-mediated endocytosis, which could be inhibited by methyl-β-cyclodextrin (MβCD) or at 4 °C. Although the exact mechanisms of endosomal escape remain unclear, DOTAP, a permanent cationic lipid highlighted in the study, may locally disrupt the membranes of LNPs and endosomes. The presence of cationic lipid–anionic lipid ion pairs has been suggested to cause disruption of the endosomal membrane by inducing a non-bilayer structure [24]. It is also believed that the membrane of PEG-coated LNPs is composed primarily of DMG-PEG, which lacks anionic phosphate groups, along with DOPE, which contains anionic phosphate groups. Thus, anionic DOPE may interact with cationic DOTAP. It is well-established that PEGylation on the surface of nanoparticles acts as a stealth polymer, suppressing aggregation, opsonization, and phagocytosis, thereby extending circulation time in drug delivery systems [29]. In this system, lipid raft-mediated endocytosis is believed to occur due to interactions with cholesterol present in the LNP membranes. Thus, PEGs did not sterically hinder their binding to the cell surface. However, membrane fusion between LNPs coated with hydrophilic PEGs and the lipid bilayer of the plasma membrane is unlikely, although it might occur depending on modifications to the PEG ends, similar to cell–cell fusion [30]. Additionally, free PEGs are commonly known to relax the lipid bilayer structure of cell membranes, thereby promoting adhesion between membranes during cell–cell fusion. Nevertheless, the endosomal escape mechanism was attributed to local membrane disruption caused by DOTAP.

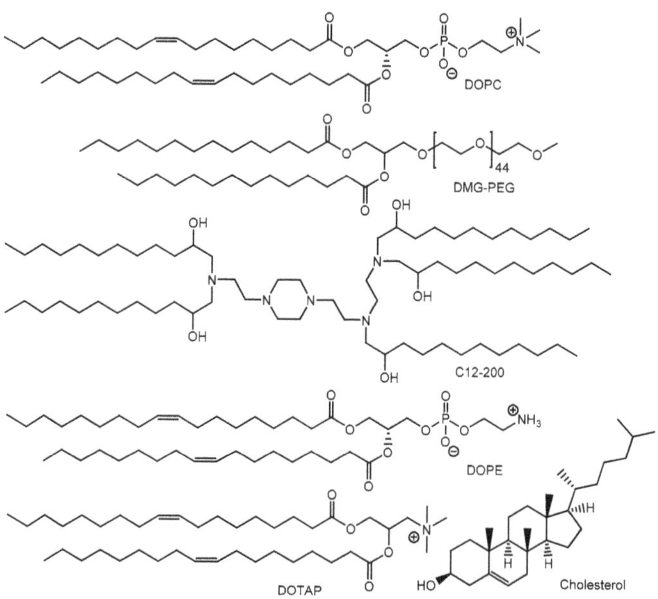

Figure 4. The structures of lipid nanoparticle components include 2,3-dioleoyl-glycero-1-phosphocholine (DOPC), 1,2-dimyristoyl-rac-glycero-3-methoxypolyethyleneglycol-2000 (DMG-PEG), cholesterol, 1,10-[[2-[4-[2-[2-[bis(2-hydroxydodecyl)amino]ethylamino]ethyl]piperazin-1-yl]ethyl]azanediyl]bis(dodecan-2-ol) (C12-200), 1,2-dioleoyl-sn-glycero-3-phosphoethanolamine (DOPE), and 1,2-dioleoyl-3-trimethylammonium-propane (DOTAP).

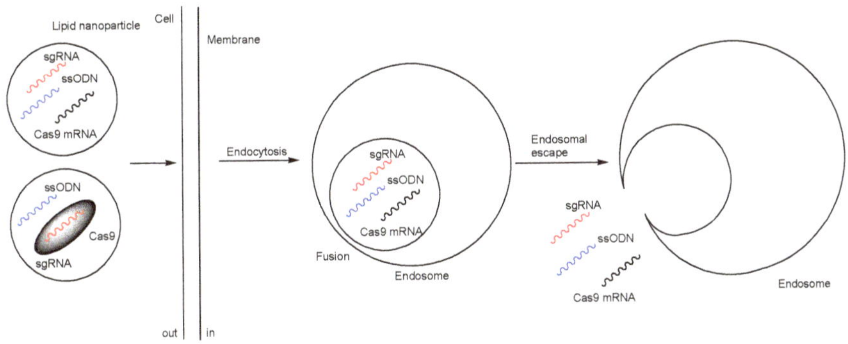

Figure 5. The outline of a lipid nanoparticle (LNP)-based Cas9 RNP delivery system uses optimally designed single-stranded oligonucleotides (ssODNs). The mechanism of endosomal escape is unknown; however, it is likely due to membrane fusion, although this remains uncertain.

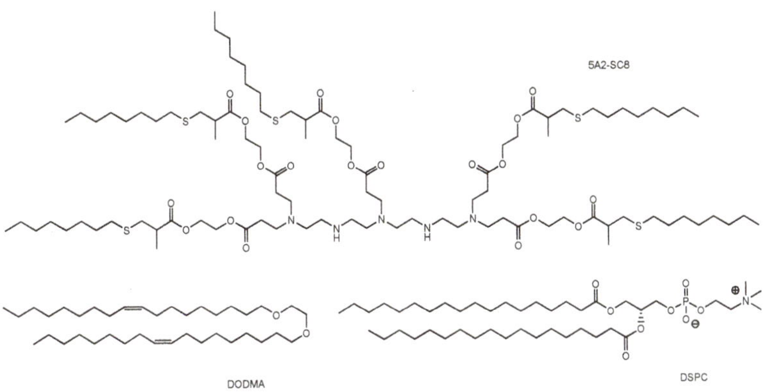

Figure 6. The structures of lipid nanoparticle components such as 5A2-SC8, 1,2-dioleyloxy-3-dimethylaminopropane (DODMA), and 1,2-distearoyl-sn-glycero-3-phosphocholine (DSPC).

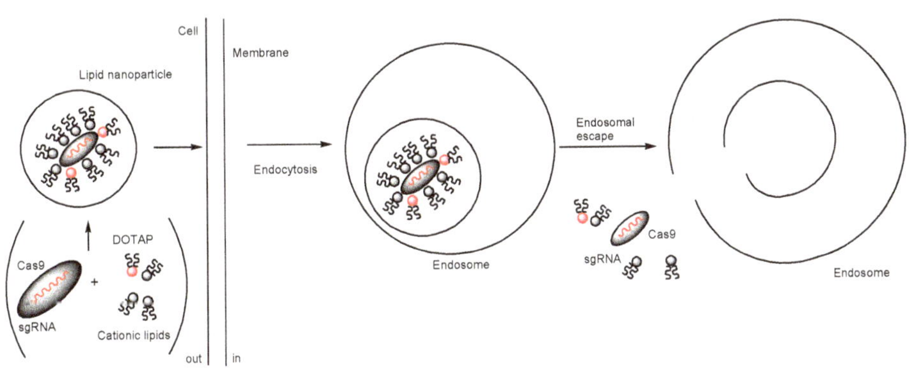

Figure 7. The pathway of LNPs encapsulating Cas9 ribonucleoprotein (Cas9 RNP), surrounded by cationic lipids such as 5A2-SC8 and 1,2-dioleoyl-3-trimethylammonium-propane (DOTAP) (depicted by a red circle), is described. Cas9 RNP is a complex of Cas9 protein and single-guide RNA (sgRNA). The LNPs are internalized into cells via lipid raft-mediated endocytosis. Although the mechanisms of endosomal escape remain unknown, DOTAP, as a permanent cationic lipid, may disrupt the membranes of both LNPs and endosomes.

(iv) Nanoassemblies composed of Cas9 RNP with oligo (20) glutamic acid tags (Cas9E20) targeting *PTEN* and gold nanoparticles, such as Arg-AuNPs, were developed using a linker that enabled electrostatic interactions between glutamic acid and arginine residues (Figure 8). Following tail-vein injection into BALB/c mice, these nanoassemblies exhibited significantly higher distribution in macrophages compared to T cells and B cells in the liver and spleen. This nanoparticle selectivity between macrophages and immune cells like T cells may be attributed to the reticuloendothelial system, which plays a critical role in host defense against infectious agents. The average gold content in sorted CD11+ macrophages was higher in the spleen compared to the liver. Nanoassemblies (285 ± 23 nm in diameter) administered via tail-vein injection into mice achieved gene editing efficiencies of over 8% in liver macrophages and over 4% in spleen macrophages [31]. The internalization mechanism of Cas9E20 into cells was not receptor-mediated endocytosis but membrane fusion facilitated by nanoassemblies, whose nanoparticle membrane consisted of linoleic acid. The uptake of nanoassemblies was not detected through fluorescence microscopy analysis of cell fluorescence. Instead, during the process of membrane fusion, the nanoassemblies integrated into the membrane and released Cas9E20 into the cytosol. Cas9E20 subsequently entered the nucleus, guided by a nuclear localization signal (PKKKRKV) [32]. However, this process might also involve macrophage phagocytosis, given the observed cell selectivity between macrophages and T cells.

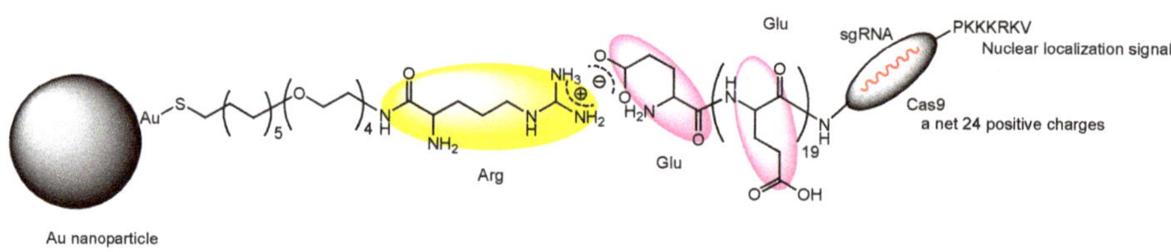

Figure 8. The structure of the nanoassembly is composed of Cas9 ribonucleoproteins (RNPs) with an oligo (20) Glu tag (Cas9E20) and a gold nanoparticle, linked through electrostatic connections between Glu (purple) and Arg (yellow).

(v) Methoxy-poly(ethylene glycol)-b-poly [2-(azepan-1-yl)ethyl methacrylate] (mPEG-PC7A) contains ionizable PC7A units with both cationic and hydrophobic side groups. Consequently, Cas9 RNPs with ssODNs, facilitating genome editing via HDR, spontaneously form nanoparticles with mPEG-PC7A polymers due to hydrophobic interactions (termed HDR-NP) when the amino groups are not protonated. Conversely, they self-disassemble when the amino groups are protonated during endosomal acidification (Figure 9). Similarly, Cas9 RNPs facilitating genome editing via NHEJ spontaneously form nanoparticles with mPEG-PC7A polymers (termed NHEJ-NP). The Cas9 RNPs targeted a STOP cassette consisting of three SV40 polyA sequences to block transcription of the downstream tdTomato gene. NHEJ-NP (29.4 nm in diameter), administered intravenously, intratracheally, or intramuscularly, induced efficient gene editing in the liver, lungs, and skeletal muscle of mice, respectively. Intravenous administration of NHEJ-NP likely resulted in its distribution to the liver via the reticuloendothelial system. The site of administration plays a crucial role. Moreover, intramuscular injection of HDR-NP (33.3 nm in diameter) demonstrated muscle strength recovery in a Duchenne muscular dystrophy mouse model. Clathrin-dependent endocytosis (with an endosomal diameter of 85–150 nm) is the most prominent mode of endocytosis, although several mechanisms exist. The endocytosis of NHEJ-NPs in HEK293 cells was inhibited by chlorpromazine (an inhibitor of clathrin-dependent endocytosis), methyl-β-cyclodextrin, and genistein (both inhibitors of caveolin-dependent endocytosis).

This indicates that clathrin-dependent and caveolin-dependent endocytosis occur simultaneously. The size of the nanoparticles is appropriate for cellular uptake through these pathways, whereas the Cas9 protein has a diameter of 7.5 nm, the sgRNA has a diameter of 5.5 nm. After endocytosis, HDR-NPs containing Cas9 RNPs with ssODNs as the payload and mPEG-PC7A polymers dissociated from the positively charged protonated polymers in endosomes and escaped into the cytosol via the proton sponge effect [33]. Interestingly, CPP-NHEJ-NPs, prepared using 50% CPP-PEG-PC7A and 50% mPEG-PC7A, were not internalized via endocytosis in HEK293 cells, unlike NHEJ-NPs. Their internalization was not inhibited by chlorpromazine or genistein. TAT (CYGRKKRRQRRR) was used as the CPP. Typically, positively charged TAT induces receptor-mediated endocytosis by utilizing negatively charged heparan sulfate proteoglycans (HSPGs) as receptors. It is unlikely that CPP-NHEJ-NPs (approximately 33.3 nm in diameter) cross the membrane via passive diffusion or direct translocation due to their overall molecular size, although both endocytosis and direct translocation are widely recognized as mechanisms for CPP internalization into cells. Thus, the internalization mechanisms of CPP-NHEJ-NPs remain uncertain. Possible mechanisms could include macropinocytosis, other less understood forms of endocytosis, or membrane disruption. While it is true that CPP-NHEJ-NPs were not internalized via endocytosis, the trajectories of fluorophore-labeled compounds often differ from those of their original counterparts due to alterations in physical properties. Furthermore, the in vivo distribution of HDR-NPs and CPP-NHEJ-NPs may vary after administration due to interactions between the positively charged TAT and endogenous materials, such as serum albumin.

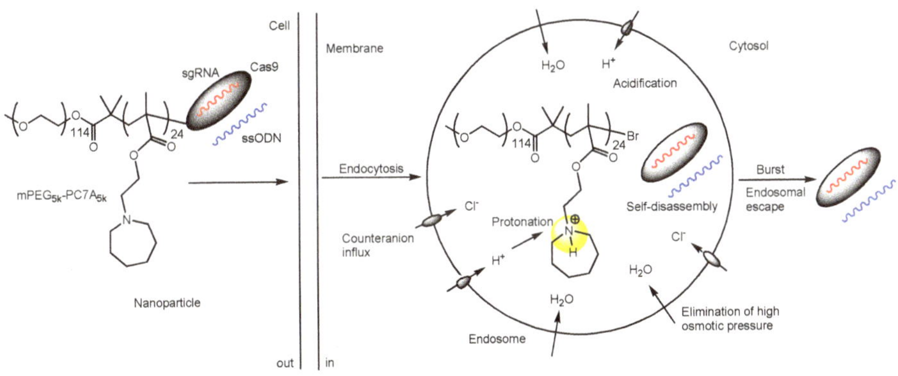

Figure 9. The pathway of nanoparticles, composed of the Cas9 ribonucleoprotein (RNP) formed by Cas9 and single-guide RNA (sgRNA), and the single-stranded oligonucleotide (ssODN) self-assembled with mPEG-PC7A through hydrophobic interactions, involves the release of the Cas9 RNP and ssODN complex into the cytosol via endocytosis. This process is followed by self-disassembly due to acidification in the endosomes and endosomal escape based on the proton sponge effect. As indicated in yellow, proton acceptance by amines in the endosomes triggers an influx of chloride ions and water, leading to osmotic rupture of the endosomal/lysosomal membrane.

(vi) Endosomal escape via the proton sponge effect was achieved using CPPs such as the Lys- and His-rich amphipathic peptide LAH5 (KKALLALALHHLAHLAHHLALALKKA). Nanocomplexes, measuring 200–400 nm in diameter and formed electrostatically by combining Cas9 RNP with cationic LAH5 peptides, were internalized via endocytosis, likely mediated by anionic HSPGs acting as receptors for receptor-mediated endocytosis. Upon acidification of the endosomes, the LAH5 peptides became protonated, causing an endosomal burst through the proton sponge effect and facilitating the release of the nanocomplexes into the cytosol. The nuclear localization signal (NLS) sequences on Cas9 directed it into the nucleus. Nanocomplexes targeting *CCR5* successfully achieved gene editing across the membrane in various cell lines [34]. Although this LAH5-based delivery strategy for CRISPR-Cas genetic engineering shows promise for in vitro and ex vivo applications, the current nanocomplex formulation is unlikely to be stable for in vivo applications. However, Cas9 RNP-loaded nanoparticles coated with LAH5 could be practical for in vivo use.

(vii) Exosomes are lipid bilayer membrane vesicles (40–160 nm in diameter) and a type of natural nanoparticle released by cells to facilitate cell-to-cell communication. It is well-established that cancer cells release exosomes containing miRNAs to prepare a favorable microenvironment for metastasis [35]. This discovery suggests that natural exosomes could be utilized for tissue-selective drug delivery. In fact, exosomes derived from hepatic stellate cells have demonstrated selectivity toward hepatic cells. Exosome^{RNP} nanocomplexes, created by loading Cas9 RNPs into purified exosomes isolated from hepatic stellate cells via electroporation, were successfully taken up by hepatocytes. Tail vein administration of Exosome^{RNP} targeting *PUMA* alleviated acute liver injury caused by a single overdose of acetaminophen. The p53 up-regulated modulator of apoptosis (PUMA) protein was significantly elevated in the livers of acetaminophen-treated mice. Similarly, tail vein administration of exosome^{RNP} targeting *CcnE1* alleviated chronic liver fibrosis caused by three intraperitoneal injections of CCl_4 per week for five weeks. Cyclin E1 (CcnE1), a member of the cyclin-dependent protein kinase family, promotes the proliferation of hepatic stellate cells. Additionally, tail vein administration of exosome^{RNP} targeting *KAT5* improved orthotopic hepatocellular carcinoma (HCC) transplanted via hepatic portal injection. K (lysine) acetyltransferase 5 (KAT5) is essential for HCC growth. Studies revealed that exosomeRNP nanocomplexes were internalized via endocytosis, which was inhibited by chlorpromazine (a clathrin-dependent endocytosis inhibitor), genistein (a caveolin-dependent endocytosis inhibitor), or exposure to 4 °C [36]. The cargo delivery mechanisms of exosomes into cells are categorized into two modes: (a) fusion with the plasma membrane of recipient cells or (b) endosomal escape after endocytosis. The liver is a key metabolic organ. The exosomes taken up by hepatocytes may undergo degradation or follow the secretory pathway within the endo-lysosomal system, depending on the signal proteins on the surface of the exosomes or the endocytic internalization of bystander exosomes due to the high frequency of endocytosis in the liver. Nonetheless, it is believed that the endosomal escape mechanism of loaded Cas9 RNPs involves fusion between the endosome and the exosomes (Figure 10).

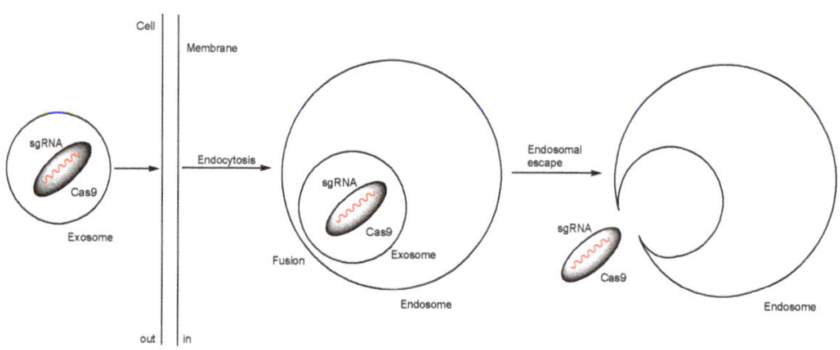

Figure 10. The pathway of exosome^{RNP} nanocomplex internalization involves endocytosis and endosomal escape of the Cas9 ribonucleoprotein (RNP), composed of Cas9 and single-guide RNA (sgRNA), with these components acting as their cargos via membrane fusion.

2.2.2. Restricted Distribution to the Target Sites via Local Injections

Simple Cas9 RNP-loaded nanoparticles lack cell selectivity, leading to off-target side effects. To improve bioavailability and minimize such side effects, local injections can restrict distribution to specific target sites, compared to oral or intravenous administration, although intravenous administration is technically a form of injection. Local injections have been implemented in the eye, bone marrow cavity, and cranial regions. The blood–retinal barrier (BRB) and the blood–aqueous barrier (BAB) pose challenges for delivering drugs to the eyes. Therefore, local injections directly into the eye are effective. The cornea of the eye consists of three cell layers: the epithelial cell layer, the stromal layer, and the endothelial layer.

(viii) Intrastromal injection of LNP-encapsulated CRISPR targeting tdTomato in the loxP-3xStop-loxP-tdTomato reporter system resulted in successful transfection not only of stromal cells but also of endothelial cells in all wild-type eyes of 3-month-old hybrid B6129F1-loxP-3xStop-loxP-tdTomato mice. tdTomato is an exceptionally bright red fluorescent protein, capable of producing fluorescence detectable up to a depth of 1 cm from the epidermis [37]. Several Cas9 RNPs were tested for their ability to disrupt stop signals and activate tdTomato expression. Although the corneal stroma is separated from the single-cell-thick endothelium by a thin, acellular matrix known as Descemet's membrane, the injected LNPs successfully crossed this barrier, as evidenced by tdTomato fluorescence. Moreover, Cas9 RNP-loaded LNPs corrected the Sey variant and restored FLAG-tagged PAX6 expression in 38% of ex vivo cortical neurons derived from a mouse model of aniridia, known as Small eye (Sey). This model carries the nonsense c.580G>T (p.G194X) variant in the *Pax6* gene [38]. The LNPs used in these experiments were composed of 1,2-dioleyloxy-3-dimethylaminopropane (DODMA), DOPE, 1,2-distearoyl-sn-glycero-3-phosphocholine (DSPC), and cholesterol (Figure 6). This delivery strategy, utilizing LNPs encapsulating RNP and DNA template mixtures via intrastromal injection, holds potential for CRISPR/Cas9-based genome editing therapies for corneal diseases.

Leukemia stem cells (LSCs) sustain the progression of acute myeloid leukemia (AML), a type of cancer that originates in the blood-forming cells of the bone marrow, and contribute to disease relapse.

(ix) Mesenchymal stem cell membrane-coated nanofibrils (MSCM-NFs), loaded with (a) bioreducible LNPs encapsulating Cas9 RNP targeting the critical gene interleukin-1 receptor accessory protein (*IL1RAP*) in human LSCs and (b) CXCL12α, a ligand of the C-X-C chemokine receptor 4 (CXCR4) (Figure 11), were injected into the bone marrow cavity of mice. This approach successfully demonstrated that IL1RAP knockout reduced the colony-forming capacity of LSCs and decreased the leukemic burden [39]. CXCL12α facilitated the recruitment of leukemia cells. Consequently, vascular cell adhesion molecule 1 (VCAM-1), derived from the mesenchymal stem cell membrane (173.8 $\pm$ 46.4 nm in thickness) on nanofibrils (1.0 $\pm$ 0.2 μm in diameter and 22.0 $\pm$ 10.1 μm in length), bound to integrin α4 (VLA-4) on the surface of leukemia stem cells. LNPs (73.3–162.2 nm in diameter) encapsulating Cas9 RNPs, which separated from MSCM-NFs, were internalized into leukemia stem cells via endocytosis. They induced efficient gene editing after endosomal escape, likely through membrane fusion [40]. The endosomal diameter in clathrin-dependent endocytosis or clathrin- and caveolae-independent endocytosis is approximately 85–150 nm and around 90 nm, respectively, although the specific type of endocytosis that occurs remains uncertain. A single LNP could fit within an endosome. The SpCas9 protein has a molecular weight of 160 kDa and a hydrodynamic diameter of approximately 7.5 nm, while sgRNA has a molecular weight of around 31 kDa and a hydrodynamic diameter of about 5.5 nm [41]. Therefore, due to their size, several Cas9 RNP molecules could be encapsulated within one LNP. Since the target *IL1RAP* gene is located in the cell nucleus, efficient gene editing is successfully achieved.

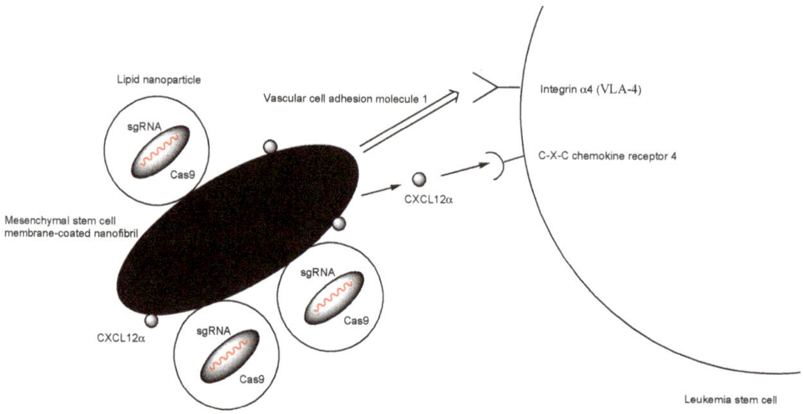

Figure 11. The interaction between leukemia stem cells and injected mesenchymal stem cell membrane-coated nanofibrils (MSCM-NFs) loaded with LNPs encapsulating Cas9 RNP targeting the critical genes interleukin-1 receptor accessory protein (IL1RAP) and CXCL12α was studied. CXCL12α, a ligand for C-X-C chemokine receptor 4 (CXCR4), recruited leukemia stem cells. As a result, vascular cell adhesion molecule 1 (VCAM-1) derived from the mesenchymal stem cell membrane bound integrin α4 (VLA-4) on the surface of leukemia stem cells. Endocytosed LNPs encapsulating Cas9 RNP induced efficient gene editing in leukemia stem cells following endosomal escape.

The BBB significantly hinders drug delivery to the brain. Consequently, local injections into the brain serve as an alternative and urgent method for direct delivery. The potential of locally injecting Cas9 RNP into the brain was evaluated to assess genome editing using the tdTomato reporter system, similar to the intrastromal injections into the cornea described earlier. Although the tdTomato locus is relatively large, the expression of the tdTomato protein in a mouse model indicates the actual genome editing efficiency and the distribution of Cas9 RNP. Four repeats of the positively charged Simian Vacuolating Virus 40 nuclear localization sequences (SV40-NLS) were fused to the *N*-terminus, along with two repeats fused to the *C*-terminus of Cas9, enabling it to spontaneously move to the cell nucleus. In total, six SV40-NLS sequences were fused to Cas9 (4x-Cas9-2x). Cell-penetrant Cas9 RNPs were injected into the striatum of lox-stop-lox Ai9 mice. As a result, these cell-penetrant Cas9 RNPs edited more neurons in the region near the injection site, as indicated by tdTomato fluorescence. In contrast, transient RNP complexes delivered using AAV serotype 9 (Cas9-AAV) exhibited better diffusion throughout the brain. However, brains treated with Cas9-AAV showed significantly elevated *Cd3e* gene expression, indicative of an adaptive immune response. Meanwhile, brains treated with Cas9 RNPs experienced acute microglial activation, which could be mitigated by reducing endotoxin levels. Consequently, Cas9 RNPs demonstrated superior safety compared to Cas9-AAV [42]. The restricted distribution of Cas9 RNPs, along with their safety profile, could potentially be enhanced through the use of nanoparticles.

(x) Non-viral, biodegradable PEGylated nanocapsules (NCs) encapsulating Cas9 RNPs (less than 35 nm in diameter) were evaluated for genome editing in neurons following intracerebral injection. The optimal formulation utilized a molar ratio of acrylic acid (AA)/*N*-(3-aminopropyl)methacrylamide hydrochloride (APMA)/1-vinylimidazole

(VI)/N,N'-bis(acryloyl)cystamine (BACA)/acrylate-polyethylene glycol (Ac-PEG)/RNP at 927/927/244/231/64/1. The small size of the NCs may facilitate their diffusion within the brain. Indeed, NCs loaded with RNPs containing anti-Ai14 sgRNA, administered via intracerebral injection into the healthy mouse striatum, demonstrated successful genome editing of striatal neurons as determined by the tdTomato reporter system. The Cas9 RNP targeting sites were located within each of the three SV40 polyA sequences in the STOP cassette. The removal of two SV40 polyA cassettes led to the expression of tdTomato. Genome editing was predominantly observed in medium spiny neurons (>80%), with occasional editing in cholinergic, calretinin, and parvalbumin interneurons [43]. The conjugation of neuron-specific ligands, such as the rabies virus glycoprotein-derived peptide (RVG peptide: YTIWMPENPRPGTPCDIFTNSRGKRASNG), a ligand for the α-7 subunit of nicotinic acetylcholine receptors (AchRs) in neuronal cells [44], could enhance specificity for neuron-targeted delivery. Moreover, conjugation of cell-penetrating peptides (CPPs) such as TAT may enhance cellular uptake. However, contrary to expectations, the addition of either CPP or RVG to the nanocapsules (NCs) did not significantly affect the neuronal editing efficiency or the size of the edited brain area in this assay system. The Cas9 RNP-loaded NCs facilitated endosomal escape via imidazole-containing monomers (VIs) and amine-containing monomers (APMAs), likely due to the proton sponge effect. Subsequently, the Cas9 RNPs were separated from the NCs through cleavage of the glutathione-responsive cross-linker in the cytosol.

(xi) CRISPR–gold targeting the metabotropic glutamate receptor 5 (*mGluR5*) gene efficiently reduced local mGluR5 protein levels in the striatum by 40–50% following intracranial injection into the brains of *Fmr1* knockout mice, a model of fragile X syndrome (FXS). This reduction in mGluR5 protein levels ultimately led to a decrease in hyperlocomotor activities, such as excessive digging and jumping behavior, in the Fmr1 knockout mice [45]. FXS [46] is an inherited genetic disorder caused by mutations in the fragile X mental retardation 1 (*FMR1*) gene. The FMR1 gene contains a CGG repeat in the 5′-untranslated region. Silencing the FMR1 gene prevents the expression of the FMR protein (FMRP). Low-molecular-weight metabotropic glutamate receptor 5 (mGluR5) antagonists have been developed for the treatment of FXS [47]. mGluR5 is an excitatory G-protein-coupled receptor. According to the mGluR theory of FXS, glutamatergic signaling through mGluR5 is enhanced in the absence of FMRP, thereby inducing various neurological and neuropsychiatric features of FXS. At present, low-molecular-weight mGluR5 antagonists are unlikely to be clinically available for the treatment of FXS. Local injections into the brain, which bypass the BBB, may be a significant strategy for treating inherited genetic diseases, including FXS.

(xii) The development and progression of psoriasis are associated with NLRP3 inflammasomes, which are large protein complexes that play a crucial role in regulating the innate immune response and inflammation. Therefore, genome editing of the NLRP3 inflammasome could alleviate the inflammatory symptoms of psoriasis. Polyamidoamine nanocomplexes (named GBLA-22), coated with phenylboronate and lipoic acid moieties as ligands and containing Cas9 RNP targeting *NLRP3*, achieved a 33.7% gene disruption at the NLRP3 locus in psoriatic skin tissues in an in vivo assay. This was accomplished via subcutaneous injection into the psoriatic skin on the back through two-point injections at the upper and lower parts of the area, using an imiquimod-induced psoriatic mouse model. The treatment led to a reduction in the levels of IL-1β, IL-17, IL-18, TNF-α, and IL-12/23p40 [48]. Psoriatic model mice were generated by applying imiquimod cream daily to a shaved 2 cm $\times$ 2 cm area of skin for seven consecutive days. Phenylboronate and lipoic acid moieties were identified as crucial ligands for protein binding, endocytosis, endosomal escape, and intracellular protein release. The cellular uptake mechanism was determined to be lipid raft-mediated endocytosis, as it was inhibited by MβCD (a cholesterol-depleting reagent) or at 4 °C but remained unaffected by cytochalasin D (a macropinocytosis inhibitor), genistein (a caveolae-dependent endocytosis inhibitor), or chlorpromazine (a clathrin-dependent endocytosis inhibitor). Furthermore, the endosomal escape mechanism was suggested to occur via membrane disruption. Following subcutaneous injection, GBLA-22-Cy5/RNP complexes were detected in the skin and liver from days 1 to 5, as indicated by Cy5 fluorescence, suggesting that these organs serve as the primary sites of nanoparticle metabolism.

(xiii) Overexpressed microRNA-21 (miR-21) in human hepatocellular carcinoma inhibits the expression of tumor suppressor genes. Therefore, miR-21-sensitive nanodelivery systems can selectively induce apoptosis in cancer cells. Nanocapsules composed of nanoassembled, engineered DNAzyme shells encasing Cas9 RNP targeting the *MIR-21* gene—referred to as miR-21-responsive DNAzyme-functionalized nanocapsules (R-DN)—demonstrated significant tumor growth inhibition of up to 75.94% in an in vivo assay. This was achieved through intratumoral injection in a HepG2 tumor mouse model, which was established by subcutaneous armpit injection of 2×10^6 cells in a PBS/Matrigel (1:1, *v*/*v*) mixture [49]. In general, DNAzyme is a catalytic single-stranded DNA that binds to and cleaves target RNA in a sequence-specific manner. R-DN entered cancer cells via endocytosis, which was inhibited by MβCD (a cholesterol-depleting reagent) and chlorpromazine (a clathrin-dependent endocytosis inhibitor). After endosomal escape—through mechanisms that remain unclear—cytosolic miR-21 was believed to interact with R-DN. It was discovered that as little as 1.7 pM of miR-21 could trigger the on-demand release of Cas9 RNP from R-DN through a conformational change in the DNAzyme upon miR-21 binding. The released Cas9 RNP then edited the miR-21 gene, leading to the upregulation of tumor suppressor genes that induce apoptosis. This process established a positive feedback-driven autonomous catabolic cycle between miR-21 and R-DN.

Nonetheless, local injections cause physical pain to patients, although Cas9 RNPs are reliably delivered to the target sites and remain in the restricted area, such as the cranial cavity. Therefore, non-invasive administration methods should be developed.

2.2.3. Selective and Effective Distribution Through Non-Invasive Methods

Selective and effective distribution through non-invasive methods, combining active targeting with a deliberate approach and passive targeting with a laissez-faire approach, is anticipated to minimize off-target side effects. Nevertheless, differences between target sites and other healthy regions persist due to inherent biological mechanisms rooted in structural principles. Cell surfaces are generally covered with negatively charged HSPGs, which interact with CPPs to facilitate receptor-mediated endocytosis. However, the surface of cancer cells exhibits a more negatively charged characteristic due to their unique metabolic processes, which produce lactate anions as part of the Warburg effect [50]. The secretion of lactate anions displaces cations such as H^+ and Na^+ from the cancer cell surface, enhancing the negative charge. The Warburg effect also creates a mildly acidic extracellular environment (pH 6.6–6.8) in cancer tissues [51]. Furthermore, phospholipids such as phosphatidylethanolamine and phosphatidylserine flip to the outer surface of cancer cells [52]. Nonetheless, more pronounced differences may be necessary for effective targeting. The potential for achieving cell selectivity between cancer and normal cells remains an area for discovery. Notably, RGD peptides are well-known for inducing receptor-mediated endocytosis by utilizing $\alpha v \beta_3$ integrins as receptors on the cancer cell surface [53]. Cell-selective internalization was achieved in cancer drug delivery systems using iRGD (c(CRGDKGPDC)) as a tumor-homing ligand.

(xiv) Nanoparticles coated with iRGD were used for the co-delivery of Cas9 RNP targeting the *Nrf2* gene and the antitumor photosensitizer chlorin e6 (Ce6). These nanoparticles, with diameters ranging from 84 to 92 nm, were administered intravenously and internalized into cells via receptor-mediated endocytosis in a mouse tumor model. In this model, cellosaurus CNE-2 cells, derived from nasopharyngeal carcinoma, were subcutaneously injected into the mice's armpits. Near-infrared irradiation, administered 12 h after nanoparticle delivery, induced the generation of reactive oxygen species (ROS) by Ce6. These ROS destabilized lysosomal membranes, enabling the nanoparticles to escape the lysosome. This facilitated genome editing, thereby increasing tumor sensitivity to ROS [54]. The ROS generated by Ce6 exhibited a synergistic antitumor effect, inducing apoptosis in addition to destabilizing lysosomal membranes. Furthermore, the Nrf2 protein is implicated in promoting angiogenesis. Hypoxia-inducible factor 1α (HIF1α) and vascular endothelial growth factor-A (VEGF-A) are key angiogenic factors. Treatment with Ce6 and Cas9 RNP-encapsulated nanoparticles led to a reduction in HIF1α and VEGF-A levels by 50.0% and 44.6%, respectively, compared to levels observed with Ce6-encapsulated nanoparticles without Cas9 RNP. Consequently, intravenously administered nanoparticles encapsulating both Ce6 and Cas9 RNP demonstrated a more significant antitumor effect than those lacking iRGD.

(xv) Intravenously administered glutathione-responsive silica nanocapsules (SNCs) conjugated with glucose and an RVG peptide, encapsulating Cas9 mRNA along with either App^{659} sgRNA or Th sgRNA, achieved up to 6.1% editing efficiency of the amyloid precursor protein (*App*) gene in the thalamus/hypothalamus (resulting in a 19.1% reduction in intact APP expression levels) and up to 3.9% editing efficiency of the tyrosine hydroxylase (*Th*) gene in the thalamus/hypothalamus (resulting in a 30.3% reduction in Th expression levels), respectively, in C57BL/6J wild-type mice. These SNCs crossed the BBB via receptor-mediated transcytosis, utilizing glucose transporter-1 (GLUT1) and the α-7 AchR of endothelial cells. Subsequently, the SNCs were internalized into neurons via receptor-mediated endocytosis, utilizing the α-7 AchR of neuronal cells. The GSH-responsive SNCs released their cargo into the cytosol following endo/lysosomal escape, likely facilitated by the proton sponge effect of the imidazole-containing chains [55]. Overall, the strategy of using GSH-responsive, imidazole-rich SNCs conjugated with glucose and the RVG peptide was highly effective in achieving targeted delivery. However, the precise mechanisms underlying endosomal escape remain unclear.

(xvi) Intravenously injected, thin disulfide-cross-linked polymeric shell nanocapsules decorated with angiopep-2 peptide and encapsulating Cas9 RNP (approximately 30 nm in diameter) achieved high *PLK1* gene editing efficiency in brain tumors, reaching up to 38.1% in orthotopic CSC2-Luc GSC tumor-bearing mice [56]. The angiopep-2 peptide (TFFYGGSRGKRNNFKTEEY) is a ligand that binds to low-density lipoprotein receptor-related protein-1 (LRP-1), which is highly expressed on both BBB endothelial cells and glioblastoma (GBM) cells. These nanocapsules crossed the BBB via receptor-mediated transcytosis using LRP-1 as a receptor, as well as through passive diffusion induced by BBB disruption caused by GBM. Subsequently, the nanocapsules were internalized into the neurons via receptor-mediated endocytosis, utilizing LRP-1.

Similarly, (xvii) a brain-targeted CRISPR/Cas9-based nanomedicine was developed by fabricating angiopep-2-decorated, guanidinium- and fluorine-functionalized polymeric nanoparticles that loaded Cas9/gRNA RNP for the treatment of GBM. These nanoparticles exhibited approximately 32% gene knockout and a 67% reduction in the protein levels of the targeted proto-oncogene, polo-like kinase 1 (PLK1), in in vitro tests using U87MG cells. This gene knockout was sufficient to prolong the median survival time of mice bearing orthotopic glioblastoma to 40 days, without serious side effects or off-target effects [57].

Table 1. Summary of the Cas9 ribonucleoprotein (RNP)-encapsulated nanoparticles discussed in this review.

#	Formulations	Administration	Diseases	Targeting Gene	Results	Status	Refs.
i	NTLA-2001 (LNPs encapsulating Cas9 mRNA and sgRNA targeting TTR	Intravenous injection	TTR amyloidosis	*TTR*	Preclinical studies showed that at day 28 the mean reduction in serum TTR protein concentration was 52% in the group that received a dose of 0.1 mg per kilogram and was 87% in the group that received a dose of 0.3 mg per kilogram.	Phase 1 clinical trial (NCT046 01051)	[17]
ii	LNP-based Cas RNP delivery system using optimally designed ssODNs	Intravenous injection	TTR amyloidosis	*TTR*	RNP-loaded LNPs with anti-TTR sgRNA (sgTTR-G269) and designed ssODNs with a complementation rate of 50% (calculated Tm: 30 °C) suppressed TTR the highest (34%) as TTR protein levels in serum, quantified 1 week after intravenous administration in mice.	Basic research	[21]
iii	5A2-DOT-5 LNPs encapsulating Cas9/sgP53/sgPTEN/ sgRB1 RNPs	Intravenous injection	Induction of cancer	*P53*, *PTEN*, or *RB1*	Generation of visible tumors on the liver in adult C57BL/6 mice.	Basic research	[28]
iv	Nanoassemblies composed of Cas9 RNP with oligo (20) Glu tags (Cas9E20) and Arg-AuNPs electrostatically connected between Glu and Arg	Intravenous injection	-	*PTEN*	>8% and >4% gene editing efficiency in macrophages of the liver and the spleen, respectively.	Basic research	[31]
v	Nanoparticles composed of Cas9 RNPs with ssODNs and mPEG-PC7A	Intravenous, intratracheal, or intramuscular injection	-	A STOP cassette that consists of three SV40 polyA sequences to prevent transcription of the downstreamtd-Tomato	Intravenously, intratracheally, and intramuscularly injected NHEJ-NP (29.4 nm in diameter) induced efficient gene editing in mouse liver, lung, and skeletal muscle, respectively.	Basic research	[33]

Table 1. *Cont.*

#	Formulations	Administration	Diseases	Targeting Gene	Results	Status	Refs.
vi	Nanocomplexes (200–400 nm in diameter) electrostatically composed of a Cas9 RNP and cationic LAH5 peptides	-	-	*CCR5*	Nanocomplexes targeting CCR5 exhibited the gene editing across the membrane in diverse cell lines.	Basic research	[34]
vii	Exosome^{RNP} nanocomplexes, prepared by loading Cas9 RNPs into purified exosomes isolated from hepatic stellate cells through electroporation	Intravenous injection	liver injury, chronic liver fibrosis, or hepatocellular carcinoma	*PUMA*, *CcnE1*, or *KAT5*	Exosome^{RNP} targeting *PUMA* ameliorated acute liver injury. Exosome^{RNP} targeting *CcnE1* ameliorated chronic liver fibrosis. Exosome^{RNP} targeting *KAT5* administered by tail vein injections ameliorated orthotopic hepatocellular carcinoma (HCC).	Basic research	[36]
viii	LNP-encapsulated tdTomato-targeted CRISPR strategy in the loxP-3xStop-loxP-tdTomato reporter system	Intrastromal injection	corneal diseases	A STOP cassette that consists of three SV40 polyA sequences to prevent transcription of the downstream tdTomato	Transfection not only of the stromal cells but also of the endothelial cells in all wild-type eyes of 3-month-old hybrid B6129F1-loxP-3xStop-loxP-tdTomato wild-type mice.	Basic research	[37]
ix	MSCM-NFs loading (a) bioreducible LNPs encapsulating Cas9 RNP targeting the critical gene IL1RAP in human LSCs and (b) CXCL12α that is a ligand of CXCR4	Injection into the bone marrow cavity	acute myeloid leukemia	*IL1RAP* or *CXCL12α*	*IL1RAP* knockout reduced LSC colony-forming capacity and leukemic burden.	Basic research	[39]
x	PEGylated nanocapsules encapsulating Cas9 RNPs	Intracerebral injection	CNS diseases	A STOP cassette that consists of three SV40 polyA sequences to prevent transcription of the downstream tdTomato	Genome editing of striatal neurons, which was determined by the tdTomato reporter system.	Basic research	[43]

Table 1. *Cont.*

#	Formulations	Administration	Diseases	Targeting Gene	Results	Status	Refs.
xi	CRISPR–gold targeting the mGluR5 gene	Intracranial injection	fragile X syndrome	*mGluR5*	Reduction in local mGluR5 protein levels in the striatum by 40–50% after an intracranial injection into the brains of Fmr1 knockout mice and eventual reduction in hyperlocomotor activities such as excessive digging and jumping behavior.	Basic research	[45]
xii	Nanocomplexes coated with phenylboronate and lipoic acid and containing Cas9 RNP	Subcutaneous injection	psoriasis	*NLRP3*	33.7% gene disruption at the NLRP3 locus in psoriatic skin tissues.	Basic research	[48]
xiii	Nanocapsules composed of nanoassembled, engineered DNAzyme shells encasing Cas9 RNP	Intratumoral injection	cancer	*MIR-21*	Tumor growth inhibition of up to 75.94%.	Basic research	[49]
xiv	Nanoparticles encapsulating Cas9 RNP targeting *Nrf2* gene and antitumor photosensitizer chlorin e6 (Ce6) and covered with iRGD	Intravenous injection	cancer	*Nrf2*	The Nrf2 protein is associated with angiogenesis promotion. Hypoxia-inducible factor 1α (HIF1α) and vascular endothelial growth factor-A (VEGF-A) are representative angiogenetic factors. Ce6 and Cas9 RNP-encapsulated NP treatment resulted in 50.0 and 44.6% reduction in HIF1α and VEGF-A levels, respectively.	Basic research	[54]
xv	GSH-responsive SNCs conjugated with glucose and RVG peptide and encapsulating Cas9 mRNA and App659 sgRNA or Th sgRNA	Intravenous injection	CNS diseases including Alzheimer's disease	*App659* or *Th*	Up to 6.1% amyloid precursor protein (*App*) gene editing efficiency for thalamus/hypothalamus (resulting in 19.1% reduction in the expression level of intact APP) or up to 3.9% tyrosine hydroxylase (*Th*) gene editing efficiency for thalamus/hypothalamus (resulting in 30.3% reduction in the expression level of Th), respectively.	Basic research	[55]

Table 1. *Cont.*

#	Formulations	Administration	Diseases	Targeting Gene	Results	Status	Refs.
xvi	Disulfide-cross-linked polymeric shell nanocapsules decorated with angiopep-2 peptide, encapsulating Cas9 RNP (approx. 30 nm in diameter)	Intravenous injection	glioblastoma	*PLK1*	High PLK1 gene editing efficiency in a brain tumor (up to 38.1%) in orthotopic CSC2-Luc GSC tumor-bearing mice.	Basic research	[56]
xvii	CRISPR/Cas9-based nanomedicine by fabricating an angiopep-2 decorated, guanidinium and fluorine functionalized polymeric NPs loading Cas9/gRNA RNP	-	glioblastoma	*PLK1*	32% gene knockout and 67% protein reduction in the proto-oncogene polo-like kinase 1 (PLK1) (in vitro).	Basic research	[57]
xviii	LNPs and polymer Pluronic F_{127}-encapsulated CRISPR/Cas plasmid	Oral administration	colitis-associated colorectal cancer	*CD98*	Reduction in CD98 expression and demonstration of therapeutic efficacy against CAC.	Basic research	[58]
xix	Cas9 RNP-encapsulated NPs covered with anti-TfR and anti-α-7 AchR bispecific mAbs	Intravenous injection	CNS diseases	Certain proteins involved in CNS diseases	Under analysis in Tashima lab.	Basic research	-

Abbreviations. refs; references, LNP; lipid nanoparticle, mRNA; messenger ribonucleic acid, TTR; transthyretin, ssODNs; single stranded oligonucleotides, MSCM-NFs; mesenchymal stem cell membrane–coated nanofibrils, IL1RAP; interleukin-1 receptor accessory protein, CXCR4; C-X-C chemokine receptor 4, CNS; central nervous system, mGluR5; metabotropic glutamate receptor 5, SNC; silica nanocapsule, Th; tyrosine hydroxylase.

(xviii) Anti-colon disease drugs can be administered orally if they are resistant to degradation by stomach acid or proteases and are not absorbed in the small intestine. The progression of ulcerative colitis (UC) and colitis-associated colorectal cancer (CAC) is linked to the overexpression of CD98. P_{127}M@pCD98s, composed of negatively charged LNPs derived from mulberry leaves and the U.S. Food and Drug Administration (FDA)-approved polymer Pluronic F_{127} (P127), encapsulate a CRISPR/Cas plasmid targeting *CD98* (pCD98) with an approximate diameter of 267.2 nm. In an in vivo assay using a CAC mouse model, oral administration of P127M@pCD98s significantly reduced CD98 expression and demonstrated therapeutic efficacy against CAC [58]. The CRISPR/Cas plasmid encodes both sgRNA and the Cas9 protein. P127M@pCD98s penetrated the mucus layer, accumulated in inflamed colon tissues, and were internalized by colon epithelial cells and macrophages via galactose receptor-mediated endocytosis, facilitated by the galactose end groups. After endosomal/lysosomal escape, pCD98 was released and became active. Macrophages are known to express CD98. However, since LNPs with a diameter of 267.2 nm are relatively large, they may not fit into endosomes. Therefore, LNPs might undergo size reduction during transport to facilitate internalization via galactose receptor-mediated endocytosis in colon epithelial cells and macrophages. Alternatively, the tight junctions between colon epithelial cells in CAC might be partially disrupted, enabling passive diffusion of LNPs. Subsequently, LNPs might enter macrophages via macropinocytosis, likely occurring within the reticuloendothelial system.

Therefore, a selective and effective distribution of Cas9 RNP-encapsulated nanoparticles in a non-invasive manner has been developed.

2.2.4. Promising Delivery of Intravenously Administered Cas9 RNP-Encapsulated Nanoparticles to the Brain

It is well established that clustering induces endocytosis [59–61]. Therefore, a multiligand strategy on nanoparticles is more likely to promote cluster formation. For instance, SNCs conjugated with glucose as a GLUT1 ligand and RVG peptide as an α-7 AchR ligand were designed using a multiligand approach to target capillary endothelial cells [43]. The expression of receptors involved in endocytosis is often not limited to target cells. For example, HSPGs, which serve as receptors for CPPs like TAT, are expressed ubiquitously. A single-ligand strategy may result in off-target side effects. In contrast, a multiligand strategy is more likely to stochastically bind to specific regions that act as entry points to target sites, especially within systemic circulation.

The ligand–receptor selectivity of monoclonal antibodies is higher than that of CPPs or THPs. Consequently, Cas9 RNP-encapsulated nanoparticles coated with monoclonal antibodies can bind tightly and selectively to receptors that facilitate endocytosis or transcytosis. Notably, the clinically approved J-Brain Cargo® system [62] enables antibody–drug conjugates to cross the BBB through receptor-mediated transcytosis via the transferrin receptor in capillary endothelial cells. Thus, Cas9 RNP-encapsulated nanoparticles coated with monoclonal antibodies targeting the transferrin receptor can cross the BBB via receptor-mediated transcytosis. Additionally, nanoparticles coated with bispecific monoclonal antibodies targeting both the transferrin receptor and the α-7 AchR can subsequently enter neurons after transendothelial migration.

3. Conclusions

The CRISPR/Cas9 gene editing technique is a powerful tool for precisely cutting and modifying DNA. However, delivering Cas9 RNPs within living organisms presents challenges, including low cell selectivity and bioavailability. While internalization into cells such as fertilized eggs, CAR-T cells, or MSCs can be achieved using methods like lipofection (cationic lipid-mediated transfection), electroporation (electric pulse methods), viral vector-based gene delivery, or microinjection, these approaches have limitations. Utilizing carrier nanoparticles coated with vectors such as monoclonal antibodies, CPPs, or THPs offers a promising solution to these challenges [Table 1]. Nanoparticles protect Cas9 RNPs from enzymatic degradation, while the vectors on nanoparticles facilitate receptor-mediated endocytosis or transcytosis by targeting tissue-specific receptors. Endosomal escape, enabling the release of Cas9 RNPs, is achieved through mechanisms such as membrane disruption via the proton sponge effect [33], ionic lipid-mediated local membrane disruption [24], fusion between endosomes and LNPs [23], or other strategies. Since Cas9 RNPs cannot cross membranes via passive diffusion, carefully designed Cas9 RNP-encapsulated nanoparticles coated with appropriate vectors can effectively enter target cells through receptor-mediated endocytosis, followed by successful gene editing after endosomal escape, leveraging the biological machinery system. Intravenously administered Cas9 RNP-encapsulated nanoparticles, coated with suitable vectors such as glucose and an RVG peptide, successfully crossed the BBB via receptor-mediated transcytosis utilizing GLUT1 and α-7 AchR in capillary endothelial cells, respectively. These nanoparticles were subsequently internalized into cells, including neurons, through receptor-mediated endocytosis via α-7 AchR. Ultimately, they facilitated gene editing for central nervous system (CNS) diseases following endosomal escape [43]. Conversely, intravenously administered Cas9 RNP-encapsulated nanoparticles, coated with suitable vectors such as iRGD, entered cancer cells via receptor-mediated endocytosis mediated by αvβ3 integrins and ultimately facilitated gene editing following endosomal escape [54].

Intelligent nanoparticles coated with vectors, such as specific membrane receptor ligands that induce endocytosis/transcytosis, could address scalability, reproducibility, and safety concerns in clinical translation. Notably, NTLA-2001 (NCT04601051) is an intravenously administered lipid nanoparticle formulation encapsulating Cas9 mRNA and sgRNA targeting *TTR*. NTLA-2001 forms complexes with APOE, enabling receptor-mediated endocytosis via LDLR in the liver, ultimately leading to reduced serum TTR protein expression through *TTR* gene knockout (Figure 3) [17]. Cas9 RNP-encapsulated nanoparticles coated with appropriate vectors are expected to be clinically introduced soon. However, human germline editing remains a topic of debate, as such modifications are inherited by future generations [63]. Therefore, the use of Cas9 RNP-encapsulated nanoparticles should be strictly limited to disease treatment. Additionally, Casgevy® was clinically approved by the FDA in December 2023 for the treatment of sickle cell disease. This therapy involves using modified hematopoietic (blood) stem cells taken from the patient's bone marrow. It is the first FDA-approved treatment utilizing CRISPR/Cas9, although it does not rely on a nanodelivery system [64].

Regarding scalability [65], large-scale production must be established. Developing scalable methods for nanoparticle production with high reproducibility and consistency is essential for translating these technologies into clinical and industrial applications [66–70]. This scalability issue can be addressed by focusing on (a) establishing robust production processes that ensure nanoparticles are consistently produced at a larger scale, (b) optimizing nanoparticle synthesis to maintain high-quality standards while increasing production capacity, (c) implementing standardization and quality control measures at each stage of production, (d) automating and integrating manufacturing steps to maintain reproducibility and consistency, (e) ensuring regulatory compliance, which is crucial for successful translation into clinical and industrial applications, and (f) achieving cost efficiency for large-scale commercialization.

Regarding reproducibility [66–70], efficient and stable encapsulation methods need to be developed. Achieving effective encapsulation of RNPs while maintaining their stability remains a significant challenge. Ensuring the structural and functional integrity of RNPs during formulation and storage is crucial for their effectiveness. The optimization of encapsulation techniques is based on (a) the selection of suitable carriers, such as LNPs, and (b) encapsulation methods.

Regarding safety [71], toxicity and biocompatibility are important considerations. Certain materials used for nanoparticle functionalization, such as ligands, may induce toxicity or trigger adverse immune responses, thereby limiting their clinical applicability. (a) Utilizing highly selective ligands, such as monoclonal antibodies, CPPs, or THPs, can help prevent off-target effects caused by incorrect distribution. Additionally, (b) PEG modification on the surface of nanoparticles is known to reduce immune responses.

Nonetheless, the CRISPR/Cas9 system has the potential to create new biological machinery through precise molecular design. Medicinal chemists and pharmaceutical scientists are expected to develop innovative drugs based on Cas9 RNP-encapsulated nanoparticles to treat patients suffering from genetic disorders.

Funding: This research received no external funding.

Institutional Review Board Statement: Not applicable.

Informed Consent Statement: Not applicable.

Data Availability Statement: Data available in a publicly accessible repository. The data presented in this study are openly available in References below. ClinicalTrials.gov Identifier can be found at https://clinicaltrials.gov/ (accessed on 1 January 2025).

Acknowledgments: This review is just my opinion based on or inferred from available published articles and public knowledge. Thus, the intellectual property rights are not infringed upon.

Conflicts of Interest: The author declares no conflicts of interest.

References

1. Ran, F.A.; Hsu, P.D.; Wright, J.; Agarwala, V.; Scott, D.A.; Zhang, F. Genome engineering using the CRISPR-Cas9 system. *Nat. Protoc.* **2013**, *8*, 2281–2308. [CrossRef] [PubMed]
2. Arora, L.; Narula, A. Gene Editing and Crop Improvement Using CRISPR-Cas9 System. *Front. Plant Sci.* **2017**, *8*, 1932. [CrossRef] [PubMed]
3. Carroll, D. Genome engineering with zinc-finger nucleases. *Genetics* **2011**, *188*, 773–782. [CrossRef] [PubMed]
4. Wright, D.A.; Li, T.; Yang, B.; Spalding, M.H. TALEN-mediated genome editing: Prospects and perspectives. *Biochem. J.* **2014**, *462*, 15–24. [CrossRef]
5. Mir, A.; Edraki, A.; Lee, J.; Sontheimer, E.J. Type II-C CRISPR-Cas9 biology, mechanism, and application. *ACS Chem. Biol.* **2018**, *13*, 357–365. [CrossRef]
6. Le Rhun, A.; Escalera-Maurer, A.; Bratovič, M.; Charpentier, E. CRISPR-Cas in Streptococcus pyogenes. *RNA Biol.* **2019**, *16*, 380–389. [CrossRef]
7. Gehrke, F.; Schindele, A.; Puchta, H. Nonhomologous end joining as key to CRISPR/Cas-mediated plant chromosome engineering. *Plant Physiol.* **2022**, *188*, 1769–1779. [CrossRef] [PubMed]
8. Yang, H.; Ren, S.; Yu, S.; Pan, H.; Li, T.; Ge, S.; Zhang, J.; Xia, N. Methods favoring homology-directed repair choice in response to CRISPR/Cas9 induced-double strand breaks. *Int. J. Mol. Sci.* **2020**, *21*, 6461. [CrossRef]
9. Waddington, S.N.; Privolizzi, R.; Karda, R.; O'Neill, H.C. A broad overview and review of CRISPR-Cas technology and stem cells. *Curr. Stem Cell Rep.* **2016**, *2*, 9–20. [CrossRef]
10. Tashima, T. Smart Strategies for Therapeutic Agent Delivery into Brain across the Blood–Brain Barrier Using Receptor-Mediated Transcytosis. *Chem. Pharm. Bull.* **2020**, *68*, 316–325. [CrossRef]
11. Tashima, T. Effective cancer therapy based on selective drug delivery into cells across their membrane using receptor-mediated endocytosis. *Bioorg. Med. Chem. Lett.* **2018**, *28*, 3015–3024. [CrossRef] [PubMed]

12. Kadry, H.; Noorani, B.; Cucullo, L. A blood–brain barrier overview on structure, function, impairment, and biomarkers of integrity. *Fluids Barriers CNS.* **2020**, *17*, 69. [CrossRef] [PubMed]
13. Emran, T.B.; Shahriar, A.; Mahmud, A.; Rahman, T.; Abir, M.H.; Siddiquee, M.F.; Ahmed, H.; Rahman, N.; Nainu, F.; Wahyudin, E.; et al. Multidrug Resistance in Cancer: Understanding Molecular Mechanisms, Immunoprevention and Therapeutic Approaches. *Front. Oncol.* **2022**, *12*, 891652. [CrossRef]
14. Laughlin, C.D.; D'Aquili, E.G. *Biogenetic Structuralism*; Columbia University Press: New York, NY, USA, 1974.
15. Leavy, S.A. Biogenetic Structuralism. *Yale J. Biol. Med.* **1976**, *49*, 420–421.
16. Chen, Y.; Cai, Y.; Xu, Q.; Chen, Z.W. Atomic force bio-analytics of polymerization and aggregation of phycoerythrin-conjugated immunoglobulin G molecules. *Mol. Immunol.* **2004**, *41*, 1247–1252. [CrossRef] [PubMed]
17. Gillmore, J.D.; Gane, E.; Taubel, J.; Kao, J.; Fontana, M.; Maitland, M.L.; Seitzer, J.; O'Connell, D.; Walsh, K.R.; Wood, K.; et al. CRISPR-Cas9 In Vivo Gene Editing for Transthyretin Amyloidosis. *N. Engl. J. Med.* **2021**, *385*, 493–502. [CrossRef]
18. Vieira, M.; Saraiva, M.J. Transthyretin: A multifaceted protein. *Biomol. Concepts* **2014**, *5*, 45–54. [CrossRef] [PubMed]
19. Liu, Z.; Li, Z.; Li, B. Nonviral Delivery of CRISPR/Cas Systems in mRNA Format. *Adv. Nanobiomed. Res.* **2022**, *2*, 2200082. [CrossRef]
20. Klösgen, B.; Helfrich, W. Cryo-Transmission Electron Microscopy of a Superstructure of Fluid Dioleoylphosphatidylcholine (DOPC) Membranes. *Biophys J.* **1997**, *73*, 3016–3029. [CrossRef]
21. Walther, J.; Porenta, D.; Wilbie, D.; Seinen, C.; Benne, N.; Yang, Q.; de Jong, O.G.; Lei, Z.; Mastrobattista, E. Comparative analysis of lipid Nanoparticle-Mediated delivery of CRISPR-Cas9 RNP versus mRNA/sgRNA for gene editing in vitro and in vivo. *Eur. J. Pharm. Biopharm.* **2024**, *196*, 114207. [CrossRef] [PubMed]
22. Zhang, X.; Li, T.; Ou, J.; Huang, J.; Liang, P. Homology-based repair induced by CRISPR-Cas nucleases in mammalian embryo genome editing. *Protein Cell* **2022**, *13*, 316–335. [CrossRef]
23. Caracciolo, G.; Caminiti, R. Efficient Escape from Endosomes Determines the Superior Efficiency of Multicomponent Lipoplexes. *J. Phys. Chem. B* **2009**, *113*, 4995–4997. [CrossRef] [PubMed]
24. Hafez, I.M.; Maurer, N.; Cullis, P.R. On the mechanism whereby cationic lipids promote intracellular delivery of polynucleic acids. *Gene Ther.* **2001**, *8*, 1188–1196. [CrossRef] [PubMed]
25. Zou, Y.; Ito, S.; Yoshino, F.; Suzuki, Y.; Zhao, L.; Komatsu, N. Polyglycerol Grafting Shields Nanoparticles from Protein Corona Formation to Avoid Macrophage Uptake. *ACS Nano.* **2020**, *14*, 7216–7226. [CrossRef]
26. Torrice, M. Does nanomedicine have a delivery problem? *ACS Cent. Sci.* **2016**, *2*, 434–437. [CrossRef] [PubMed]
27. Kobayashi, H.; Watanabe, R.; Choyke, P.L. Improving Conventional Enhanced Permeability and Retention (EPR) Effects; What Is the Appropriate Target? *Theranostics* **2014**, *4*, 81–89. [CrossRef] [PubMed]
28. Wei, T.; Cheng, Q.; Min, Y.L.; Olson, E.N.; Siegwart, D.J. Systemic nanoparticle delivery of CRISPR-Cas9 ribonucleoproteins for effective tissue specific genome editing. *Nat. Commun.* **2020**, *11*, 3232. [CrossRef] [PubMed]
29. Suk, J.S.; Xu, Q.; Kim, N.; Hanes, J.; Ensign, L.M. PEGylation as a strategy for improving nanoparticle-based drug and gene delivery. *Adv. Drug Deliv. Rev.* **2016**, *99 Pt A*, 28–51. [CrossRef] [PubMed]
30. Yoshihara, A.; Watanabe, S.; Goel, I.; Ishihara, K.; Ekdahl, K.N.; Nilsson, B.; Teramura, Y. Promotion of cell membrane fusion by cell-cell attachment through cell surface modification with functional peptide-PEG-lipids. *Biomaterials* **2020**, *253*, 120113. [CrossRef]
31. Lee, Y.W.; Mout, R.; Luther, D.C.; Liu, Y.; Castellanos-García, L.; Burnside, A.S.; Ray, M.; Tonga, G.Y.; Hardie, J.; Nagaraj, H.; et al. In Vivo Editing of Macrophages through Systemic Delivery of CRISPR-Cas9-Ribonucleoprotein-Nanoparticle Nanoassemblies. *Adv. Ther.* **2019**, *2*, 1900041. [CrossRef]
32. Mout, R.; Ray, M.; Yesilbag Tonga, G.; Lee, Y.-W.; Tay, T.; Sasaki, K.; Rotello, V.M. Efficient Gene Editing through Direct Cytosolic Delivery of CRISPR/Cas9-Ribonucleoprotein. *ACS Nano* **2017**, *11*, 2452–2458. [CrossRef] [PubMed]
33. Xie, R.; Wang, X.; Wang, Y.; Ye, M.; Zhao, Y.; Yandell, B.S.; Gong, S. pH-Responsive Polymer Nanoparticles for Efficient Delivery of Cas9 Ribonucleoprotein With or Without Donor DNA. *Adv. Mater.* **2022**, *34*, e2110618. [CrossRef]
34. Öktem, M.; Mastrobattista, E.; de Jong, O.G. Amphipathic Cell-Penetrating Peptide-Aided Delivery of Cas9 RNP for In Vitro Gene Editing and Correction. *Pharmaceutics* **2023**, *15*, 2500. [CrossRef]
35. Lin, Q.; Zhou, C.R.; Bai, M.J.; Zhu, D.; Chen, J.W.; Wang, H.F.; Li, M.A.; Wu, C.; Li, Z.R.; Huang, M.S. Exosome-mediated miRNA delivery promotes liver cancer EMT and metastasis. *Am. J. Transl. Res.* **2020**, *12*, 1080–1095.
36. Wan, T.; Zhong, J.; Pan, Q.; Zhou, T.; Ping, Y.; Liu, X. Exosome-mediated delivery of Cas9 ribonucleoprotein complexes for tissue-specific gene therapy of liver diseases. *Sci. Adv.* **2022**, *8*, eabp9435. [CrossRef] [PubMed]
37. Winnard, P.T., Jr.; Kluth, J.B.; Raman, V. Noninvasive Optical Tracking of Red Fluorescent Protein–Expressing Cancer Cells in a Model of Metastatic Breast Cancer. *Neoplasia* **2006**, *8*, 796–806. [CrossRef] [PubMed]

38. Mirjalili Mohanna, S.Z.; Djaksigulova, D.; Hill, A.M.; Wagner, P.K.; Simpson, E.M.; Leavitt, B.R. LNP-mediated delivery of CRISPR RNP for wide-spread in vivo genome editing in mouse cornea. *J. Control Release* **2022**, *350*, 401–413. [CrossRef] [PubMed]
39. Ho, T.C.; Kim, H.S.; Chen, Y.; Li, Y.; LaMere, M.W.; Chen, C.; Wang, H.; Gong, J.; Palumbo, C.D.; Ashton, J.M.; et al. Scaffold-mediated CRISPR-Cas9 delivery system for acute myeloid leukemia therapy. *Sci. Adv.* **2021**, *7*, eabg3217. [CrossRef] [PubMed]
40. Zheng, L.; Bandara, S.R.; Tan, Z.; Leal, C. Lipid nanoparticle topology regulates endosomal escape and delivery of RNA to the cytoplasm. *Proc. Natl. Acad. Sci. USA* **2023**, *120*, e2301067120. [CrossRef] [PubMed]
41. Mout, R.; Ray, M.; Lee, Y.W.; Scaletti, F.; Rotello, V.M. In Vivo Delivery of CRISPR/Cas9 for Therapeutic Gene Editing: Progress and Challenges. *Bioconjug. Chem.* **2017**, *28*, 880–884. [CrossRef] [PubMed]
42. Stahl, E.C.; Sabo, J.K.; Kang, M.H.; Allen, R.; Applegate, E.; Kim, S.E.; Kwon, Y.; Seth, A.; Lemus, N.; Salinas-Rios, V.; et al. Genome editing in the mouse brain with minimally immunogenic Cas9 RNPs. *Mol. Ther.* **2023**, *31*, 2422–2438. [CrossRef] [PubMed]
43. Metzger, J.M.; Wang, Y.; Neuman, S.S.; Snow, K.J.; Murray, S.A.; Lutz, C.M.; Bondarenko, V.; Felton, J.; Gimse, K.; Xie, R.; et al. Efficient in vivo neuronal genome editing in the mouse brain using nanocapsules containing CRISPR-Cas9 ribonucleoproteins. *Biomaterials* **2023**, *293*, 121959. [CrossRef]
44. Sajjanar, B.; Saxena, S.; Bisht, D.; Singh, A.K.; Manjunatha Reddy, G.B.; Singh, R.; Singh, R.P.; Kumar, S. Effect of nicotinic acetylcholine receptor alpha 1 (nAChRα1) peptides on rabies virus infection in neuronal cells. *Neuropeptides* **2016**, *57*, 59–64. [CrossRef]
45. Lee, B.; Lee, K.; Panda, S.; Gonzales-Rojas, R.; Chong, A.; Bugay, V.; Park, H.M.; Brenner, R.; Murthy, N.; Lee, H.Y. Nanoparticle delivery of CRISPR into the brain rescues a mouse model of fragile X syndrome from exaggerated repetitive behaviours. *Nat. Biomed. Eng.* **2018**, *2*, 497–507. [CrossRef]
46. Hagerman, R.J.; Berry-Kravis, E.; Hazlett, H.C.; Bailey, D.B., Jr.; Moine, H.; Kooy, R.F.; Tassone, F.; Gantois, I.; Sonenberg, N.; Mandel, J.L.; et al. Fragile X syndrome. *Nat. Rev. Dis. Primers* **2017**, *3*, 17065. [CrossRef] [PubMed]
47. Pop, A.S.; Gomez-Mancilla, B.; Neri, G.; Willemsen, R.; Gasparini, F. Fragile X syndrome: A preclinical review on metabotropic glutamate receptor 5 (mGluR5) antagonists and drug development. *Psychopharmacology* **2014**, *231*, 1217–1226. [CrossRef]
48. Tan, E.; Wan, T.; Pan, Q.; Duan, J.; Zhang, S.; Wang, R.; Gao, P.; Lv, J.; Wang, H.; Li, D.; et al. Dual-responsive nanocarriers for efficient cytosolic protein delivery and CRISPR-Cas9 gene therapy of inflammatory skin disorders. *Sci. Adv.* **2024**, *10*, eadl4336. [CrossRef] [PubMed]
49. Tang, X.; Chen, Y.; Wang, B.; Luo, D.; Wang, J.; He, Y.; Feng, L.; Xu, Y.; Xie, S.; Chen, M.; et al. Autonomous Feedback-Driven Engineered DNAzyme-Coated Trojan Horse-like Nanocapsules for On-Demand CRISPR/Cas9 Delivery. *ACS Nano* **2024**, *18*, 13950–13965. [CrossRef] [PubMed]
50. Le, W.; Chen, B.; Cui, Z.; Shi, D. Detection of cancer cells based on glycolytic-regulated surface electrical charges. *Biophys. Rep.* **2019**, *5*, 10–18. [CrossRef]
51. Sun, H.; Chen, L.; Cao, S.; Liang, Y.; Xu, Y. Warburg Effects in Cancer and Normal Proliferating Cells: Two Tales of the Same Name. *Genom. Proteom. Bioinform.* **2019**, *17*, 273–286. [CrossRef]
52. Preta, G. New Insights Into Targeting Membrane Lipids for Cancer Therapy. *Front. Cell Dev. Biol.* **2020**, *8*, 571237. [CrossRef] [PubMed]
53. Cressman, S.; Sun, Y.; Maxwell, E.J.; Fang, N.; Chen, D.D.Y.; Cullis, P.R. Binding and Uptake of RGD-Containing Ligands to Cellular $a_v\beta_3$ Integrins. *Int. J. Pept. Res. Ther.* **2009**, *15*, 49–59. [CrossRef]
54. Deng, S.; Li, X.; Liu, S.; Chen, J.; Li, M.; Chew, S.Y.; Leong, K.W.; Cheng, D. Codelivery of CRISPR-Cas9 and chlorin e6 for spatially controlled tumor-specific gene editing with synergistic drug effects. *Sci. Adv.* **2020**, *6*, eabb4005. [CrossRef] [PubMed]
55. Wang, Y.; Wang, X.; Xie, R.; Burger, J.C.; Tong, Y.; Gong, S. Overcoming the Blood–Brain Barrier for Gene Therapy via Systemic Administration of GSH-Responsive Silica Nanocapsules. *Adv. Mater.* **2023**, *35*, e2208018. [CrossRef] [PubMed]
56. Zou, Y.; Sun, X.; Yang, Q.; Zheng, M.; Shimoni, O.; Ruan, W.; Wang, Y.; Zhang, D.; Yin, J.; Huang, X.; et al. Blood-brain barrier–penetrating single CRISPR-Cas9 nanocapsules for effective and safe glioblastoma gene therapy. *Sci. Adv.* **2022**, *8*, eabm8011. [CrossRef] [PubMed]
57. Ruan, W.; Jiao, M.; Xu, S.; Ismail, M.; Xie, X.; An, Y.; Guo, H.; Qian, R.; Shi, B.; Zheng, M. Brain-targeted CRISPR/Cas9 nanomedicine for effective glioblastoma therapy. *J. Control Release* **2022**, *351*, 739–751. [CrossRef]
58. Ma, L.; Ma, Y.; Gao, Q.; Liu, S.; Zhu, Z.; Shi, X.; Dai, F.; Reis, R.L.; Kundu, S.C.; Cai, K.; et al. Mulberry Leaf Lipid Nanoparticles: A Naturally Targeted CRISPR/Cas9 Oral Delivery Platform for Alleviation of Colon Diseases. *Small* **2024**, *20*, 2307247. [CrossRef] [PubMed]
59. Liu, A.P.; Aguet, F.; Danuser, G.; Schmid, S.L. Local clustering of transferrin receptors promotes clathrin-coated pit initiation. *J. Cell Biol.* **2010**, *191*, 1381–1393. [CrossRef]

60. Cureton, D.K.; Harbison, C.E.; Cocucci, E.; Parrish, C.R.; Kirchhausen, T. Limited transferrin receptor clustering allows rapid diffusion of canine parvovirus into clathrin endocytic structures. *J. Virol.* **2012**, *86*, 5330–5340. [CrossRef] [PubMed]
61. Kibbey, R.G.; Rizo, J.; Gierasch, L.M.; Anderson, R.G. The LDL Receptor Clustering Motif Interacts with the Clathrin Terminal Domain in a Reverse Turn Conformation. *J. Cell Biol.* **1998**, *142*, 59–67. [CrossRef] [PubMed]
62. Kida, S.; Kinoshita, M.; Tanaka, S.; Okumura, M.; Koshimura, Y.; Morimoto, H. Non-clinical evaluation of a blood-brain barrier-penetrating enzyme for the treatment of mucopolysaccharidosis type I. *Mol Genet Metab.* **2019**, *126*, S83–S84. [CrossRef]
63. Wiley, L.; Cheek, M.; LaFar, E.; Ma, X.; Sekowski, J.; Tanguturi, N.; Iltis, A. The Ethics of Human Embryo Editing via CRISPR-Cas9 Technology: A Systematic Review of Ethical Arguments, Reasons, and Concerns. *HEC Forum*, 2024; *Online ahead of print*. [CrossRef]
64. Singh, A.; Irfan, H.; Fatima, E.; Nazir, Z.; Verma, A.; Akilimali, A. Revolutionary breakthrough: FDA approves CASGEVY, the first CRISPR/Cas9 gene therapy for sickle cell disease. *Ann. Med. Surg.* **2024**, *86*, 4555–4559. [CrossRef]
65. The U.S. Food and Drug Administration, Human Gene Therapy Products Incorporating Human Genome Editing, Guidance for Industry, January 2024. Available online: https://www.fda.gov/media/156894/download (accessed on 1 January 2025).
66. Khairnar, S.V.; Pagare, P.; Thakre, A.; Nambiar, A.R.; Junnuthula, V.; Abraham, M.C.; Kolimi, P.; Nyavanandi, D.; Dyawanapelly, S. Review on the Scale-Up Methods for the Preparation of Solid Lipid Nanoparticles. *Pharmaceutics* **2022**, *14*, 886. [CrossRef]
67. Shen, Y.; Gwak, H.; Han, B. Advanced manufacturing of nanoparticle formulations of drugs and biologics using microfluidics. *Analyst* **2024**, *49*, 614–637. [CrossRef]
68. Hussain, M.; Binici, B.; O'Connor, L.; Perrie, Y. Production of mRNA lipid nanoparticles using advanced crossflow micromixing. *J. Pharm. Pharmacol.* **2024**, *76*, 1572–1583. [CrossRef] [PubMed]
69. Saffie-Siebert, S.; Torabi-Pour, N.; Gibson, A.; Sutera, F.M.; Dehsorkhi, A.; Baran-Rachwalska, P.; Quinn, S. Toward a large-batch manufacturing process for silicon-stabilized lipid nanoparticles: A highly customizable RNA delivery platform. *Mol. Ther. Methods Clin. Dev.* **2024**, *32*, 101299. [CrossRef]
70. Webb, C.; Forbes, N.; Roces, C.B.; Anderluzzi, G.; Lou, G.; Abraham, S.; Ingalls, L.; Marshall, K.; Leaver, T.J.; Watts, J.A.; et al. Using microfluidics for scalable manufacturing of nanomedicines from bench to GMP: A case study using protein-loaded liposomes. *Int. J. Pharm.* **2020**, *582*, 119266. [CrossRef] [PubMed]
71. Pharmaceuticals and Medical Devices Agency (PMDA), White-Paper for Quality and Safety for Gene Therapy Products Using Gene Editing Technology. Available online: https://www.pmda.go.jp/files/000237636.pdf (accessed on 1 January 2025).

pharmaceutics

MDPI

Article

NLC-Based Rifampicin Delivery System: Development and Characterization for Improved Drug Performance Against *Staphylococcus aureus*

Javiera Carrasco-Rojas [1], Felipe I. Sandoval [2], Christina M. A. P. Schuh [2], Carlos F. Lagos [3,4], Javier O. Morales [1,5,6], Francisco Arriagada [1,*] and Andrea C. Ortiz [3,*]

1 Departamento de Ciencias y Tecnología Farmacéutica, Facultad de Ciencias Químicas y Farmacéuticas, Universidad de Chile, Santiago 8380494, Chile; javiera.carrasco@uchile.cl (J.C.-R.)
2 Centro de Medicina Regenerativa, Facultad de Medicina, Clínica Alemana-Universidad del Desarrollo, Santiago 7550000, Chile
3 Escuela de Química y Farmacia, Facultad de Ciencias, Universidad San Sebastián, Campus Los Leones, Lota 2465, Providencia, Santiago 7510157, Chile
4 Centro Basal Ciencia & Vida, Fundación Ciencia & Vida, Av. del Valle Norte 725, Huechuraba 8580702, Chile
5 Centro de Nuevos Fármacos para Hipertensión e Insuficiencia Cardíaca, CENDHY, Santiago 8380494, Chile
6 Centro Avanzado de Enfermedades Crónicas, ACCDiS, Santiago 8380494, Chile
* Correspondence: francisco.arriagada@ciq.uchile.cl (F.A.); andrea.ortizo@uss.cl (A.C.O.)

Abstract: Background/Objectives: Rifampicin is a typical antibiotic used for the treatment of *Staphylococcus aureus* (*S. aureus*) infections; however, its clinical utility is limited by poor aqueous solubility, chemical instability, and increasing bacterial resistance. Nanostructured lipid carriers (NLCs) offer a promising strategy to improve drug solubility, stability, and antimicrobial performance. **Methods**: In this study, rifampicin-loaded NLC (NLC-RIF) was developed using a hot homogenization with a low energy method and characterized in terms of particle size, polydispersity index, zeta potential, encapsulation efficiency, colloidal stability, and drug loading. **Results:** In vitro release studies under sink conditions demonstrated a biphasic release pattern, best described by the Korsmeyer–Peppas model, suggesting a combination of diffusion and matrix erosion mechanisms. Antimicrobial activity against *S. aureus* revealed a substantial increase in potency for NLC-RIF, with an IC_{50} of 0.46 ng/mL, approximately threefold lower than that of free rifampicin. Cytotoxicity assays in HepG2 cells confirmed over 90% cell viability across all tested concentrations. **Conclusions**: These findings highlight the potential of NLC-RIF as a biocompatible and effective nanocarrier system for enhancing rifampicin delivery and antibacterial activity.

Keywords: nanostructured lipid carrier; antibiotic; rifampicin; drug release; *Staphylococcus aureus*; HepG2 cell

1. Introduction

Global concern about multidrug resistance to antimicrobials has increased in recent years. The World Health Organization has emphasized the need to develop new strategies to combat antimicrobial resistance [1]. Conventional antibiotics still suffer from limitations such as low bioavailability, rapid clearance, and poor penetration into infection sites, often aggravated by biofilm formation. These shortcomings contribute to therapeutic failures, rising resistance levels, and patient discontinuation of treatment due to lack of efficacy. To overcome these barriers, nanoscale delivery systems can be employed to encapsulate, target, and release drugs at specific sites, thereby improving treatment efficacy, reducing dosage, and enhancing patient adherence [2–4].

In this context, two main categories of nanoantibiotics have been described [5]. The first corresponds to nanosystems that have antibacterial activity in themselves, such as silver, zinc, and gold nanosystems. The second group is nanoparticles capable of conjugating antibiotics on the surface or incorporating them into the matrix of the nanosystem. Nanotechnology is an appropriate strategy to circumvent bacterial resistance mechanisms because it circumvents the mechanisms that bacteria have developed for drugs. Additionally, improves key physicochemical and biopharmaceutical properties of drugs, such as water solubility and half-life [2,6]. Furthermore, strategies can be developed to design modified releases that could reduce the frequency and dose of administration [7], and may enable synergistic effects by co-delivering multiple antibiotics in a single formulation [8,9].

Among the various nanocarrier types, solid lipid nanoparticles (SLN) and nanostructured lipid carriers (NLCs) are particularly attractive due to their low toxicity, high stability, biocompatibility, biodegradability, and ability to encapsulate both hydrophilic and lipophilic drugs [10]. Moreover, their manufacturing is cost-effective and readily scalable [11].

Specifically, NLCs mitigate burst release during storage by their reduced recrystallization rate. This arises from a less-ordered lipid matrix formed by mixing solid and liquid lipids, which creates additional spaces for drug accommodation [12]. Accordingly, NLCs have emerged as versatile platforms to optimize the delivery of various therapies, including antibiotics [13]. Moreover, these nanoparticles can be fabricated using a combination of lipids and polymer-lipid conjugates, including polyethylene glycols (PEGs) of varying lengths and end-moieties or functionalized with different polymers [14,15]. This approach enables multiple functionalizations with various agents to direct the nanosystem to specific targets [16]. Therefore, they combine excellent biocompatibility and long-term drug retention with surface properties amenable to targeting ligands, while their nanoscale dimensions facilitate enhanced interactions with bacterial membranes [17,18].

Rifampicin is a hydrophobic, zwitterionic antibiotic characterized by both acidic and basic functional groups, pKa 1.7 and 7.9, respectively. It is regarded as one of the most potent bactericidal agents and is widely used to treat infections caused by surface-adherent microorganisms [19]. Its therapeutic efficacy is attributed to its broad-spectrum antimicrobial activity, which includes most Gram-positive bacteria and select Gram-negative species. However, rifampicin is not exempt from pharmacokinetic drawbacks, including variable solubility and instability in different conditions of pH, which may limit their efficacy against resistant strains and intracellular pathogens [20]. Rifampicin remains a cornerstone of tuberculosis therapy; in 2012, an estimated 450,000 cases of multidrug-resistant tuberculosis were reported globally, 10% of which were extensively drug-resistant [21]. Resistance to rifampicin in *Staphylococcus aureus* remains relatively low (~6.2%), yet vigilance is warranted given its potential to increase over time [22]. Therefore, strategies to reduce the emergence of antimicrobial resistance are urgently needed.

The high lipophilicity of rifampicin (LogP 2.83) favors its encapsulation within lipid matrices [10], while its ionizable groups support the controlled release into aqueous environments. Although rifampicin-loaded nanosystems have been evaluated against *Mycobacterium tuberculosis* [23], *Escherichia coli* [24], *Pseudomonas aeruginosa* [25], *Klebsiella pneumoniae* [26], and *Staphylococcus aureus* [27]. To our knowledge, no study has yet reported rifampicin encapsulation in NLCs specifically against *Staphylococcus aureus* [28–30].

In this study, we developed nanostructured lipid carriers (NLCs) for the encapsulation and controlled release of rifampicin, with a primary focus on evaluating their biological performance. We conducted a comprehensive biological assessment, including antimicrobial activity against *Staphylococcus aureus* and cytotoxicity using a hepatic cell model to investigate biocompatibility. Additionally, physicochemical characterization and colloidal stability of the nanosystems, along with in vitro release kinetics, were evaluated to support the biological findings. Our results contribute to the growing evidence supporting NLCs as promising platforms for advanced antibiotic delivery, offering a potential route to enhance drug performance and address the ongoing challenge of bacterial resistance.

2. Materials and Methods

2.1. Materials

Gelucire® 44/14 was donated by Gattefosse. Miglyol® 812, Tween® 80, rifampicin (≥95% HPLC), dimethyl sulfoxide (DMSO, Molecular Biology), 3-(4,5-dimethylthiazol-2-yl)-2,5-diphenyltetrazolium (MTT, ≥97.5%), Amicon® Ultra 10 KDa, and triple sugar iron medium (TSI) were purchased from Merck (Merck KGaA, Darmstadt, Germany). SnakeSkinTM Dialysis Tubing of cellulose membrane with a MWCO 10 kDa was purchased from Thermo Fisher Scientific (Thermo Fisher Scientific, Carlsbad, CA, USA). Acetic acid (HPLC grade) and acetonitrile (HPLC grade) were purchased from PanReac AppliChem ITW Reagents (Darmstadt, Germany). Dulbecco's Modified Eagle Medium (DMEM high-glucose), phosphate-buffered saline (PBS, without calcium and magnesium), fetal bovine serum (FBS), and penicillin/streptomycin solution were obtained from Cytiva. Trypsin-EDTA solution (0.25%) was purchased from Gibco. Ultra-pure water (18.2 MΩcm) was produced using a Simplicity System from Millipore. All materials were used as received without any further purification.

2.2. Preparation of NLCs

The fabrication process was described in a previous report [31]. The NLC is composed of Gelucire® 44/14, Miglyol® 812, Tween® 80, and water. Briefly, for a 100% formulation, the lipid phase consisted of 4% *w*/*w* Gelucire® 44/14, 1% *w*/*w* Miglyol® 812, and 2% *w*/*w* Tween® 80, which were mixed and heated to 85 °C. Meanwhile, the aqueous phase, comprising 93% *w*/*w* water heated to 85 °C, was slowly added to the lipid phase under constant stirring at 400 rpm. Then the mixture was then cooled rapidly at 4 °C for at least 30 min without stirring. To prepare rifampicin-loaded NLC, rifampicin (50 mg) was incorporated into the lipid phase, and the process continued as described. We previously determined the optimal amount of rifampicin loaded into the nanocarrier. The resulting formulation was designated NLC-RIF.

2.3. Incorporation Efficiency and Drug Loading

To quantify the different drugs incorporated in NLC, the formulations were ultracentrifuged in 10,000 Da MWCO Amicon® Ultra for 15 min at 6900 rcf. Then, the filtered section was separated and quantified through HPLC.

The rifampicin method was adapted from Goutal and collaborators [32]. Chromatographic analysis was performed using a Shimadzu Prominence HPLC system (Kyoto, Japan). The separation was achieved using an Inertsil ODS-3 C18 column (GL Science Inc., Tokyo, Japan) (4.6 mm × 250 mm, 5 μm). The mobile phase was composed of water/acetic acid 0.01% (phase A) and acetonitrile/acetic acid 0.01% (phase B). The gradient elution mode considers a change of phase B to phase A increased linearly from 30% to 60% from 0 to 10 min. The flow was set at 1.2 mL/min while the injection volume was 50 μL. Rifampin was detected at 7.3 min during a total run time of 13 min at a wavelength of 335 nm.

The drug loading and incorporation efficiency percentages were determined according to the following Equations (1) and (2), respectively:

$$\% \text{ Incorporation efficiency} = \frac{\text{Rifampicin}_{\text{total}} - \text{Rifampicin}_{\text{free}}}{\text{Rifampicin}_{\text{total}}} * 100 \quad (1)$$

$$\% \text{ Drug loading} = \frac{\text{Rifampicin}_{\text{total}} - \text{Rifampicin}_{\text{free}}}{\text{Solid lipid} + \left(\text{Rifampicin}_{\text{total}} - \text{Rifampicin}_{\text{free}}\right)} * 100 \quad (2)$$

where $\text{Rifampicin}_{\text{total}}$ and $\text{Rifampicin}_{\text{free}}$ are initial drug amounts, incorporated into the NLC-RIF preparation. The unloaded drug amount corresponds to the evaluated by the corresponding HPLC method.

2.4. Physicochemical Characterization

To determine hydrodynamic diameter (HD), polydispersity index (PdI), and zeta potential (ZP), a Malvern Zetasizer Nano ZS equipment (Malvern Panalytical Ltd., Malvern, UK) was used. HD and PdI were measured by dynamic light scattering (DLS) and ZP was determined by Doppler laser microelectrophoresis. Samples were measured in an aqueous medium with a dilution of 1:10 and each determination was measured three times corresponding to the average of eleven determinations. The condition of the medium presents a viscosity of 0.08872 cP, refractive index of 1.330, and sample refractive index of 1.333 with a wavelength of 633 nm with a detection angle of 173° and equilibration time of 120 s. Disposable cuvettes (model DTS0012) and disposable folded capillary cells (model DTS1060) were used for HD and ZP measurements, respectively.

On the other hand, to determine the formulation concentration, a nanoparticle tracking analysis (NTA) NanoSight NS300 (Malvern Panalytical, Malvern, UK) was used. The diluted samples were injected with sterile syringes into the sample chamber equipped with a 532 nm diode laser (green). All samples were measured in a single shutter and gain mode for 90 s with manual shutter, gain, brightness, and threshold adjustments at room temperature.

2.5. Colloidal Stability Assessment

The colloidal stability of all types of nanoparticles was determined over 1, 2, 3, 4, 7, and 8 weeks at 4 °C. In each case, HD, PdI, and ZP were registered.

2.6. Morphology of Nanoparticles

To visualize the NLC and NLC-RIF morphology, transmission electron microscopy was performed. Briefly, NLC and NLC-RIF were diluted 100 times with water, deposited on film-coated copper grids, stained with 1% phosphotungstic acid for 2 min, washed with water, and dried for 48 h at room temperature. The samples were evaluated on an Inspect F50 Scanning Transmission Electron Microscope (FEI, Hillsboro, OR, USA).

2.7. Drug Release Studies

Release of rifampicin-loaded NLC and free rifampicin was measured using the dialysis bag method. SnakeSkin ™ dialysis bag of 10 kDa MWCO was embedded in 10 mL of PBS (pH 7.4) containing 0.1% of Tween® 80 at 37 °C. Samples (1 mL) were withdrawn at selected time intervals for 72 h and replaced with 1 mL of fresh medium. The amount of release drugs was quantified in triplicate by HPLC methods described previously. Once the data were obtained, kinetic models were used to fit the release profile of studies and the coefficient of determination (R^2) and Akaike information criterion (AIC) were used to select the model with the best fit.

2.8. Bacterial Strain and Culture Conditions

The reference strain *Staphylococcus aureus* ATCC 25923 (American Type Culture Collection, Manassas, VA, USA) was used for all microbiological assays. Bacterial stocks were maintained at −80 °C and routinely cultured in TSI at 37 °C under aerobic conditions.

2.9. Antibacterial Activity Assay

To evaluate the antibacterial activity of NLC, NLC-RIF, and free rifampicin, a fresh bacterial suspension was prepared on Falcon® 5 mL Round Bottom Polystyrene Test Tube at a turbidity equivalent to 0.5 McFarland standard, using a DensiCheck Plus turbidimeter (BioMérieux, Marcy-l'Étoile, France). An inoculum corresponding to 1.5×10^7 colony-forming units (CFUs) was dispensed into each well of a sterile 96-well microplate, followed by the addition of treatment formulations at final concentrations ranging from 0.5 ng/mL to 1 µg/mL in a total volume of 200 µL of TSI per well.

Plates were incubated for 18 h at 37 °C under normoxic conditions. Bacterial growth was quantified by measuring absorbance at 600 nm using a microplate reader (Multiskan Sky, Thermo Scientific, Waltham, MA, USA). The percentage of growth inhibition relative to untreated controls was calculated for each concentration, and dose–response curves were fitted to determine the half-maximal inhibitory concentration (IC_{50}) values. All experiments were performed independently at least three times, each in biological quintuplicates.

2.10. Metabolic Activity Assay

To evaluate the biosafety of rifampicin, a human hepatocellular carcinoma cell line (HepG2, ATCC® HB-806), kindly provided by Dr. Carlos F. Lagos, was used to assess cell viability. Cells were cultured under standard conditions (37 °C, 5% CO_2, humidified atmosphere) in DMEM high-glucose supplemented with 10% fetal FBS and penicillin/streptomycin solution. When cultures reached approximately 90% confluence, cells were washed twice with 1X PBS (0.0067 M) and detached by incubation with 0.25% trypsin-EDTA solution at 300× *g* for 5 min. Subsequently, cells were seeded at a density of 1×10^4 cells per well in 96-well plates. Treatments were applied, including NLC, NLC-RIF, and free rifampicin, at increasing concentrations ranging from 0.5 ng/mL to 1 µg/mL.

After 24 h of incubation, cell viability was assessed using the MTT assay at a final concentration of 0.5 mg/mL. Following 1.5 h of incubation at 37 °C, the resulting formazan crystals were dissolved in DMSO, and absorbance was measured at 570 nm using a SYNERGY H1 microplate reader (BioTek Inc., Winooski, VT, USA) with Gen5 v3.16 software. Statistical analysis was performed using two-way ANOVA with multiple comparisons in GraphPad Prism (10.0), and differences were considered statistically significant at $p < 0.05$.

2.11. Data Analysis and Statistics

The experiments were performed at least in triplicate, and the data were described as the mean value ± standard deviation (SD). The software used for statistical data analysis was GraphPad Prism software version 10.4.2. Statistical analysis was performed using two-way ANOVA with multiple comparisons and differences were considered statistically significant at $p < 0.05$. In the case of kinetic fit, the DDSolver add-Inn Microsoft Excel program was used.

3. Results and Discussion

3.1. NLC Physicochemical Properties

In this study, the biological evaluation of rifampicin-loaded nanostructured lipid carriers (NLC-RIF) was conducted, focusing on their antimicrobial activity against *Staphylococcus aureus* and cytotoxicity in HepG2 cells. Physicochemical characterizations were performed to evidence the suitable characteristics of the nanocarrier and its colloidal stability, alongside an assessment of release kinetics, to support the biological findings. NLC and NLC-RIF were successfully prepared by a low-energy hot-emulsion method. The results show that both nanomaterials exhibit nanometric mean hydrodynamic diameter (Table 1), and low polydispersity indices, indicative of a highly monodisperse distribution. Antibiotic loading produced a modest increase in size, yet both systems remained well below 250 nm. It is well established that nanoparticles smaller than ~280 nm can more readily navigate biological barriers—such as mucosal layers and tight endothelial junctions—to access infection sites [33]. However, achieving effective therapeutic delivery requires more than an optimal size; colloidal stability, surface charge, and the ability to protect the cargo until reaching the target are equally critical determinants of in vivo performance [34]. In this work, the low-energy fabrication yielded particles with hydrodynamic diameters below 130 nm and zeta potentials around −4 to −6 mV, balancing steric and electrostatic stabilization without compromising biocompatibility. Moreover, these dimensions compare favorably with other rifampicin-loaded lipid carriers, which often exceed 200 nm in diameter and may face clearance or penetration limitations [35]. This scalable, energy-efficient approach therefore produces NLCs ideally sized for systemic administration and targeted antibiotic release.

Table 1. Characterization of NLC and NLC-RIF through hydrodynamic diameter (nm ± SD), PdI (±SD), and zeta potential (mV ± SD) by dynamic light scattering (DLS) and particle concentration by nanoparticle tracking analysis (NTA).

Nanomaterial	Hydrodynamic Size ± SD (nm)	Polydispersity Index ± SD	Zeta Potential ± SD (mV)	Particle Concentration ± SD (Particle/mL)
NLC	98.6 ± 2.2	0.06 ± 0.03	−4.9 ± 0.1	$2.1 \times 10^{17} \pm 1.8 \times 10^{16}$
NLC-RIF	121.5 ± 2.0	0.18 ± 0.03	−3.4 ± 1.6	$8.7 \times 10^{16} \pm 1.2 \times 10^{15}$

On the other hand, NLCs exhibited near-neutral zeta potentials, which alone could compromise colloidal stability by reducing electrostatic repulsion and promoting particle aggregation. However, Gelucire® 44/14 and Tween® 80 in the lipid matrix provide robust steric stabilization: the polyethylene glycol moieties of Gelucire® 44/14 coat the particle surface, while Tween® 80 lowers interfacial tension, together preventing agglomeration. Upon rifampicin loading, the zeta potential became slightly less negative, a change attributable to the drug's protonated aliphatic amine groups at pH 5.5, which partially neutralized the surface charge (pKa 7.9). Consequently, despite low electrostatic zeta values, the dual steric–electrostatic stabilization mechanism maintains the colloidal integrity of both NLC and NLC-RIF.

According to Guo and colleagues, one of the main limitations in the application of nanomedicine is the lack of pure and well-characterized nanoparticles. In this regard, accurate population size determination is known to be relevant and crucial for drug delivery systems [36]. Nanoparticle tracking analysis (NTA) measurements revealed that empty NLC and rifampicin-loaded NLC (NLC-RIF) differ markedly in both mean size and distribution width. Empty NLC exhibited a mean hydrodynamic diameter of

84.3 ± 1.1 nm, a mode of 76.5 ± 1.0 nm, and a standard deviation of 21.4 ± 3.0 nm. Its D10, D50, and D90 values indicated a relatively narrow, monomodal distribution (Table 2). Upon incorporation of rifampicin, the mean diameter increased to 110.2 ± 3.9 nm with a mode of 89.1 ± 5.5 nm and a larger standard deviation of 33.9 ± 2.8 nm. Correspondingly, D10, D50, and D90 all shifted toward higher values, and the D90–D10 span grew from ~32 nm to ~84 nm, reflecting a broader distribution and a pronounced tail of larger particles. These findings indicate that rifampicin loading leads to particle enlargement and increased polydispersity, likely due to interfacial localization of the drug and reorganization of the lipid–surfactant shell.

Table 2. Analysis of NLC and NLC-RIF size distributions using NTA.

Nanosystem	Nanoparticle Tracking Analysis (nm ± SD)		
	D10	D50	D90
NLC	84.3 ± 1.1	80.0 ± 0.2	100.4 ± 2.6
NLC-RIF	75.3 ± 2.2	101.7 ± 4.6	159.1 ± 8.9

On the other hand, the colloidal stability studies demonstrate that the hydrodynamic size and ζ-potential persist without significant changes during at least eight weeks at 4 °C (Figure 1). Blank NLC maintained a mean diameter of 140–145 nm throughout the study (140 ± 2 nm at week 1 vs. 144 ± 2 nm at week 8), whereas NLC-RIF exhibited a modest increase from 185 ± 3 nm to 205 ± 5 nm over the same period. ζ-Potential values for blank NLC remained consistently around −11.5 ± 0.5 mV at week 1 and −12.0 ± 0.6 mV at week 8, indicative of stable electrostatic repulsion, while NLC-RIF showed a slightly less negative surface charge, consistent with partial neutralization by rifampicin at the lipid–water interface. The polydispersity index of both formulations remained below 0.2 throughout the entire eight-week evaluation at 4 °C, demonstrating a consistently narrow size distribution. These results indicate that both NLC and NLC-RIF maintain satisfactory colloidal stability under refrigerated storage.

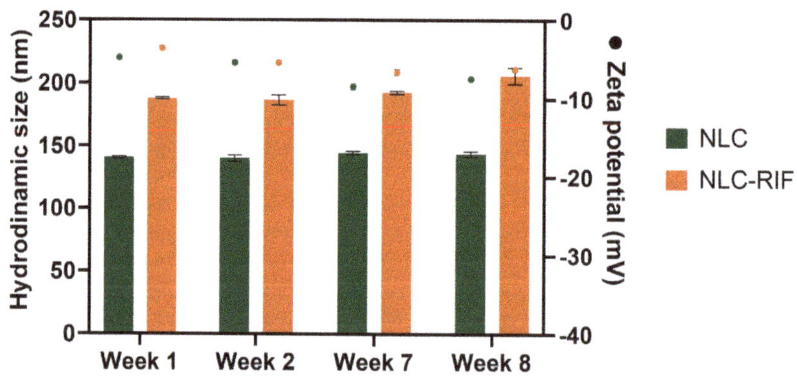

Figure 1. Colloidal stability of NLC and NLC-RIF at 4 °C assessed by hydrodynamic diameter and zeta potential (mean ± SD). n = 3.

To investigate the shape and morphology of NLC and NLC-RIF, transmission electron microscopy (TEM) was employed. Both samples exhibited a spherical shape with a monodisperse size distribution, demonstrating the reproducibility of the preparation process (Figure 2). A slight discrepancy was observed between the sizes measured by dynamic light scattering (DLS) and TEM. This difference arises because TEM images reflect the size of the nanoparticles in their solid state after drying, which can lead to a flattened shape and slight deformation at the matrix edges.

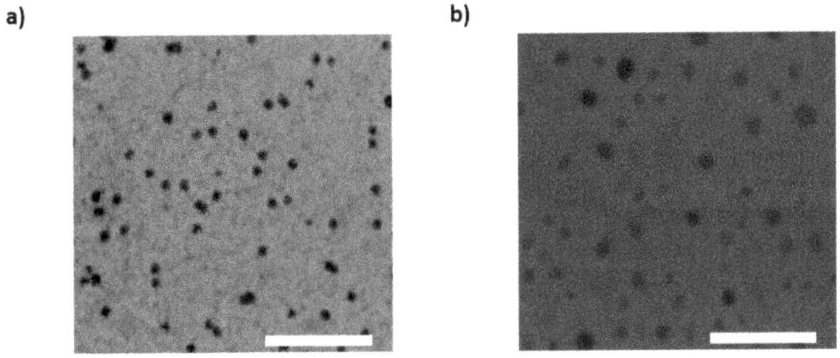

Figure 2. Transmission electron microscopy (TEM) micrograph of (**a**) NLC and (**b**) NLC-RIF. Scale bar: 2 μm.

3.2. Incorporation Efficiency and Drug Loading

Rifampicin is a molecule with pH-dependent low water solubility, reported as 0.31, 0.87, and 1.4 mg/mL at pH 4.0, 7.0, and 9.0, respectively [37]. Additionally, rifampicin exhibits a high log p value, which explains its high lipophilicity. However, it contains functional groups such as -OH groups (phenols), an -NH group, and ketone/quinone functions, which give it a certain capacity to form hydrogen bonds, so it is not completely insoluble in water. Therefore, it can behave as an amphipathic molecule in certain environments, which is useful for its incorporation into systems such as lipid nanoparticles (NLC, SLN, liposomes, etc.).

In this case, rifampicin was readily incorporated into the NLCs without complications, yielding an encapsulation efficiency of 98.1 ± 0.6% and a drug loading of 4.5 ± 0.3%. These results indicate that 59.6 μmol of rifampicin is encapsulated, ensuring that 1.64 mg of rifampicin can be available per 1 mL of formulation, independent of the pH of the medium. Consequently, nearly all of the rifampicin introduced during formulation became encapsulated, eliminating the need for organic solvents to enhance loading. While some reports employ organic solvents to improve rifampicin incorporation [38,39], such strategies necessitate additional solvent-removal steps that may leave residual traces and adversely affect in vitro or in vivo performance. In contrast, our method produced a solvent-free NLC-RIF formulation with high encapsulation efficiency.

Regarding the location of drugs within such nanostructures, the literature highlights several key considerations. Firstly, a correlation has been reported between the proportion of the liquid lipid used in the formulation and the drug's location within the matrix. Specifically, it has been demonstrated that when less than 10% liquid lipid is employed, it forms a distinct drug-loaded phase located at the edges of the nanostructure matrix [12]. Secondly, the emulsification and cooling temperatures significantly affect the initial drug incorporation and its retention within the matrix during storage. During emulsification, typically conducted at approximately 85 °C, rifampicin's aqueous solubility may slightly increase. As the formulation cools, rifampicin rapidly loses aqueous solubility, leading to enhanced interactions with the lipid phase and subsequent entrapment within the lipid matrix [40]. Additionally, the literature indicates that the cooling process can influence the nanostructure's conformation. In this context, slow cooling rates (<1–2 °C/min) promote the formation of a β conformation, resulting in a more stable nanostructure and reduced likelihood of drug expulsion from the matrix [41]. Therefore, in this case, rifampicin is expected to be predominantly dispersed within the liquid lipid at the edges of the matrix, which should exhibit a β conformation.

3.3. Drug Release and Kinetic Fitting

The release of rifampicin loaded in NLC and free rifampicin at pH 7.4 and 37 °C is presented in Figure 3. Free rifampicin achieves complete diffusion in approximately 2 h. In contrast, rifampicin release from NLC reaches about 35% within 72 h, corresponding to approximately 1.94 mg of rifampicin. The results show that there is no abrupt release of rifampicin; this is attributed to the drug being incorporated into the lipid matrix and remaining there. This responds to the conformation of an NLC since by containing a liquid lipid, it forms an imperfect matrix that allows for a less structured conformation inside, thus preventing the abrupt expulsion of the drug from the interior of the matrix [12]. In this case, the type of NLC corresponds to type I, since they contain mono-, di-, and triglycerides from Gelucire® 44/14, which generate multiple spaces providing a more suitable environment for the incorporation of the drug [16]. In general, there is little background in the literature on rifampicin loaded into lipid systems, so it is difficult to find direct comparisons; however, there are some precedents. For example, the literature shows that rifampicin loaded in a solid lipid nanoparticle (SLN) presents a burst release, possibly because the nanosystem design only includes solid lipids, therefore the internal conformation is completely compacted and rapidly expels drug molecules to the outside [42]. In another case, in the work conducted by Aliyazdi, et al., they fabricated lipid nanocapsules loaded with rifampicin; and release studies showed a release similar to that obtained in our study, but in a shorter period of time [43]. Meanwhile, the release of rifampicin from polyelectrolyte nanoparticles has been evaluated and in that case, it reaches about 60% at pH 7.4 [19]. Rifampicin is an oxygen-sensitive molecule; therefore, various articles have been described in the literature that use an antioxidant during release tests. Additionally, it has been shown that rifampicin undergoes oxidation reactions usually at neutral or basic pH more rapidly than in acidic media. However, it has been shown that there is no significant difference in the antimicrobial activity of free rifampicin and its oxidized form [44].

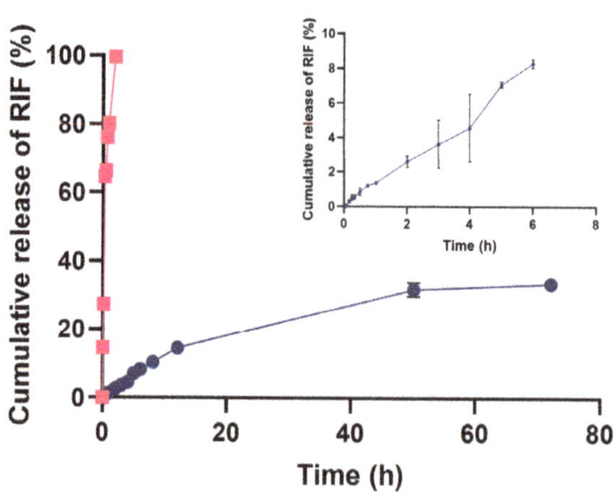

Figure 3. Cumulative release of free rifampicin (magenta line) and rifampicin-loaded NLC (blue line) at 37 °C and pH 7.4 over 72 h (mean ± SD). Inset: release of rifampicin-loaded NLC up to 6 h. n = 3.

The kinetic release data for NLC-RIF were fitted to zero-order, first-order, and Korsmeyer–Peppas models Table 3. The results show that the Korsmeyer–Peppas model provided the best overall description of the release profile (k_{KP} = 4.07; n = 0.99; R^2 = 0.986; AIC = 56.6), as indicated by its markedly lower AIC value compared with the other models. The diffusional exponent (n = 0.99) corresponds to case II transport, consistent with a release mechanism governed primarily by matrix relaxation and/or erosion rather than simple Fickian diffusion. These results demonstrate that rifampicin release from the NLCs follows an ostensibly zero-order regime in its principal phase, controlled by lipid matrix dynamics.

Table 3. Kinetic fit evaluation to NLC-RIF at 37 °C. In equations, Q_0 and Q_t are the initial amount of rifampicin and the amount of rifampicin dissolved at time t, respectively. Q_t/Q_∞ is the fractional release of rifampicin, K_0 is the zero-order constant, K_1 is the first-order constant, k_{KP} is the Korsmeyer–Peppas constant and n is the diffusional exponent.

	Zero-Order			First Order			Korsmeyer–Peppas			
	k_0 (%h^{-1})	R^2	AIC	k_1 (h^{-1})	R^2	AIC	k_{KP}	*n*	R^2	AIC
NLC-RIF	1.64	0.879	130.2	0.04	0.995	75.3	4.07	0.99	0.986	56.6

3.4. Antimicrobial Activity

The antimicrobial activity of NLC-RIF against *Staphylococcus aureus* (*S. aureus*) was evaluated through a non-linear regression analysis using a four-parameter logistic (4-PL) model without constraints (Figure 4a). The fitted curve demonstrated a high goodness of fit, with an R^2 value of 0.9791. The IC_{50} value obtained was 0.0004638 µg/mL, corresponding to a $LogIC_{50}$ of −3.334. The Hill slope was −4.730, indicating inhibitory activity with moderate cooperativity and a relatively steep response curve. The maximum viability observed (Top) was 90.81%, while the minimum viability achieved (Bottom) was 4.215%, resulting in a span of 86.59%. Confidence intervals (95% CI) were narrow for most parameters, supporting the robustness of the model. Altogether, these results suggest that the NLC-RIF nanosystem exhibits potent and efficient antibacterial activity against *S. aureus*, achieving substantial bacterial inhibition at very low concentrations. On the other hand, rifampicin shows a fit yielded an IC_{50} of 0.0019 µg/mL with a Hill coefficient of −19.06, suggesting extremely rapid and cooperative inhibition, characterized by an almost abrupt response curve (Figure 4b). In addition to its markedly lower IC_{50} and steeper Hill slope, the NLC-RIF formulation exhibits a substantially narrower dynamic range than free rifampicin: the transition from ~80% to ~20% bacterial viability occurs over only ~0.2 log µg/mL for NLC-RIF, versus ~0.4 log µg/mL for the unencapsulated drug. This narrow window underscores a pronounced threshold-type response induced by lipid encapsulation, which may help minimize the gap between subtherapeutic and potentially toxic doses allowing it to be useful for applications where effective inhibition is desired but with some modulation. Moreover, both curves display small standard errors at their upper and lower plateaus, attesting to the high reproducibility of the assay, and both NLC-RIF and free rifampicin achieve virtually complete bactericidal activity ($\leq$5% viability) at concentrations $\geq 10^{-2}$ mg/mL. These findings reinforce that NLC encapsulation not only enhances potency but also enhances dose–response control and experimental consistency.

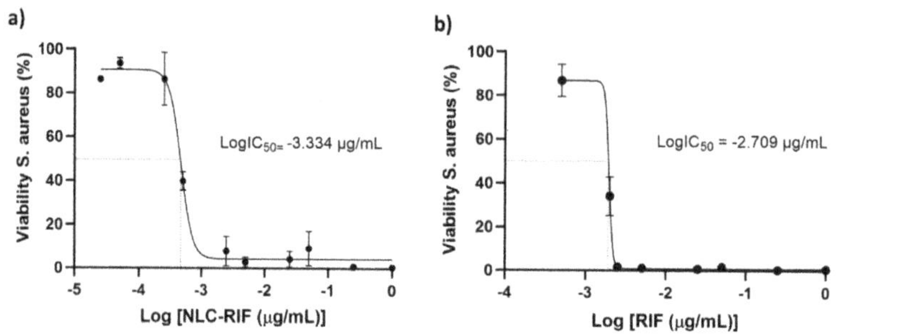

Figure 4. Viability study of *Staphylococcus aureus* at concentrations of 1.0 to 0.0005 μg/mL of rifampicin. (**a**) NLC-RIF and (**b**) RIF. n = 5.

Although a complete inhibition of bacterial growth (0% viability) was not achieved under the experimental conditions, the concentration corresponding to the minimal observed bacterial viability (~4%) was estimated at 1.0 μg/mL, suggesting a near-MIC behavior for NLC-RIF nanosystems.

This outcome may be attributed to two factors: intrinsic rifampicin resistance in *S. aureus* and/or the ability of NLCs to overcome this resistance, thereby enhancing antimicrobial efficacy. Rifampicin has been shown to inhibit transcription by interfering with the beta subunit of RNA polymerase; therefore, the mechanism of rifampin resistance in *S. aureus* is determined by mutations in the RpoB gene, which encodes the B subunit of RNA polymerase [45]. The most common mutations are those that cause changes in the amino acid sequence of the RpoB protein [46]. In terms of the interaction of the nanosystem with the microorganism, this result is consistent with what has been reported in the literature, since Gram-positive bacteria have a thin layer of peptidoglycan with teichoic acid and pores, which allows easy penetration of NLC and cause cell membrane damage and cell death [34,47]. In addition, the increase in antimicrobial activity is attributed to the NLC lipids diffusing inward through the thick peptidoglycan layer causing a disruption in the bacteria and enhancing the effect of the antibiotic [48]. For this reason, a lower concentration of rifampicin is required when loaded into the NLC to achieve the same effect as free rifampicin.

In relation to the data obtained in the literature, the group of Tran in 2018 evaluated non-lamellar lyotropic liquid crystalline nanoparticles loaded with rifampicin and obtained an IC_{50} between 0.01 and 0.03 μg/mL [49]. Therefore, the system proposed by our group requires a lower concentration of loaded rifampicin to decrease the IC_{50} by two orders of magnitude. There are more reports in the literature of rifampicin-loaded nanosystems; however, these are nanoparticles that are tested on strains other than *S. aureus* [50,51], or of a different nature than inorganic nanoparticles [52,53], or to evaluations of intracellular infection [54,55], for different purposes [56–60], so they are not comparable with this research.

3.5. *Cytotoxicity Outcomes*

To assess the potential cytotoxicity of the NLC-RIF formulation, a cell viability assay was performed using HepG2 cells exposed to various concentrations of free rifampicin, blank NLCs, and NLC-RIF in a range of 1 to 0.0005 μg/mL, expressed as rifampicin concentration (Figure 5). The results showed that cell viability remained above 90% under all tested conditions, with no significant differences observed between treatments. Neither free rifampicin, blank NLCs, nor NLC-RIF negatively affected the viability of HepG2 cells, even at the highest tested concentration (1 µg/mL), indicating good cellular tolerability of the nanostructured formulation.

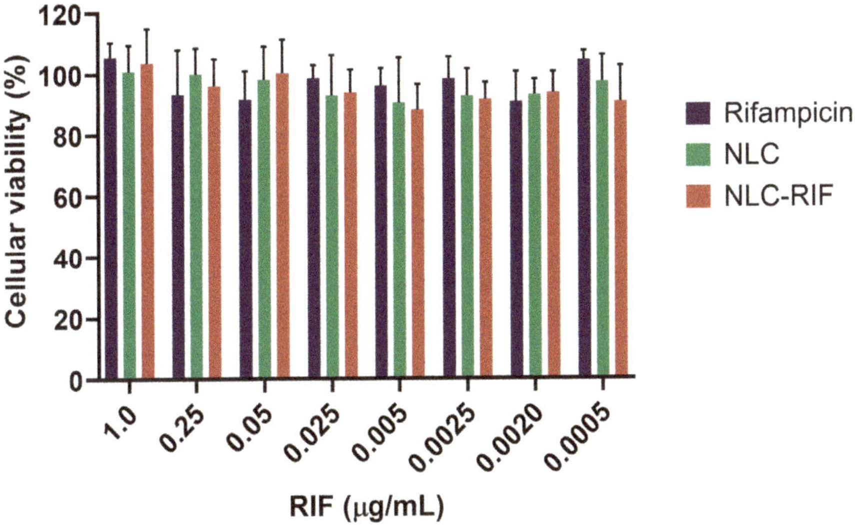

Figure 5. Metabolic effect of rifampicin and NLC formulations treatment on HepG2 cells. Cell viability of HepG2 cells after 24 h of exposure to increasing concentrations of rifampicin (RIF), blank nanostructured lipid carriers (NLCs), and rifampicin-loaded NLC (NLC-RIF). Cells were treated with concentrations ranging from 0.5 ng/mL to 1 µg/mL. Cell viability was assessed using the MTT assay and expressed as a percentage relative to untreated controls. No statistically significant differences were observed among treatments at any concentration tested. Data are presented as mean ± standard deviation (n = 4).

The excellent cell viability observed in HepG2 cells following exposure to NLC-RIF indicates that the formulation does not exert significant cytotoxic effects at concentrations up to 1 µg/mL. This result is consistent with the NLC formulation, as Gelucire® 44/14 has been recognized as GRAS (generally recognized as safe) [61]; Miglyol® 812 has been studied as a safe excipient [62,63]. In the case of Tween® 80, its surfactant properties can disrupt cell membranes [64,65]; however, this material interacts with the NLC surface, and at its present concentration, it is not sufficient to cause destabilization and subsequent cell death.

The comparable viability values across free rifampicin, blank NLCs, and NLC-RIF suggest that encapsulation of rifampicin within the lipid matrix does not introduce additional cytotoxicity. These findings support the suitability of NLC-RIF for further development as a safe delivery system, particularly for therapeutic applications requiring systemic exposure or long-term treatment.

A study evaluating an NLC against HepG2 cells is reported in the literature; however, the objective of this study was to investigate an anticancer NLC. Despite this, based on the published results, it can be concluded that NLC cytotoxicity is not predominant [66]. Gundogdu and colleagues developed an NLC with two types of Gelucire®, oleic acid, compritol, Span® 80, Tween® 80, and lipoid derivatives; therefore, the materials are similar to those used by our group. In this case, the nanosystem was studied in a gastric adenocarcinoma cell line. Although the authors do not show the cytotoxicity of NLC, all nanocarriers showed cell viability above 80% [67]. On the other hand, the Chandan group manufactured an NLC composed of glyceryl monostearate, castor oil, and Pluronic F68.

The authors evaluated cytotoxicity using MTT in Caco-2 cells and determined that the unloaded NLCs exhibited cell viability greater than 80% at all concentrations tested [68].

4. Conclusions

The development of nanostructured lipid carriers loaded with rifampicin resulted in a stable and biocompatible nanosystem with enhanced antibacterial activity against *Staphylococcus aureus*. The NLC-RIF formulation exhibited favorable physicochemical characteristics, sustained drug release, and a significantly lower IC_{50} compared to free rifampicin, indicating improved potency. In addition, the system maintained excellent colloidal stability over extended storage and demonstrated high cell viability in hepatic cells, supporting its safety for biomedical use. These findings position NLC-RIF as a promising platform for targeted antibiotic delivery. Future work should explore the in vivo pharmacokinetics, biodistribution, and therapeutic efficacy of this nanosystem, as well as its applicability against other clinically relevant bacterial strains and biofilm-associated infections.

Author Contributions: Conceptualization, A.C.O.; methodology, A.C.O., F.A., J.C.-R. and F.I.S.; software, A.C.O., F.A. and J.C.-R.; validation, A.C.O., F.A. and J.C.-R.; formal analysis, A.C.O. and J.C.-R.; investigation, A.C.O., F.A., J.C.-R. and F.I.S.; resources, A.C.O., F.A., J.O.M., C.M.A.P.S. and C.F.L.; data curation, A.C.O., F.A. and J.C.-R.; writing—original draft preparation, A.C.O., F.A. and J.C.-R.; writing—review and editing, A.C.O., F.A., J.C.-R., J.O.M., C.M.A.P.S. and C.F.L.; visualization, A.C.O.; supervision, A.C.O.; project administration, A.C.O.; funding acquisition, A.C.O., J.O.M., C.M.A.P.S. and C.F.L. All authors have read and agreed to the published version of the manuscript.

Funding: This research was funded by FONDECYT Project 11240466 (A.C.O.), Regular FONDECYT Project 1231154 (J.O.M.), FONDEF Project ID24I10171 (J.O.M.), PIA/ANID ACT240058 (J.O.M.), FONDAP Project 15130011 (J.O.M.), FONDEQUIP EQM160157, regular FONDECYT Project 1220803 (C.M.A.P.S.), FONDEF project ID22I10099 (C.M.A.P.S.), Centro Basal Ciencia & Vida, FB210008, Financiamiento Basal para Centros Científicos y Tecnológicos de Excelencia de ANID (C.F.L.).

Institutional Review Board Statement: Not applicable.

Informed Consent Statement: Not applicable.

Data Availability Statement: The manuscript contains the reported data. Additional relevant data can be obtained upon request from the corresponding author.

Conflicts of Interest: The authors declare no conflicts of interest.

References

1. Antimicrobial Resistance. Available online: https://www.who.int/news-room/fact-sheets/detail/antimicrobial-resistance (accessed on 28 March 2025).
2. Hetta, H.F.; Ahmed, E.A.; Hemdan, A.G.; El-Deek, H.E.; Abd-Elregal, S.; Abd Ellah, N.H. Modulation of Rifampicin-Induced Hepatotoxicity Using Poly(Lactic-Co-Glycolic Acid) Nanoparticles: A Study on Rat and Cell Culture Models. *Nanomedicine* **2020**, *15*, 1375–1390. [CrossRef] [PubMed]
3. Mubeen, B.; Ansar, A.N.; Rasool, R.; Ullah, I.; Imam, S.S.; Alshehri, S.; Ghoneim, M.M.; Alzarea, S.I.; Nadeem, M.S.; Kazmi, I. Nanotechnology as a Novel Approach in Combating Microbes Providing an Alternative to Antibiotics. *Antibiotics* **2021**, *10*, 1473. [CrossRef] [PubMed]
4. Zara, G.P.; Bargoni, A.; Cavalli, R.; Fundarò, A.; Vighetto, D.; Gasco, M.R. Pharmacokinetics and Tissue Distribution of Idarubicin-Loaded Solid Lipid Nanoparticles After Duodenal Administration to Rats. *J. Pharm. Sci.* **2002**, *91*, 1324–1333. [CrossRef]
5. Mamun, M.M.; Sorinolu, A.J.; Munir, M.; Vejerano, E.P. Nanoantibiotics: Functions and Properties at the Nanoscale to Combat Antibiotic Resistance. *Front. Chem.* **2021**, *9*, 687660. [CrossRef]
6. Wang, D.-Y.; van der Mei, H.C.; Ren, Y.; Busscher, H.J.; Shi, L. Lipid-Based Antimicrobial Delivery-Systems for the Treatment of Bacterial Infections. *Front. Chem.* **2020**, *7*, 872. [CrossRef]

7. Pandey, R.; Sharma, S.; Khuller, G.K. Oral Solid Lipid Nanoparticle-Based Antitubercular Chemotherapy. *Tuberculosis* **2005**, *85*, 415–420. [CrossRef]
8. Guo, Q.; Guo, H.; Lan, T.; Chen, Y.; Chen, X.; Feng, Y.; Luo, Y.; Yao, Y.; Li, Y.; Pan, X.; et al. Co-Delivery of Antibiotic and Baicalein by Using Different Polymeric Nanoparticle Cargos with Enhanced Synergistic Antibacterial Activity. *Int. J. Pharm.* **2021**, *599*, 120419. [CrossRef] [PubMed]
9. Ahsan, A.; Thomas, N.; Barnes, T.J.; Subramaniam, S.; Loh, T.C.; Joyce, P.; Prestidge, C.A. Lipid Nanocarriers-Enabled Delivery of Antibiotics and Antimicrobial Adjuvants to Overcome Bacterial Biofilms. *Pharmaceutics* **2024**, *16*, 396. [CrossRef]
10. Arana, L.; Gallego, L.; Alkorta, I. Incorporation of Antibiotics into Solid Lipid Nanoparticles: A Promising Approach to Reduce Antibiotic Resistance Emergence. *Nanomaterials* **2021**, *11*, 1251. [CrossRef]
11. Shegokar, R.; Nakach, M. Chapter 4—Large-Scale Manufacturing of Nanoparticles—An Industrial Outlook. In *Drug Delivery Aspects*; Shegokar, R., Ed.; Elsevier: Amsterdam, The Netherlands, 2020; pp. 57–77, ISBN 978-0-12-821222-6.
12. Gordillo-Galeano, A.; Mora-Huertas, C.E. Solid Lipid Nanoparticles and Nanostructured Lipid Carriers: A Review Emphasizing on Particle Structure and Drug Release. *Eur. J. Pharm. Biopharm.* **2018**, *133*, 285–308. [CrossRef]
13. Bazán Henostroza, M.A.; Diniz Tavares, G.; Nishitani Yukuyama, M.; De Souza, A.; José Barbosa, E.; Carlos Avino, V.; dos Santos Neto, E.; Rebello Lourenço, F.; Löbenberg, R.; Araci Bou-Chacra, N. Antibiotic-Loaded Lipid-Based Nanocarrier: A Promising Strategy to Overcome Bacterial Infection. *Int. J. Pharm.* **2022**, *621*, 121782. [CrossRef] [PubMed]
14. Sadeghzadeh, F.; Motavalizadehkakhky, A.; Mehrzad, J.; Zhiani, R.; Homayouni Tabrizi, M. Folic Acid Conjugated-Chitosan Modified Nanostructured Lipid Carriers as Promising Carriers for Delivery of Umbelliprenin to Cancer Cells: In Vivo and in Vitro. *Eur. Polym. J.* **2023**, *186*, 111849. [CrossRef]
15. Muheem, A.; Wasim, M.; Aldosari, E.; Baboota, S.; Ali, J. Fabrication of TPGS Decorated Etravirine Loaded Lipidic Nanocarriers as a Neoteric Oral Bioavailability Enhancer for Lymphatic Targeting. *Discov. Nano* **2024**, *19*, 5. [CrossRef] [PubMed]
16. Viegas, C.; Patrício, A.B.; Prata, J.M.; Nadhman, A.; Chintamaneni, P.K.; Fonte, P. Solid Lipid Nanoparticles vs. Nanostructured Lipid Carriers: A Comparative Review. *Pharmaceutics* **2023**, *15*, 1593. [CrossRef]
17. Dhiman, N.; Awasthi, R.; Sharma, B.; Kharkwal, H.; Kulkarni, G.T. Lipid Nanoparticles as Carriers for Bioactive Delivery. *Front. Chem.* **2021**, *9*, 580118. [CrossRef]
18. Shirodkar, R.K.; Kumar, L.; Mutalik, S.; Lewis, S. Solid Lipid Nanoparticles and Nanostructured Lipid Carriers: Emerging Lipid Based Drug Delivery Systems. *Pharm. Chem. J.* **2019**, *53*, 440–453. [CrossRef]
19. Motiei, M.; Pleno de Gouveia, L.; Šopík, T.; Vícha, R.; Škoda, D.; Císař, J.; Khalili, R.; Domincová Bergerová, E.; Münster, L.; Fei, H.; et al. Nanoparticle-Based Rifampicin Delivery System Development. *Molecules* **2021**, *26*, 2067. [CrossRef] [PubMed]
20. Chen, W.; Glackin, C.A.; Horwitz, M.A.; Zink, J.I. Nanomachines and Other Caps on Mesoporous Silica Nanoparticles for Drug Delivery. *Acc. Chem. Res.* **2019**, *52*, 1531–1542. [CrossRef]
21. Koch, A.; Mizrahi, V.; Warner, D.F. The Impact of Drug Resistance on Mycobacterium Tuberculosis Physiology: What Can We Learn from Rifampicin? *Emerg. Microbes Infect.* **2014**, *3*, 1–11. [CrossRef]
22. Yılmaz, E.Ş.; Aslantaş, Ö. Antimicrobial Resistance and Underlying Mechanisms in *Staphylococcus aureus* Isolates. *Asian Pac. J. Trop. Med.* **2017**, *10*, 1059–1064. [CrossRef]
23. Ellis, T.; Chiappi, M.; García-Trenco, A.; Al-Ejji, M.; Sarkar, S.; Georgiou, T.K.; Shaffer, M.S.P.; Tetley, T.D.; Schwander, S.; Ryan, M.P.; et al. Multimetallic Microparticles Increase the Potency of Rifampicin against Intracellular *Mycobacterium tuberculosis*. *ACS Nano* **2018**, *12*, 5228–5240. [CrossRef] [PubMed]
24. Wu, S.; Huang, Y.; Yan, J.; Li, Y.; Wang, J.; Yang, Y.Y.; Yuan, P.; Ding, X. Bacterial Outer Membrane-Coated Mesoporous Silica Nanoparticles for Targeted Delivery of Antibiotic Rifampicin against Gram-Negative Bacterial Infection In Vivo. *Adv. Funct. Mater.* **2021**, *31*, 2103442. [CrossRef]
25. Chaudhary, N.; Aggarwal, B.; Saini, V.; Yavvari, P.S.; Sharma, P.; Srivastava, A.; Bajaj, A. Polyaspartate-Derived Synthetic Antimicrobial Polymer Enhances the Activity of Rifampicin against Multidrug-Resistant Pseudomonas Aeruginosa Infections. *Biomater. Sci.* **2022**, *10*, 5158–5171. [CrossRef] [PubMed]
26. Farooq, U.; Ahmad, T.; Khan, A.; Sarwar, R.; Shafiq, J.; Raza, Y.; Ahmed, A.; Ullah, S.; Ur Rehman, N.; Al-Harrasi, A. Rifampicin Conjugated Silver Nanoparticles: A New Arena for Development of Antibiofilm Potential against Methicillin Resistant *Staphylococcus aureus* and Klebsiella Pneumoniae. *Int. J. Nanomed.* **2019**, *14*, 3983–3993. [CrossRef]
27. Scolari, I.R.; Páez, P.L.; Musri, M.M.; Petiti, J.P.; Torres, A.; Granero, G.E. Rifampicin Loaded in Alginate/Chitosan Nanoparticles as a Promising Pulmonary Carrier against *Staphylococcus aureus*. *Drug Deliv. Transl. Res.* **2020**, *10*, 1403–1417. [CrossRef]
28. Banerjee, S.; Roy, S.; Bhaumik, K.N.; Pillai, J. Mechanisms of the Effectiveness of Lipid Nanoparticle Formulations Loaded with Anti-Tubercular Drugs Combinations toward Overcoming Drug Bioavailability in Tuberculosis. *J. Drug Target.* **2020**, *28*, 55–69. [CrossRef]

29. Pinheiro, M.; Ribeiro, R.; Vieira, A.; Andrade, F.; Reis, S. Design of a Nanostructured Lipid Carrier Intended to Improve the Treatment of Tuberculosis. *Drug Des. Dev. Ther.* **2016**, *10*, 2467–2475. [CrossRef]
30. Vieira, A.C.; Magalhaes, J.; Rocha, S.; Cardoso, M.S.; Santos, S.G.; Borges, M.; Pinheiro, M.; Reis, S. Targeted Macrophages Delivery of Rifampicin-Loaded Lipid Nanoparticles to Improve Tuberculosis Treatment. *Nanomedicine* **2017**, *12*, 2721–2736. [CrossRef]
31. Ortiz, A.C.; Yañez, O.; Salas-Huenuleo, E.; Morales, J.O. Development of a Nanostructured Lipid Carrier (NLC) by a Low-Energy Method, Comparison of Release Kinetics and Molecular Dynamics Simulation. *Pharmaceutics* **2021**, *13*, 531. [CrossRef]
32. Goutal, S.; Auvity, S.; Legrand, T.; Hauquier, F.; Cisternino, S.; Chapy, H.; Saba, W.; Tournier, N. Validation of a Simple HPLC-UV Method for Rifampicin Determination in Plasma: Application to the Study of Rifampicin Arteriovenous Concentration Gradient. *J. Pharm. Biomed. Anal.* **2016**, *123*, 173–178. [CrossRef]
33. Huh, A.J.; Kwon, Y.J. "Nanoantibiotics": A New Paradigm for Treating Infectious Diseases Using Nanomaterials in the Antibiotics Resistant Era. *J. Control. Release* **2011**, *156*, 128–145. [CrossRef] [PubMed]
34. Wang, L.; Hu, C.; Shao, L. The Antimicrobial Activity of Nanoparticles: Present Situation and Prospects for the Future. *Int. J. Nanomed.* **2017**, *12*, 1227–1249. [CrossRef] [PubMed]
35. Fazly Bazzaz, B.S.; Khameneh, B.; Zarei, H.; Golmohammadzadeh, S. Antibacterial Efficacy of Rifampin Loaded Solid Lipid Nanoparticles against *Staphylococcus epidermidis* Biofilm. *Microb. Pathog.* **2016**, *93*, 137–144. [CrossRef]
36. Carneiro, S.P.; Carvalho, K.V.; de Oliveira Aguiar Soares, R.D.; Carneiro, C.M.; de Andrade, M.H.G.; Duarte, R.S.; dos Santos, O.D.H. Functionalized Rifampicin-Loaded Nanostructured Lipid Carriers Enhance Macrophages Uptake and Antimycobacterial Activity. *Colloids Surf. B Biointerfaces* **2019**, *175*, 306–313. [CrossRef]
37. Anjani, Q.K.; Domínguez-Robles, J.; Utomo, E.; Font, M.; Martínez-Ohárriz, M.C.; Permana, A.D.; Cárcamo-Martínez, Á.; Larrañeta, E.; Donnelly, R.F. Inclusion Complexes of Rifampicin with Native and Derivatized Cyclodextrins: In Silico Modeling, Formulation, and Characterization. *Pharmaceuticals* **2022**, *15*, 20. [CrossRef]
38. Li, Y.Y.; Chen, X.G.; Zhang, J.; Liu, C.S.; Xue, Y.P.; Sun, G.Z.; Zhang, W.F. In Vitro Release of Rifampicin and Biocompatibility of Oleoylchitosan Nanoparticles. *J. Appl. Polym. Sci.* **2009**, *111*, 2269–2274. [CrossRef]
39. Mistry, N.; Bandyopadhyaya, R.; Mehra, S. Enhancement of Antimycobacterial Activity of Rifampicin Using Mannose-Anchored Lipid Nanoparticles against Intramacrophage Mycobacteria. *ACS Appl. Bio Mater.* **2022**, *5*, 5779–5789. [CrossRef]
40. Bunjes, H. Structural Properties of Solid Lipid Based Colloidal Drug Delivery Systems. *Curr. Opin. Colloid Interface Sci.* **2011**, *16*, 405–411. [CrossRef]
41. Cholakova, D.; Denkov, N. Polymorphic Phase Transitions in Triglycerides and Their Mixtures Studied by SAXS/WAXS Techniques: In Bulk and in Emulsions. *Adv. Colloid Interface Sci.* **2024**, *323*, 103071. [CrossRef]
42. Aboutaleb, E.; Noori, M.; Gandomi, N.; Atyabi, F.; Fazeli, M.R.; Jamalifar, H.; Dinarvand, R. Improved Antimycobacterial Activity of Rifampin Using Solid Lipid Nanoparticles. *Int. Nano Lett.* **2012**, *2*, 33. [CrossRef]
43. Aliyazdi, S.; Frisch, S.; Neu, T.; Veldung, B.; Karande, P.; Schaefer, U.F.; Loretz, B.; Vogt, T.; Lehr, C.-M. A Novel 3D Printed Model of Infected Human Hair Follicles to Demonstrate Targeted Delivery of Nanoantibiotics. *ACS Biomater. Sci. Eng.* **2024**, *10*, 4947–4957. [CrossRef] [PubMed]
44. Woźniak-Budych, M.J.; Przysiecka, Ł.; Langer, K.; Peplińska, B.; Jarek, M.; Wiesner, M.; Nowaczyk, G.; Jurga, S. Green Synthesis of Rifampicin-Loaded Copper Nanoparticles with Enhanced Antimicrobial Activity. *J. Mater. Sci. Mater. Med.* **2017**, *28*, 42. [CrossRef] [PubMed]
45. Aubry-Damon, H.; Soussy, C.-J.; Courvalin, P. Characterization of Mutations in therpoB Gene That Confer Rifampin Resistance in *Staphylococcus aureus*. *Antimicrob. Agents Chemother.* **1998**, *42*, 2590–2594. [CrossRef]
46. Mlynarczyk-Bonikowska, B.; Kowalewski, C.; Krolak-Ulinska, A.; Marusza, W. Molecular Mechanisms of Drug Resistance in *Staphylococcus aureus*. *Int. J. Mol. Sci.* **2022**, *23*, 8088. [CrossRef]
47. Abedi, E.; Akhavan, H.-R.; Mohammadi, H.; Banasaz, S. Structure-Based Modifications of Nano Lipid Carriers: Comparative Review on Release Properties and Anti-Microbial Activities of Bioactive Compounds. *Food Control* **2024**, *159*, 110237. [CrossRef]
48. Dyett, B.P.; Yu, H.; Sarkar, S.; Strachan, J.B.; Drummond, C.J.; Conn, C.E. Uptake Dynamics of Cubosome Nanocarriers at Bacterial Surfaces and the Routes for Cargo Internalization. *ACS Appl. Mater. Interfaces* **2021**, *13*, 53530–53540. [CrossRef]
49. Tran, N.; Hocquet, M.; Eon, B.; Sangwan, P.; Ratcliffe, J.; Hinton, T.M.; White, J.; Ozcelik, B.; Reynolds, N.P.; Muir, B.W. Non-Lamellar Lyotropic Liquid Crystalline Nanoparticles Enhance the Antibacterial Effects of Rifampicin against *Staphylococcus aureus*. *J. Colloid Interface Sci.* **2018**, *519*, 107–118. [CrossRef]
50. Ghaderkhani, J.; Yousefimashouf, R.; Arabestani, M.; Roshanaei, G.; Asl, S.S.; Abbasalipourkabir, R. Improved Antibacterial Function of Rifampicin-Loaded Solid Lipid Nanoparticles on Brucella Abortus. *Artif. Cells Nanomed. Biotechnol.* **2019**, *47*, 1181–1193. [CrossRef] [PubMed]

51. Khatak, S.; Mehta, M.; Awasthi, R.; Paudel, K.R.; Singh, S.K.; Gulati, M.; Hansbro, N.G.; Hansbro, P.M.; Dua, K.; Dureja, H. Solid Lipid Nanoparticles Containing Anti-Tubercular Drugs Attenuate the *Mycobacterium marinum* Infection. *Tuberculosis* **2020**, *125*, 102008. [CrossRef]
52. Patra, J.K.; Das, G.; Baek, K.-H. Phyto-Mediated Biosynthesis of Silver Nanoparticles Using the Rind Extract of Watermelon (*Citrullus lanatus*) under Photo-Catalyzed Condition and Investigation of Its Antibacterial, Anticandidal and Antioxidant Efficacy. *J. Photochem. Photobiol. B Biol.* **2016**, *161*, 200–210. [CrossRef]
53. Grumezescu, A.M.; Ghitulica, C.D.; Voicu, G.; Huang, K.-S.; Yang, C.-H.; Ficai, A.; Vasile, B.S.; Grumezescu, V.; Bleotu, C.; Chifiriuc, M.C. New Silica Nanostructure for the Improved Delivery of Topical Antibiotics Used in the Treatment of Staphylococcal Cutaneous Infections. *Int. J. Pharm.* **2014**, *463*, 170–176. [CrossRef] [PubMed]
54. Suciati, T.; Nafisa, S.; Nareswari, T.L.; Juniatik, M.; Julianti, E.; Wibowo, M.S.; Yudhistira, T.; Ihsanawati, I.; Triyani, Y.; Khairurrijal, K. ArtinM Grafted Phospholipid Nanoparticles for Enhancing Antibiotic Cellular Uptake Against Intracellular Infection. *Int. J. Nanomed.* **2020**, *15*, 8829–8843. [CrossRef] [PubMed]
55. Yang, X.; Tang, X.; Yi, S.; Guo, T.; Liao, Y.; Wang, Y.; Zhang, X. Maltodextrin-Derived Nanoparticles Resensitize Intracellular Dormant *Staphylococcus aureus* to Rifampicin. *Carbohydr. Polym.* **2025**, *348*, 122843. [CrossRef]
56. Aguilar-Colomer, A.; Colilla, M.; Izquierdo-Barba, I.; Jiménez-Jiménez, C.; Mahillo, I.; Esteban, J.; Vallet-Regí, M. Impact of the Antibiotic-Cargo from MSNs on Gram-Positive and Gram-Negative Bacterial Biofilms. *Microporous Mesoporous Mater.* **2021**, *311*, 110681. [CrossRef]
57. Amarnath Praphakar, R.; Sumathra, M.; Sam Ebenezer, R.; Vignesh, S.; Shakila, H.; Rajan, M. Fabrication of Bioactive Rifampicin Loaded κ-Car-MA-INH/Nano Hydroxyapatite Composite for Tuberculosis Osteomyelitis Infected Tissue Regeneration. *Int. J. Pharm.* **2019**, *565*, 543–556. [CrossRef]
58. Jia, D.; Zou, Y.; Cheng, J.; Zhang, Y.; Zhang, H.; Lu, K.; Chen, H.; Zhang, Y.; Yu, Q. A Multifunctional Nanoplatform with "Disruption and Killing" Function to Improve the Efficiency of Conventional Antibiotics for Biofilm Eradication. *J. Mater. Sci. Technol.* **2025**, *205*, 98–108. [CrossRef]
59. Topsakal, A.; Ekren, N.; Kilic, O.; Oktar, F.N.; Mahirogullari, M.; Ozkan, O.; Sasmazel, H.T.; Turk, M.; Bogdan, I.M.; Stan, G.E.; et al. Synthesis and Characterization of Antibacterial Drug Loaded β-Tricalcium Phosphate Powders for Bone Engineering Applications. *J. Mater. Sci. Mater. Med.* **2020**, *31*, 16. [CrossRef] [PubMed]
60. Kranthi Kiran, A.S.; Kizhakeyil, A.; Ramalingam, R.; Verma, N.K.; Lakshminarayanan, R.; Kumar, T.S.S.; Doble, M.; Ramakrishna, S. Drug Loaded Electrospun Polymer/Ceramic Composite Nanofibrous Coatings on Titanium for Implant Related Infections. *Ceram. Int.* **2019**, *45*, 18710–18720. [CrossRef]
61. Chambin, O.; and Jannin, V. Interest of Multifunctional Lipid Excipients: Case of Gelucire® 44/14. *Drug Dev. Ind. Pharm.* **2005**, *31*, 527–534. [CrossRef]
62. Traul, K.A.; Driedger, A.; Ingle, D.L.; Nakhasi, D. Review of the Toxicologic Properties of Medium-Chain Triglycerides. *Food Chem. Toxicol.* **2000**, *38*, 79–98. [CrossRef]
63. Buss, N.; Ryan, P.; Baughman, T.; Roy, D.; Patterson, C.; Gordon, C.; Dixit, R. Nonclinical Safety and Pharmacokinetics of Miglyol 812: A Medium Chain Triglyceride in Exenatide Once Weekly Suspension. *J. Appl. Toxicol.* **2018**, *38*, 1293–1301. [CrossRef] [PubMed]
64. Xia, W.J.; Onyuksel, H. Mechanistic Studies on Surfactant-Induced Membrane Permeability Enhancement. *Pharm. Res.* **2000**, *17*, 612–618. [CrossRef] [PubMed]
65. Rege, B.D.; Kao, J.P.Y.; Polli, J.E. Effects of Nonionic Surfactants on Membrane Transporters in Caco-2 Cell Monolayers. *Eur. J. Pharm. Sci.* **2002**, *16*, 237–246. [CrossRef] [PubMed]
66. Yan, Z.; Yang, K.; Tang, X.; Bi, Y.; Ding, Y.; Deng, M.; Xia, D.; Zhao, Y.; Chen, T. Norcantharidin Nanostructured Lipid Carrier (NCTD-NLC) Suppresses the Viability of Human Hepatocellular Carcinoma HepG2 Cells and Accelerates the Apoptosis. *J. Immunol. Res.* **2022**, *2022*, 3851604. [CrossRef] [PubMed]
67. Gundogdu, E.; Demir, E.-S.; Ekinci, M.; Ozgenc, E.; Ilem-Ozdemir, D.; Senyigit, Z.; Gonzalez-Alvarez, I.; Bermejo, M. An Innovative Formulation Based on Nanostructured Lipid Carriers for Imatinib Delivery: Pre-Formulation, Cellular Uptake and Cytotoxicity Studies. *Nanomaterials* **2022**, *12*, 250. [CrossRef]
68. Chandan, C.; Phani Kumar, G.; Jawahar, N.; Sushma, B.V.; Amachawadi, R.G.; Shati, A.A.; Alfaifi, M.Y.; Elbehairi, S.E.I.; Prasad, S.K.; Shivamallu, C.; et al. Design, Development and Characterization of Papain-Loaded Nanostructured Lipid Carriers for Enhanced Stability and Bio-Accessibility in Acidic Environments. *Results Chem.* **2024**, *8*, 101571. [CrossRef]

pharmaceutics

MDPI

Review

Tapinarof Nanogels as a Promising Therapeutic Approach

Barbara Balogh [1,2], Ágota Pető [1], Pálma Fehér [1], Zoltán Ujhelyi [3] and Ildikó Bácskay [1,4,*]

1 Department of Pharmaceutical Technology, Faculty of Pharmacy, University of Debrecen, Rex Ferenc Utca 1, H-4002 Debrecen, Hungary; balogh.barbara@pharm.unideb.hu (B.B.); peto.agota@pharm.unideb.hu (Á.P.); feher.palma@pharm.unideb.hu (P.F.)

2 Doctoral School of Pharmaceutical Sciences, University of Debrecen, Nagyerdei Körút 98, H-4032 Debrecen, Hungary

3 Department of Industrial Pharmaceutical Technology, Faculty of Pharmacy, University of Debrecen, Rex Ferenc Utca 1, H-4002 Debrecen, Hungary; ujhelyi.zoltan@pharm.unideb.hu

4 Institute of Healthcare Industry, University of Debrecen, Rex Ferenc Utca 1, H-4002 Debrecen, Hungary

* Correspondence: bacskay.ildiko@pharm.unideb.hu; Tel.: +36-52-411-717 (ext. 58421)

Abstract: Psoriasis is a chronic inflammatory skin disease characterised by increased oxidative stress, the overproliferation of keratinocytes, the accumulation of inflammatory mediators, and skin barrier damage. Although a number of therapeutic options are available, finding long-term treatments that are well-tolerated and patient-friendly treatments remains a challenge. Tapinarof is a new type of aryl hydrocarbon receptor (AhR) modulator that has recently attracted attention as a promising non-steroidal alternative. However, its application may be limited by its poor water solubility and low degree of skin penetration. Nanotechnology-based drug carriers, specially nanogels, offer new opportunities to overcome these limitations by combining the advantages of targeted drug delivery and enhanced skin penetration. Furthermore, nanogel formulations can improve skin hydration and support the restoration of skin barrier function, which are important in the treatment of psoriasis. This review focuses on current and emerging therapeutic approaches, with particular emphasis on the potential of incorporating tapinarof into nanogel formulations as a novel alternative to topical psoriasis treatment.

Keywords: tapinarof; non-steroid; aryl hydrocarbon receptor; dermatological therapies; nanotechnology for psoriasis

1. Introduction

Although several reviews have already discussed the pathophysiology and treatment of psoriasis, most have focused on either systemic biological therapies or conventional topical treatments. However, a significant number of patients with psoriasis still try to avoid the use of biologic agents and corticosteroids, and there is still a strong need for well-tolerated topical treatments that can be used in the long term. Despite this, few publications have explored the formulation-specific challenges and opportunities associated with newer topical agents.

Tapinarof is a recently approved topical aryl hydrocarbon receptor (AhR) agonist that has emerged as a promising therapeutic option due to its anti-inflammatory, antioxidant, and barrier-restoring properties. However, despite its growing clinical relevance, there is a lack of comprehensive discussion on how its efficacy and safety could be further optimised. This review aims to bridge that gap by combining current knowledge on tapinarof with pharmaceutical strategies designed to enhance its performance.

Among these strategies, nanogels deserve particular attention for their ability to improve skin penetration, offer controlled drug release, and reduce systemic side effects, properties that are beneficial for chronic inflammatory skin diseases such as psoriasis. While nanotechnology-based drug delivery systems are increasingly being explored in dermatology, the incorporation of tapinarof into nanogel formulations is a novel and unexplored area. This review summarises current progress and emerging trends, highlighting the potential of advanced carrier platforms to improve the therapeutic profile of tapinarof. This review also offers a unique perspective to help readers better understand where emerging therapies, such as nanogel-based approaches, fit into the current context of psoriasis treatment.

2. Methodology

This review was conducted using a structured search of peer-reviewed, English-language scientific articles, clinical trial reports, and regulatory documents related to psoriasis, tapinarof, and nanogel-based drug delivery systems. Only publications from 2014 to 2025 that were written in English were considered. Databases searched included PubMed, Scopus, Web of Science, PubPharm, and Google Scholar including the following terms: "tapinarof", "nanogel", "psoriasis", "topical treatment", "AhR agonist", "drug delivery systems", "nanotechnology in psoriasis", "dermal drug delivery", and "tapinarof in dermatology". In addition, clinical trials related to tapinarof were identified through www.clinicaltrials.gov (accessed on 9 March 2025). Articles were selected based on their relevance to the topic of the review, with priority given to recent studies and those involving regulatory approvals, particularly by the FDA.

3. Structure and Barrier Function of the Epidermis

The skin is the largest organ in the human body, operating as a multifunctional defence system and maintaining homeostasis. The structure of the skin is extremely complex and divided into three major layers: the epidermis, the dermis, and the hypodermis. Each has different anatomical and functional characteristics [1–3].

The epidermis is composed of five different layers: the uppermost layer is the stratum corneum, followed by the stratum lucidum, stratum granulosum, stratum spinosum, and stratum basale. As the epidermis is the most external layer of the skin, it consists of exfoliating epithelial cells, called keratinocyte cells, that are constantly renewing themselves, ensuring the integrity of the outer protective layer. When exposed to UV radiation, the keratinocytes of the epidermis synthesise vitamin D, which is essential for calcium metabolism and the proper functioning of the immune system. It also contains specific cell types such as melanocytes, Langerhans cells, and Merkel cells, which are responsible for pigmentation, immune defence, and sensory function. In addition, the sebaceous and sweat glands of the skin perform exocrine functions, helping to keep the skin hydrated and eliminating toxins. The functions of the sweat glands is also crucial for thermoregulation, as they reduce body temperature by evaporating sweat. The stratum corneum prevents water loss and the penetration of pathogens due to the presence of a hydrophobic lipid layer [1,2].

Overall, the upper layer of the skin not only provides physical and chemical protection, but it is also an active participant in the metabolic processes of the body [2].

4. General Characteristics of Psoriasis

Psoriasis is a chronic immune-mediated skin disease that affects 1–3% of the world's population, more than 100 million people [4–7]. It can emerge at any age, but it typically first appears between the ages of 15 and 25, with a second peak in emergence observed between the ages of 50 and 69 [8,9].

Genetic, environmental, and immunopathological factors all influence the pathogenesis of this disease [6,9–11]. Familial accumulation and specific gene variants (e.g., genes involved in the regulation of immune response and keratinocyte proliferation) may also contribute to its development. The initial presentation of psoriasis and subsequent flare-ups can be provoked by a variety of environmental factors, such as infections, stress, smoking, alcohol consumption, and cold, dry weather [1,6,9].

Both the adaptive and innate immune systems take part in the pathogenesis of psoriasis, characterised by a mixture of autoimmune and autoinflammatory mechanisms, with a central role played by antigen-presenting cells (APCs) and inflammatory cytokines (TNF-α, IL-17, and IL-23) [6,9,11–14]. These inflammatory responses result in keratinocyte hyperproliferation, which lead to a thickening of the skin and the typical plaque lesions [1,14,15].

Psoriasis is not a contagious illness, but it is associated with significant inflammation that manifests as painful, itchy, scaly plaques that show intermittent remission and exacerbation [16–19]. Skin cells normally mature and slough off in about a month, whereas in psoriasis this process takes only 3–5 days, leading to an accumulation of cells [20].

There is currently no permanent cure, but symptoms can be controlled and the quality of life can be improved with the right therapy [8,19]. The National Psoriasis Foundation categorises psoriasis as mild (<3%), moderate (3–10%), and severe (>10%) according to the extent of the disease and the proportion of the affected body surface. The severity of psoriasis is measured by calculating a PASI (Psoriasis Area and Severity Index) score ranging from 1 to 10, where 10 indicates severe and 1 indicates good psoriatic skin condition. [6].

Psoriasis can occur in several forms, which can be classified into the following types according to the American Academy of Dermatology [18]:

- Plaque psoriasis (psoriasis vulgaris): The most common form (>80% of cases), associated with erythematous plaques covered with silvery scales [18,21,22]. This usually appears in the elbows, knees, scalp, navel, and sacrum regions [19,23].
- Guttate psoriasis: Appears mainly in children and young adults, with drop-shaped erythematous lesions, often after infections [8,23,24].
- Pustular psoriasis: A less common but more severe form with sterile, pustular blisters, which can be localised (palms, soles) or generalised (involving the whole body surface) [23–25].
- Erythrodermic psoriasis: The most severe form, affecting more than 90% of the skin and associated with severe inflammation, scaling, oedema, and even life-threatening complications [23].
- Intertriginous psoriasis (inverse psoriasis): A form that appears in the folds of the body (armpits, groin) and is associated with red, scaleless, painful plaques [8,23,25].

Psoriasis not only damages the skin but it can also trigger inflammatory processes throughout the body, which can be associated with a number of comorbidities, affecting the daily lives of patients [21,26]. These inflammatory processes can contribute to cardiovascular disease, metabolic syndrome, obesity, and inflammatory bowel disease [6,19,21]. In addition, psoriatic arthritis evolves in 10–30% of cases and manifests as arthritis. It can lead to the erosion of cartilage tissue in joints and cause irreversible joint damage in the long term [12,18,19,21]. Psoriasis also increases the risk of depression and anxiety threefold, which can be a consequence of the social stigmatisation and chronic pain associated with skin symptoms [22,27]. In the early-onset form of the disease, around 75% of patients experience more severe comorbidities, while the late-onset form has a higher prevalence of other types of comorbidities [19]. Although the disease has a fluctuating progression and it is currently incurable, targeted therapies can provide significant improvements in quality of life [8].

5. Immunological and Molecular Pathogenesis of Psoriasis

The most important of the pathogenesis is uncontrolled proliferation of keratino-cytes and dysregulation of the immune system that lead to abnormal skin lesions [16,27,28]. The condition of this disease becomes autoimmune as the cells of the body activate the immune system [24].

The evolution of psoriasis can be described in two main phases: initiation and maintenance. In the initiation phase, various environmental factors, such as stress, trauma (e.g., tattooing), bacterial or viral infections (e.g., streptococcal pharyngitis in guttate psoriasis), drugs (e.g., beta-blockers, lithium), and dry, cold weather conditions, can lead to the activation of the disease [10,19,21,27]. In the maintenance phase, chronic activation of the immune system, overproduction of inflammatory mediators, and abnormal cell proliferation ensure the persistence of the condition. Several components of the immune system active participate in the maintenance of the disease [21]. Dendritic cells, T cells (particularly Th17 cells), and various other immune cells, including neutrophils, monocytes, macrophages, and mast cells, play roles in its pathogenesis [10,21,24,25]. These cellular interactions and immune mechanisms are illustrated in Figure 1.

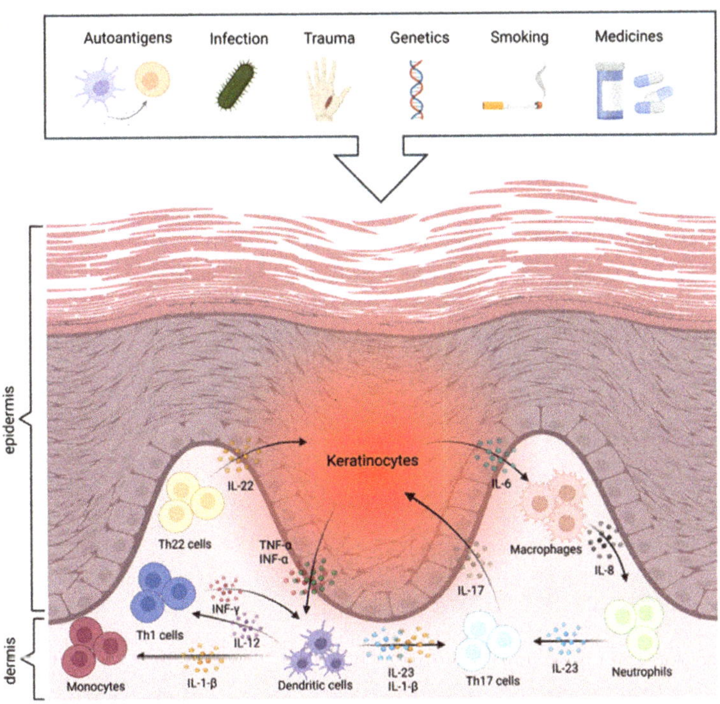

Figure 1. Aetiology and pathomechanisms of psoriasis (created with BioRender.com).

Following activation by external or internal stimuli, dendritic cells in the epidermis produce various inflammatory cytokines, including TNF-α, IL-12, and IL-23, which promote the differentiation of naive T cells into Th1 and Th17 cells [2,14,19,29]. IL-17, IL-22, and IFN-γ cytokines produced by Th17 cells contribute to inflammatory processes, stimulating keratinocyte proliferation and neutrophil recruitment to the skin, thus exacerbating the inflammatory condition [2,8,27,29]. IFN-γ produced by Th1 cells also maintains the inflammatory cascade and skin barrier dysfunction [10,21,28]. Inflammatory cytokines result in an over-proliferation of keratinocytes, which causes exfoliation. This increases the production of inflammatory mediators (TNF-β, IL-1, IL-22, and IL-6), which generate a positive feedback loop that recruits more inflammatory cells to the skin, thus increasing the intensity of the inflammatory response [10]. It also enhances angiogenesis and aggravates psoriasis symptoms by increasing the production of vascular endothelial growth factor (VEGF) to serve the increased metabolic needs of rapidly proliferating skin cells [8,24]. In general, innate immune cells, especially neutrophils, are highly important in the early stages of psoriasis development, whereas T cell-dominated adaptive inflammation is a feature of stable plaques in the later stages.

Different signalling pathways are also essential in the maintenance and exacerbation of psoriasis. The NF-κB (TNFAIP3, NFKBIA, NFKBIZ) and JAK-STAT (STAT3) signalling pathways play a prominent role in the regulation of the inflammatory environment through excessive keratinocyte growth and immune cell activation [21]. The IL-23/IL-17 axis is substantial, as IL-23 maintains the activity of Th17 cells, which produce IL-17 and IL-22, promoting inflammation and keratinocyte proliferation [10]. In addition, adjacent keratinocytes, leukocytes, and fibroblasts also communicate to enhance and maintain inflammation [2]. In short, the interaction between immune cells, cytokines, and keratinocytes creates a self-sustaining chronic inflammatory cycle that leads to the persistent nature of the condition [8].

A deeper knowledge of the pathogenesis of psoriasis is essential for identifying new therapeutic targets for the development of more effective and targeted treatment strategies [4,10].

Future goals are further exploring the molecular mechanisms of psoriasis, gaining a better understanding of the role of genetic factors, and developing personalised treatments to improve the quality of life of the patients. The application of innovative therapies, such as nanotechnology-based drug delivery systems, could open up new possibilities for treatment. These new approaches may contribute to the long-term improvement of the condition of patients with psoriasis and provide an opportunity for more effective and long-term control of the disease [10,30–32].

6. Psoriasis Therapies and Their Limitations

The treatment of psoriasis involves multiple approaches, such topical, oral, biological, and intravenous therapies, as well as phototherapy [1,23]. As psoriasis is mainly a skin disease, topical treatment is the first choice in mild and moderate cases [23]. The majority of patients with psoriasis (80%) can be effectively treated with topical therapy [33]. However, these treatments present a number of challenges, including side effects, discomfort, and low patient adherence, which are significant problems for therapeutic success.

The currently available topical treatments are corticosteroids, vitamin D analogues (calcipotriene, calcitriol), retinoids (tazarotene), and calcineurin inhibitors (pimecrolimus, tacrolimus) [14,16,34]. These agents often have side effects, which limits their long-term use; the problems associated with each are summarised in Table 1 [5,34–36]. This situation creates an urgent need for well-tolerated topical therapies with minimal side effects that are suitable for long-term use [4,28].

Combination therapy can be more effective than monotherapy as it targets several pathomechanisms at one time. The combined use of corticosteroids and vitamin D analogues not only increases therapeutic efficacy but also moderates irritation. Similarly, the combination of topical retinoids and calcineurin inhibitors can boost treatment efficacy while reducing side effects [12,14,17]. Previous studies have also indicated that salicylic acid can improve the permeation of calcineurin inhibitors, vitamin D analogues, and corticosteroids through the skin, thereby helping to improve their therapeutic profile [37].

Moderate and severe psoriasis may require systemic therapies, which include methotrexate, cyclosporine, and acitretin [1,17,23]. Although great advances have been made in newer biologic therapies and phototherapy, they are not available or appropriate for all patients due to their high cost and potential immunological side effects [31]. Systemic drugs are very effective and can relieve the patient of symptoms for a period of time, but the risk of side effects is much higher than with topical treatments; this is presented in Table 2. Thus, in moderate to severe forms of the disease, the combination of systemic and topical agents could be necessary to achieve a complete cure [22].

The priorities of psoriasis treatment are to alleviate symptoms, clear the skin, and make patients' daily life easier. The optimal therapy is selected on an individual basis, considering disease severity, plaque location, comorbidities, patient preferences, response to prior treatments, and expert opinion [9,27,31]. Maintaining patient adherence remains a major challenge in the management of psoriasis. The main reasons for discontinuing therapy include low efficacy, unsatisfactory cosmetic results, and suboptimal doctor–patient relationships [12,22]. In addition, conventional topical treatments are often uncomfortable, difficult to apply, or take too long to achieve the desired therapeutic effect [12,22,27]. More than 50% of patients are dissatisfied with the current treatment options, which results in poor patient compliance [27,38].

Although advances have been made in current therapeutic options, high patient dissatisfaction rates and low adherence suggest a potential need for a well-tolerated, non-steroidal topical therapy [12,33,39]. Novel nanocarrier-based strategies offer a promising alternative, as they improve the skin penetration of active ingredients, reduce side effects, and can provide effective therapy at lower doses, thus increasing patient satisfaction. The aim for the future is to develop therapeutic approaches that are not only more effective and safer, but also better adapted to patients' lifestyles and expectations [17].

New topical therapeutic options include tapinarof nanogels, which may offer a promising alternative to current treatments as a non-corticosteroid agent. Tapinarof is an aryl hydrocarbon receptor (AhR) agonist that exerts anti-inflammatory and antioxidant effects, thereby helping to reduce plaque formation and inflammatory processes [33,40]. Clinical trials have demonstrated that tapinarof is effective in improving skin lesions and is well-tolerated in the long term, with a more favourable side effect profile than that of conventional corticosteroids [33]. Therefore, this compound could potentially be a new and promising option for the treatment of psoriasis [40].

Table 1. Topical treatments for psoriasis: mechanisms, side effects, and examples.

Therapy	Mechanism	Adverse Effects	Examples	References
Corticosteroids	Inhibit the production of cytokines and reduce inflammatory mediators	Tachyphylaxis, atrophy, stretch marks, erythema	Clobetasol, betametazon, mometazon	[9,12,22,23,33]
Vitamin D analogues	Inhibit dendritic cell maturation, T-cell activation, and keratinocyte proliferation	Skin irritation, burning, erythema	Calcipotriene, calcitriol	[2,12,17,23,25]

Table 1. *Cont.*

Therapy	Mechanism	Adverse Effects	Examples	References
Retinoids	Inhibit keratinocyte proliferation	Erythema, peeling, skin irritation, burning, itching	Tazarotene, tretinoin	
Calcineurin inhibitors	Reduce T-cell activation and the production of inflammatory cytokines (IL-2) by inhibiting the enzyme calcineurin	Skin irritation, burning, itching	Tacrolimus, pimecrolimus	[22,23,27,33,41]
Keratolytics	Reduce the intercellular cohesion of the stratum corneum by dissolving the intercellular cementum	Frontal headache, central nervous system symptoms, metabolic acidosis, tinnitus, nausea, vomiting	Salicylic acid	[23,42]

Table 2. Systematic treatments for psoriasis: mechanisms, side effects, and examples.

Therapy	Mechanism	Adverse Effects	Examples	References
Dihydrofolate reductase inhibitors	Block the proliferation of keratinocytes and immune cells by inhibiting dihydrofolate reductase	Dry skin, hair loss, liver toxicity, risk of skin cancer, nausea, infections, bone marrow suppression	Methotrexate	
Retinoids	Inhibit keratinocyte proliferation	Nausea, hepatotoxicity, infections, xerosis, nail and hair fragmentation, teratogenicity	Acitretin	[1,9,17,23]
Calcineurin inhibitors	Reduce T-cell activation and the production of inflammatory cytokines (IL-2) by inhibiting the enzyme calcineurin	Dry skin, cardiovascular and gastrointestinal problems, gingival hyperplasia, tremor, leukopenia, hepatotoxicity, nephrotoxicity, hypertension, increased immunosuppression	Cyclosporine	[1,2,6,17,41]
Biological therapies	Inhibit cytokines or cytokine receptors (IL-12/23 inhibitors, TNF inhibitors, IL-17 inhibitors)	Expensive, safety concerns, high risk of malignant tumours and facial paralysis	Etanercept, adalimumab, ustekinumab, infliximab	[1,2,9,23]
Phototherapy	Causes cell death by apoptosis, necrosis or autophagy; reduces epidermal proliferation	Melanoma, photoaging, burning, erythema, pruritus, xerosis, pain, and discomfort	Ultraviolet B light, psoralen ultraviolet A light, photodynamic therapy	[1,2,6,23,41]

7. New Therapeutic Strategies for Psoriasis

Over the last decade, several new therapies have been developed to attack the physiological mechanisms behind inflammatory diseases. The US Food and Drug Administration (FDA) has approved multiple new agents for the treatment of psoriasis, which have led to the establishment of new therapeutic options. Notably, IL-17 and IL-23 antagonists, as well as oral TYK2 inhibitors and topical agents such as PDE4 inhibitors and AhR agonists, have shown significant clinical efficacy. Table 3 summarises recently FDA-approved agents, detailing their applications, drug classes, and indications [43–46].

Table 3. New active agents recently approved by the FDA.

Active Substance	Market Name	Application	Drug Class	Indication	FDA Approval Year	Ref.
Bimekizumab	Bimzelx	subcutaneous injection	IL-17A, IL-17F antagonist	moderate-to-severe plaque psoriasis	2023	[47–49]
Deucravacitinib	Sotyktu	oral application	TYK2 inhibitor	moderate-to-severe plaque psoriasis	2022	[43,47–50]
Roflumilast	Zoryve	topical application	PDE4 inhibitor	mild-to-severe plaque psoriasis	2022	[51–54]
Tapinarof	Vtama	topical application	AhR agonist	mild-to-severe plaque psoriasis	2022	[27,42,55]
Risankizumab	Skyrizi	subcutaneous injection	IL-23 antagonist	moderate-to-severe plaque psoriasis	2019	[43,44,56,57]
Certolizumab pegol	Cimzia	subcutaneous injection	TNF-α blocker	moderate-to-severe plaque psoriasis	2018	[58–61]
Tildrakizumab	Ilumya	subcutaneous injection	IL-23 antagonist	moderate-to-severe plaque psoriasis	2018	[62–65]
Guselkumab	Tremfya	subcutaneous injection	IL-23 antagonist	moderate-to-severe plaque psoriasis	2017	[43,66–69]
Brodalumab	Siliq	subcutaneous injection	IL-17A antagonist	moderate-to-severe plaque psoriasis	2017	[43,70–72]
Ixekizumab	Taltz	subcutaneous injection	IL-17A antagonist	moderate-to-severe plaque psoriasis, psoriatic arthritis	2016	[43,73–75]

8. Tapinarof

8.1. Structural Properties and Pharmacological Relevance of Tapinarof

The chemical properties of tapinarof are also crucial for its pharmacological application. It is a solid white powder with the chemical formula $C_{17}H_{18}O_2$ and a molecular weight of 254.32 g/mol [76,77]. Its solubility in DMSO is 2 mg/ml, and it is insoluble in water, which is a challenging issue for drug technology development [78]. Stability studies have demonstrated that it degrades in aqueous environments, and the degradation process is enhanced at higher temperatures and with exposure to UVA radiation. Degradation of about 50% in the presence of molecular oxygen was observed in a pH 7.4 aqueous solution after 48 h at laboratory temperature [79].

Tapinarof belongs to the group of polyphenols, specifically phenolic transstilbenes, which are produced by certain bacteria. It is structurally and functionally different from polyphenols of plant origin, such as resveratrol. However, tapinarof can be considered an isopropyl analogue of resveratrol, as shown in Figure 2. Tapinarof and its structurally related derivatives— resveratrol, pterostilbene, and pinocylvin—have been extensively studied [79,80]. Natural stilbenes are produced by many plants as a defence mechanism against various stresses, such as excessive UV radiation, heat stress, insect attack, and fungal or bacterial infections [79].

Although tapinarof and resveratrol are structurally similar compounds, their activities are significantly different [80]. Resveratrol is a natural phenolic compound produced by many plants (grapes, nuts, and berries) and acts as a partial agonist at the AhR. Research has shown that resveratrol is able to attenuate imikimod (IMQ)-induced psoriasis-like dermatitis in mice by reducing IL-17A and IL-19 [29]. While tapinarof directly activates the AhR pathway, resveratrol is only weakly or barely able to do so. The different origins (bacterial vs. plant) likely contribute to the differences in their activity. Profiling studies have demonstrated that the two compounds have different interactions with the AhR-ARNT complex, which is important for their biological effects. This suggests that tapinarof has a more specific and potent mechanism of action in activating AhR than resveratrol [80].

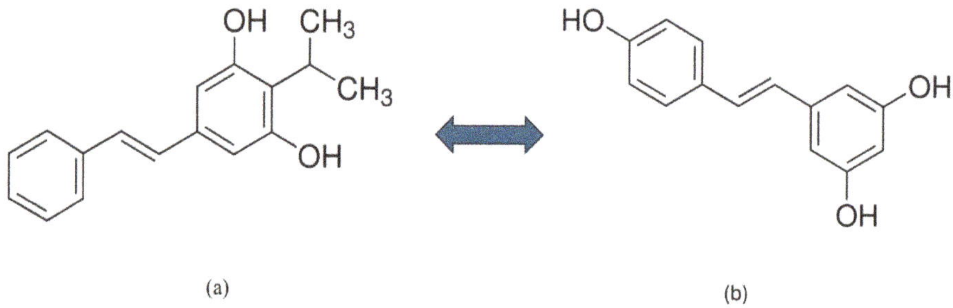

Figure 2. Chemical structure of tapinarof (**a**) and resveratrol (**b**) [77].

8.2. Tapinarof as a New Nonsteroidal AhR Modulator

Tapinarof (WBI-1001), also known as benvitimod or GSK2894512, is a non-steroidal anti-inflammatory compound that has a significant effect on the regulation of the immune response [4,10,16,81,82]. It is a novel topically applied agent that targets aryl hydrocarbon receptors, which are conductive to antioxidant activity, skin barrier protein expression, and cytokine regulation [11,12,83]. Tapinarof binds to AhRs to modulate the expression of skin barrier proteins, thereby reducing skin inflammation and promoting the recovery of skin barrier function. It also inhibits Th17 cell differentiation and reduces the expression of inflammatory cytokines such as IL-17, IL-22, and IL-23 by its mechanism of action. Furthermore, it alleviates oxidative stress through its antioxidant activity [11,83,84]. The role of AhR is crucial in this disease. Tapinarof and other AhR modulators could open up new perspectives in the treatment of psoriasis and other inflammatory skin diseases [16,83,85].

This compound was originally developed in China, and it has been evaluated for efficacy and safety in a comprehensive clinical trial programme. The formulation is optimised for twice-daily application and contains proprietary excipients to enhance its topical efficacy [11,16,86,87]. The FDA approved tapinarof 1% cream for the topical treatment of plaque psoriasis in adults in May 2022 [86,88,89].

It has shown excellent efficacy and favourable tolerability in clinical trials due to its unique mechanism [26,83,90]. It is also currently being investigated for the treatment of atopic dermatitis [33,83,91,92]. Whether marketed as benvitimod or tapinarof, despite formulation differences, this innovative molecule could become a promising alternative for the treatment of plaque psoriasis [85].

8.3. The Biological Origin of Tapinarof

Tapinarof is a natural, non-herbal polyphenol that is produced by Photorhabdus luminescens [13,81,93]. It is a bioluminescent, Gram-negative bacillus that lives in symbiosis with the nematode Heterorhabditis [26,33,82]. Nematodes harbour this bacterium in their intestinal tract and excrete it into insects. The bacterium produces metabolites that destroy the host, thus providing the optimal environment for nematode development. Observations showed that insects infected with nematodes did not decompose as quickly after death as those that were not exposed to the parasite [13,16].

The discovery of tapinarof was based on a random study of secondary metabolites of P. luminescens. The first observations date back to 1959, when Dutky and colleagues [94]. observed that insects infected with nematodes did not decompose rapidly, in contrast to the rapid decomposition observed in their absence [12,13]. This phenomenon suggested that P. luminescens may produce metabolites that have an antibiotic nature with anti-inflammatory properties, preventing rapid degradation of the infected host, and are responsible for the observed phenomenon [26,88].

Researchers isolated and identified the bioactive molecule produced by P. luminescens, called 3,5-dihydroxy-4-isopropylstilbene (also known as tapinarof) [16,90]. Subsequent research has characterised the compound in detail, which has been shown to produce immunomodulatory and anti-inflammatory effects when bound to AhR [16,81]. This led to the pharmaceutical development of the compound, which eventually became a topical therapy for the treatment of psoriasis and other skin diseases.

8.4. The Role of AhR Activation and the Therapeutic Potential of Tapinarof in the Treatment of Inflammatory Skin Diseases

The aryl hydrocarbon receptor is a cytoplasmic, ligand-dependent transcription factor that is activated by several exogenous and endogenous ligands [36,84,93]. It is widely expressed in various skin cells, including keratinocytes, fibroblasts, masocytes, and melanocytes, where it functions as a chemical sensor, converting external and internal stimuli into biological responses [14,36,89]. In healthy skin, AhR signalling is essential for maintaining skin homeostasis, regulating the immune response, and modulating inflammatory processes [36,84,93].

The expression of AhR is modified in psoriasis, influencing the pathogenesis of the disease. Activation of AhR regulates the terminal differentiation of CD4+ T-helper cells (Th17 and Th22) and the production of IL-17 and IL-22 cytokines, which are responsible for the maintenance of inflammation [22,26,33,80].

AhR can be activated by exogenous ligands such as environmental toxins (e.g., 2,3,7,8-tetrachlorodibenzo-p-dioxin) and endogenous ligands such as tryptophan derivatives [26,29,84,86]. After AhR binds to the ligand, the AhR–ligand complex migrates to the nucleus where it dimerizes with the AhR nuclear translocator (ARNT). The resulting AhR–ligand–ARNT complex binds to specific DNA recognition sites, triggering gene transcription and affecting skin barrier function, keratinocyte differentiation, and immune cell activity [26,29,87,93].

Since AhR activation is associated with the regulation of inflammatory and immunological processes in the skin, AhR agonists thus provide a promising therapeutic strategy for the treatment of inflammatory skin diseases [13,26,27,33,87]. In vitro and in vivo studies have demonstrated that AhR deficiency induces an enhanced inflammatory response, which increases the production of pro-inflammatory cytokines and makes the skin more vulnerable to inflammatory processes [16,29]. Furthermore, since keratinocyte proliferation and differentiation play a central role in the development of psoriasis, inhibiting keratinocyte hyperproliferation is also considered an effective therapeutic approach to treat the disease [36].

Through the activation of AhR, tapinarof induces anti-inflammatory and antioxidant effects, and repairs the skin and maintains its homeostasis (Table 4) [39,55,80,90]. It downregulates the production of inflammatory cytokines such as IL-17 and IL-22 [13,55,80]. These cytokines play a central role in keratinocyte proliferation, so their inhibition helps to suppress excessive cell division and restore normal skin structure. In this way, they do not only play a role in alleviating inflammatory symptoms, but also support the long-term health of the skin [29,33,90].

Table 4. Molecular pathways of tapinarof [13,27,92].

Effect	Mechanism
Anti-inflammatory effect	Reduces the expression of Th2 cytokines (IL-4, IL-5, IL-13, IL-31) and Th17 cytokines (IL-17A, IL-17F), which are involved in the pathogenesis of psoriasis.
Skin barrier enhancement	Increases the expression of skin barrier proteins (filaggrin, loricrin, involucrin) and ceramide skin lipids, helping to normalise the skin barrier.
Antioxidant effect	Activates the Nrf2 pathway and reduces reactive oxygen species levels, contributing to the reduction in oxidative stress and the antioxidant response of the skin.

In addition, tapinarof upregulates the production of skin barrier proteins with reduced expression in psoriasis, such as filaggrin and loricrin, resulting in normalising of the skin barrier [55,84,88]. It also promotes the maintenance of skin function by regulating the gene expression of immune cells and skin cells [26,29,33,80].

Tapinarof also has antioxidant properties, partly due to its own chemical structure and partly due to activation of the nuclear factor erythroid 2-related factor 2 (Nrf2) pathway [35,81,88]. The 4′-OH group plays an significant role in the antioxidant activity of stilbenols [81]. In addition, the molecule contains two phenolic groups that directly neutralise reactive oxygen species (ROS): superoxide anions and hydroxyl radicals. By activating the AhR-Nrf2 transcription factor pathway, it promotes the expression of antioxidant enzymes such as NADPH quinone oxidoreductase 1 (NQO1) and heme oxygenase-1 (HO-1), decreasing oxidative stress. This dual effect may help to moderate psoriasis-associated cell damage [83,84,86,92].

Tapinarof also has antimicrobial activity, especially against Gram-positive bacteria and fungi, which can promote the balance and condition of the microbiome of the skin [90].

8.5. Results of Clinical Trials

Tapinarof is a first-in-class AhR approved by the FDA for the treatment of psoriasis, and it is currently being investigated for the treatment of atopic dermatitis as well [88,91,92,95]. It is presently only available as a cream. Its efficacy and safety have been evaluated in several preclinical and clinical trials in recent years, particularly in the treatment of psoriasis [12,28,33,55].

8.5.1. Preclinical Studies

Preclinical studies of the mechanism of tapinarof have shown that its effects are strongly dependent on the presence of AhR and have confirmed its anti-inflammatory action, providing the basis for its clinical development [27,80].

Its effects were investigated in an IMQ-induced psoriasis mouse model. Topical application of 1% tapinarof cream alleviated psoriasis symptoms and reduced IL-17A expression in wild-type mice [96]. Tapinarof application also reduced pro-inflammatory cytokine levels and alleviated clinical signs of inflammation in mouse models of IMQ-induced psoriasis and lipopolysaccharide-induced ear oedema [27,29,80,86]. Its safety profile has also been investigated: a skin sensitisation test with 8% tapinarof cream in guinea pigs did not show significant skin irritation. Although human data on the risks of application are not available during lactation and pregnancy, animal studies have shown the presence of tapinarof in breast milk, but there is no evidence of teratogenicity [27].

Zhu et al. investigated the effect of tapinarof in two mouse models of psoriasis induced by imiquimod (IMQ) and interleukin-23 (IL-23). They observed that while tapinarof ameliorated IMQ-induced psoriasis-like dermatitis, there was also evidence of decreased keratinocyte proliferation, reduced epidermal thickness, and normalised differentiation. However, it paradoxically exacerbated inflammation in the IL-23-induced model. Tapinarof treatment in the IL-23 model led to increased epidermal thickness and the presence of differentiated epithelial dysplasia.

In contrast, Urashima et al. demonstrated that tapinarof suppressed IL-23-induced psoriasis-like dermatitis in mice, reducing ear thickness, inflammatory cell infiltration, and pro-inflammatory cytokine expression. A main difference between the two studies was in the timing of tapinarof administration. Zhu et al. applied tapinarof as a prophylactic treatment (concurrently with IL-23 injection), whereas Urashima et al. initiated treatment after inflammation had already been established. This temporal variation may explain the divergent results, suggesting that the immunomodulatory effects of tapinarof are context-dependent and may change according to the stage of the disease [14,97].

Overall, in the mouse models, tapinarof treatment attenuated keratinocyte proliferation and expression of inflammatory cytokines, leading to an alleviation of lesions characteristic of psoriasis. In addition, its antioxidant properties reduced the production of reactive oxygen species in keratinocytes, which further reduced the inflammation [29].

8.5.2. Clinical Studies

In a phase II trial, 227 adult patients (aged 18 to 65 years) with plaque psoriasis were randomised to receive tapinarof cream at a concentration of 1% or 0.5% once or twice daily and were compared to the placebo group to assess efficacy and safety [12,27,33,83]. Treatment success was determined by PGA scores (Physician Global Assessment for Psoriasis) and PASI scores [27,29]. By the end of week 12, a higher proportion of those using 1% tapinarof (65% for twice daily and 56% for once daily) had achieved a "clear" or "almost clear" PGA score compared with the control group (11% for twice daily and 5% for once daily) [33,86,88]. The PASI-50 and PASI-75 response rates were also significantly higher in the tapinarof group (PASI-50: 71–92%, PASI-75: 46–65%) compared to the placebo group (PASI-50: 10–32%, PASI-75: 5–16%) [10,12,27,33].

Two major 12-week phase III trials (PSOARING 1 and PSOARING 2) included a total of 1025 adult patients (aged 18 to 75 years) with moderate to severe psoriasis [27,55,86]. The results confirmed previous observations that patients using tapinarof had improved PGA and PASI scores compared to the control group [10,16,55,86]. In these studies, once-daily application of 1% tapinarof cream significantly reduced PASI scores [10,16,33,90]. By the end of week 12, the PGA endpoint was reached by 35.4–40.2% of patients in the tapinarof group compared to only 6.0–6.3% in the control group [13,16,29]. The PASI-75 improvement rate was 36.1–47.6% in the tapinarof group compared to only 6.9–10.2% in the control group [27,55,88]. Overall, 85.8% of patients in the trials reported that tapinarof helped them to easily manage their psoriasis and 62.9% agreed that tapinarof cleared their skin and prevented recurrence [33,55].

8.5.3. Long-Term Safety Studies

The PSOARING 3 study included 763 patients who had previously completed PSOARING 1 or 2 and were allowed to receive tapinarof for up to 52 weeks from baseline to the end of PSOARING 3 [12,55,86,91]. The aim of the 40-week phase was to evaluate long-term efficacy, safety, and ability to maintain the remission of tapinarof [12,27,33]. In this study, tapinarof 1% cream was applied in long-term, intermittent periods, and treatment provided a lasting effect for 12 weeks followed by a 4-week break [27,86,91]. A total of 40.9% of patients achieved complete skin clearance (PGA score = 0) at least once, and of those with active disease at baseline (PGA score ≥ 2), 58.2% achieved a PGA score = 0 on one or more occasions [12,27,86,88]. The results of the PSOARING 3 trial showed that tapinarof treatment not only produced durable improvement but also helped patients maintain symptom-free status for up to four months after the discontinuation of therapy. Longer-term clinical trials have shown low relapse rates with tapinarof, suggesting that tapinarof not only reduces symptoms but may also contribute to longer remission maintenance [12,27,88,91]. In addition, no tachyphylaxis was observed with prolonged use [13,27,91].

8.5.4. Pharmacokinetics

Based on pharmacokinetic studies, both 1% and 2% tapinarof cream had low transdermal absorption [33]. Peak blood concentrations were observed on the first day of treatment and decreased significantly afterwards, which led to minimal systemic exposure. This is probably because the impaired skin barrier of psoriatic lesions at the start of therapy facilitates increased absorption, which may support early therapeutic efficacy. However, tapinarof activates the AhR during therapy, which boosts the expression of skin barrier proteins such as filaggrin, loricrin, and involucrin. As a consequence, the skin barrier is progressively restored over time, further reducing systemic absorption without decreasing local efficacy [27,33,88,98].

Therefore, the risk of systemic side effects remains low and no significant drug–drug interactions or cardiovascular effects have been reported [13,33]. The drug is mainly metabolised in the liver via CYP1A2 and CYP3A4 enzymes, while oxidation, sulfation, and glucuronidation occur in hepatocytes and the drug is strongly bound to plasma proteins (~99%) [27,33,86].

8.5.5. Side Effects

During treatment, the most commonly reported side effect is folliculitis [82,88,91]. Contact dermatitis [16,28,86], upper respiratory tract infection [10,12,55], headache [10,16,88], and itching [27,55,88] have also been reported. Other less common adverse reactions are nasopharyngitis, influenza, urticaria, and drug-induced skin rash [27]. The majority of adverse reactions are mild to moderate in severity, and only a small proportion of patients discontinue treatment because of them [28,33,86,90].

Clinical studies have shown that VTAMA (tapinarof 1% cream) is generally well-tolerated in patients with psoriasis. Most adverse events have been mild to moderate in severity, and long-term use of the cream has not revealed any new safety risks [13,55,88].

Clinical trials are currently ongoing in the paediatric population (2–17 years), in which all participants receive 1% tapinarof cream once daily [99]. This is important because approximately 25% of psoriasis cases occur before the age of 18 years [27,90]. Trials have been carried out in atopic dermatitis, and the results show that tapinarof is safe for children over 2 years [100–103]. However, further research is needed to evaluate the safety of the product in pregnant patients [27,90].

Ghani et al. compared tapinarof with another new topical agent, roflumilast, in terms of safety and efficacy. Their findings showed that tapinarof was associated with more frequent but less severe side effects (folliculitis and contact dermatitis), while roflumilast showed less frequent but more severe side effects (diarrhoea, headache, and insomnia). However, further comparative clinical trials are needed to determine which agent is more effective and cost-effective in the long term [104].

However, further combination treatment options remain the subject of required trials. Research is ongoing to evaluate whether tapinarof in combination with other therapeutic modalities, such as biological agents or other topical agents, can enhance the efficacy and durability of treatment [12,33].

In parallel, studies are being carried out to assess the therapeutic equivalence of the tapinarof cream 1% developed by Teva Pharmaceuticals USA and the VTAMA tapinarof cream 1% already approved by Dermavant Sciences in adult patients with plaque psoriasis [105]. Separate studies have also investigated the efficacy and safety of the VTAMA (tapinarof) cream, 1% formulation, in patients with intertriginous psoriasis (underarm or groin area) [106].

Overall, the results of clinical trials have shown that the once-daily use of 1% tapinarof cream is safe and effective for the treatment of mild-to-moderate psoriasis for up to one year [13,91]. This cream's favourable clinical profile makes it a promising therapeutic option, particularly for those who do not respond well to other treatments or who are seeking a long-term, sustainable, steroid-free therapy [16,28,33].

9. Nanotechnology in Dermatology

Nanotechnology is a scientific field that focuses on the development, synthesis, and characterisation of materials at the nanoscale (1–100 nm), creating new opportunities for drug delivery [8,17,107]. In recent decades, there has been a rising interest in nanotechnological solutions for dermal and transdermal therapies, particularly in the development of nanoparticle drug delivery systems, which have been playing a dominant role since the 1990s [31,108–110]. The application of nanoparticles in drug delivery provided valuable opportunities to enhance the therapeutic outcomes of different types of drugs, and may offer innovative strategies in the treatment of psoriasis [30,31].

One of the most important aims of pharmaceutical nanotechnology is the application of therapeutic and biocompatible drug carriers in nanoforms [111]. Nanocarriers provide the possibility for targeted delivery of active ingredients, which increases bioavailability and thereby enhances therapeutic efficacy [8,107,112].

Nanoparticles play a significant role in transdermal drug delivery, as their large surface area promotes skin penetration, retention, and sustained release [37,107,111]. Their efficacy is highly dependent on their penetration through the skin barrier, which is influenced by parameters such as particle size, molecular weight, surface charge, and pH [31,107,111]. Conventional drug formulations, such as creams and ointments, often have limited permeation, which makes it difficult to deliver drugs through the skin [9]. In contrast, nanotechnology can evade this obstacle by reducing particle size, which improves drug penetration through the stratum corneum, offering the possibility of targeted delivery of drugs to deeper layers of the skin [18,31,111].

Furthermore, drugs encapsulated in nanocarriers are protected from degradation, increasing the half-life of the active drug and minimising systemic toxicity [18,25,41]. In addition, their large surface area promotes the solubility of active ingredients and thus their bioavailability [31,112,113].

This is especially essential for hydrophobic drugs, as a significant proportion of novel drugs are lipophilic in nature and have poor water solubility, which limits their bioavailability and drug delivery efficacy [108]. Approximately 40% of drugs on the market and nearly 90% of molecules under development have poor water solubility, which is often a major cause of therapeutic failure [108,114,115]. Nanotechnology has revolutionised this field, as nanoparticle-based drug formulations can significantly improve the dissolution rate, permeation, and therapeutic efficacy of drugs [108,114]. This is advantageous in topical drug delivery, where the stratum corneum is the major limiting factor, acting as a primary barrier to drug entry [2,111,116]. Nanotechnology-based formulations offer an excellent opportunity for targeted and effective transdermal delivery of drugs with limited water solubility [108,112,114].

The active substances used to treat psoriasis (corticosteroids, retinoids, and immunomodulators) often have limited skin penetration and can cause severe side effects [8,111]. Nano-sized drug carriers offer the potential to address these issues [8,18].

10. Nanoparticle Carriers

In recent decades, several new generations of nanoparticle carriers have emerged that offer promising opportunities for topical therapies against psoriasis. Nanoparticle carriers can be grouped into four main categories: nanoparticles, nanofibres, physical carriers, and matrix nanocarriers [8,117]. In this section, we evaluate the advantages, limitations, and applicability of the most common nanoparticle types used in experimental psoriasis therapies.

10.1. Nanoparticles

10.1.1. Vesicular Carriers

Liposomes, niosomes, transferosomes, and ethosomes are vesicular carriers [8,117].

Liposomes are nano-sized, phospholipid-based vesicular structures that form spontaneously in aqueous environments and are widely used as biocompatible, low-toxicity nanocarriers to enhance drug solubility, controlled release, and targeted skin delivery [8].

In a recent study, cyclosporin-loaded cationic liposomes developed by Walunj et al. achieved an entrapment efficiency of 93%, with a particle size of ~111 nm and zeta potential of +41 mV. In vivo application of the liposomal gel decreased psoriasis-related symptoms in an IMQ-induced mouse model. The treatment led to a 3.40-fold reduction in IL-22, a 1.47-fold reduction in IL-17, and a 1.71-fold reduction in TNF-α levels. Significant reductions were also observed in PASI scores, ear thickness, and spleen-to-body weight ratios, which confirm their therapeutic potential. Despite these advantages, challenges related to large-scale production, formulation reproducibility, and long-term physical stability hinder their broader clinical translation. Moreover, high drug loading can also lead to precipitation or particle aggregation, particularly when using thin-film hydration methods [118].

Niosomes are non-phospholipid-based bilayer vesicles composed of non-ionic surfactants and cholesterol, offering improved chemical stability and lower production costs compared to liposomes [17,117]. Their structural flexibility provides improved skin penetration [17].

A study by Abu Hashim et al. investigated an acitretin-loaded niosomal gel for topical psoriasis therapy. The optimised formulation achieved high drug entrapment (90.32 $\pm$ 3.80%), a particle size of 369.73 $\pm$ 45.45 nm, and a zeta potential of -36.33 ± 1.80 mV. Ex vivo skin permeation assays demonstrated a 3.16-fold increase in cumulative permeation and significantly enhanced drug deposition in the viable epidermis and dermis (330.86 $\pm$ 5.32 $\mu g/cm^2$), compared to a conventional acitretin gel. In vivo, the formulation showed antipsoriatic efficacy in a mouse tail model, promoting orthokeratosis and reducing epidermal thickness, with no signs of skin irritation. Despite these promising findings, its clinical relevance remains uncertain due to the lack of human data. Furthermore, long-term stability and safety were not fully carried out, as the formulation was only tested under limited storage conditions and short-term application [119].

Transfersomes are vesicular particles consisting of ultradeformable lipid bilayers and at least one internal aqueous compartment. Their flexibility allows the vesicles to deform without structural damage and effectively penetrate the skin barrier, thereby facilitating targeted delivery of the drug to deeper layers. They provide a more efficient and controlled drug delivery than conventional liposomes [8,17].

Vohra et al. developed a nano-transferosomal gel loaded with aloe vera and vitamin E, which showed high entrapment efficiency (92.29%) and a stable zeta potential (−38.5 mV). The nanoformulations demonstrated notable anti-inflammatory effects and skin compatibility during topical application and significantly outperformed a commercial vitamin E cream. This was evidenced by results obtained in a TPA-induced mouse ear oedema model. However, the study lacked in vivo psoriasis-specific efficacy data, limiting its disease-specific relevance. Thus, while the carrier shows promise for herbal-based anti-inflammatory applications, its specific benefit in psoriasis therapy remains to be demonstrated in appropriate disease models [120].

Ethosomes are also ultradeformable phospholipid-based nanoparticles containing a high concentration of ethanol (20–45%) [8]. One of the major advantages of ethosomes over other lipid-based nanocarriers such as liposomes or transferosomes is that they are more efficient in penetrating the stratum corneum due to their ethanol content [8,41,117].

An ethosomal gel based on tacrolimus and hyaluronic acid was formulated by Dadwal and co-workers. The optimised nanoscale vesicles had high drug loading efficiency and excellent physicochemical properties. Ex vivo skin permeation studies using goat skin revealed a flux of 90.22 ± 0.52 μg/cm^2/h for the ethosomal gel, compared to 72.15 ± 0.31 μg/cm^2/h for the commercial tacrolimus formulation, resulting in an enhancement ratio of 1.33. These results indicate improved transdermal absorption and suggest that the combination of ethosomal nanocarriers with hyaluronic acid can offer a promising strategy for the topical treatment of psoriasis. However, further comparative preclinical and clinical studies are needed to confirm long-term safety and efficacy and to quantitatively assess the benefit over other nanocarriers [121].

10.1.2. Lipid Nanoparticles

Examples of lipid nanoparticles include solid lipid nanoparticles (SLNs) and nanostructured lipid carriers (NLCs).

Solid lipid nanoparticles are spherical, lipid-based nanoparticles that stay solid at room and body temperature [17,117]. One of their main advantages over conventional liposomes is their more efficient drug delivery, which includes higher drug loading capacity, controlled release, and enhanced stability of the active substances [17,117].

In their study, Serini and co-workers reported that solid lipid nanoparticles containing curcumin and α-linolenic acid were able to reduce IMQ-induced inflammation in an in vitro psoriasis model, which was evidenced by decreases of 43% in IL-23, 73.7% in IL-6, and 26.5% in IL-8 expression in macrophages. In addition, the levels of two markers of ferroptosis (TFRC and MDA) were significantly reduced, suggesting an antioxidant effect. However, while these findings indicate that curcumin and α-linolenic acid-loaded SLNs may represent a promising nanocarrier-based strategy, the effects have so far only been demonstrated in vitro and further preclinical and clinical studies are needed to assess the translational potential [122].

Nanostructured lipid carriers contain physiologically compatible lipids, surfactants, and emulsifying agents. They can enhance skin hydration, strengthen the skin barrier, improve bioavailability, and provide targeted drug delivery. NLCs are particularly advantageous in the field of drug delivery systems due to their ease of production, biocompatibility, non-toxicity, and scalability [8].

For the treatment of hyperproliferative skin disorders, Llorente et al. developed riluzole-containing nanostructured lipid carriers. The optimised formulation demonstrated a mean particle size under 200 nm, a high entrapment efficiency (~87%), and a zeta potential around −25 mV. In vitro studies using HaCaT cells demonstrated that the formulation significantly inhibited cell proliferation in a dose-dependent manner ($p < 0.0001$), with effects comparable to free riluzole. In animal models, topical application of the preparation led to reduced skin thickness and visible alleviation of inflammatory signs. Although the study demonstrated promising physicochemical properties and therapeutic effects of riluzole-loaded NLCs, it did not include comprehensive in vivo toxicity data. Additionally, the formulation process is relatively complex due to the low solubility and light sensitivity of the active ingredient, requiring optimisation for stability and efficacy [123].

10.2. Nanofibres

Nanofibres are fibres or fibre-like structures made from natural or synthetic polymers that provide fluid absorption, moisture control, and gas permeability [117]. Nanofibres have an excellent surface-to-volume ratio, which allows for the efficient delivery of both hydrophilic and hydrophobic drugs [124]. They are promising in the treatment of psoriasis, where they are often used in combination with nanoparticles to promote synergistic drug release and skin regeneration [117].

Kang and coworkers investigated curcumin-filled nanofibre films using electrostatic fibre pulling. The optimised film contained curcumin at 58.7 ± 10.1 $\mu g/cm^2$ with an entrapment efficiency of 56.5 ± 9.7% and tensile strength of 4.86 ± 0.14 MPa. In vitro skin deposition studies using a psoriasis mouse model showed a >2.0-fold increase in curcumin deposition with lipid-hybridised films compared to lipid-free controls. In vivo, treatment with the hybrid film significantly reduced skin thickening, scaling, and pro-inflammatory cytokines (TNF-α and IL-6), with anti-inflammatory efficacy approaching that of a commercial corticosteroid. These advantages suggest strong potential in psoriasis therapy. However, challenges such as curcumin-induced skin staining and limited dermal penetration highlight the necessity for further optimisation [125].

10.3. Physical Carriers

These systems are hybrids of hypodermic needles and transdermal patches containing hundreds of tiny, microscale needles, which create microscopic channels through the stratum corneum, enhancing drug penetration efficiency [117,126].

Du et al. designed a microneedle patch containing hyaluronic acid and methotrexate for the topical treatment of psoriasis. In vitro studies showed that the preparation preserved the antiproliferative activity of API against HaCaT keratinocytes and achieved ~90% drug release within 1 h. In vivo, the microneedles successfully penetrated both normal and psoriatic mouse skin (~150 μm), dissolved within 10 min, and significantly reduced epidermal thickness, ear swelling, and cytokine expression (IL-17, IL-23, Ki67) in an imiquimod-induced psoriasis model. Compared to oral methotrexate at the same dose (13.8 μg), microneedle-delivered methotrexate achieved superior therapeutic outcomes with less systemic toxicity. However, disadvantages include reduced mechanical strength at higher drug loads and the lack of long-term safety evaluation [126].

10.4. Matrix Nanocarriers

Matrix-based nanocarriers such as nanoemulsions and nanogels offer promising advantages in transdermal drug delivery. Nanoemulsions are kinetically stable dispersions of immiscible liquids (typically oil and water) with droplet sizes between 10 and 200 nm, which enhances drug solubility and bioavailability [112,113,115,116]. However, their low viscosity makes their application problematic, which can be solved using nanogels [115,127]. By incorporating nanoemulsions into a gel matrix, nanogels combine the advantages of gel formulations and nanotechnology to provide a more efficient drug delivery system [115].

Nanogels are a three-dimensional, nanoscale (20 to 250 nm) network of hydrophilic polymers with viscoelastic properties [109,111,128]. They are made from polymers or through heterogeneous polymerisation of monomers, and their structure can be stabilised by cross-linkages, either physical or chemical [3,8].

Their constitution affects their biocompatibility and biodegradation, which minimises toxicity and immune response. They are able to provide stable delivery and enhanced skin penetration of different types of drugs, provide protection against degradation and environmental effects, and also increase the stability and bioavailability of drugs [25,109,117,129]. Due to their high water retention capacity, they hydrate the skin surface while providing controlled and prolonged drug release, maintaining the desired therapeutic effect over a longer period of time [8,23,117,128].

Since psoriasis treatment requires long-term commitment, patient adherence is essential [3,8]. Nanogels support this by enabling sustained drug release, reducing dosing frequency, and improving treatment comfort and compliance [8,25,41,111].

Previous studies have confirmed that these new nanoformulations may present exciting potential not only for the treatment of psoriasis but also for other skin conditions.

Chandrashekhar and co-workers evaluated a tretinoin nanogel (0.025%) versus a conventional gel in a randomised, multicentre clinical trial involving 207 acne patients. The nanogel group showed significantly higher reductions in total (72.9% vs. 65.0%; $p = 0.03$) and inflammatory lesions (78.1% vs. 66.9%; $p = 0.02$), and fewer local adverse events (13.3% vs. 24.7%; $p = 0.04$) compared to the conventional formulation [130].

Avasatthi et al. developed a methotrexate-loaded nanostructured lipid carrier nanogel that achieved 47.32% drug release at 48 h, compared to 94.23% from conventional methotrexate gel, indicating sustained release. In a psoriasis mouse model, the preparation reduced PASI scores and restored skin histology, while the conventional gel resulted in persistent parakeratosis [131].

Kakade et al. formulated a tacrolimus-loaded nanogel based on nanostructured lipid carriers, which demonstrated over 90% sustained drug release over 24 h, excellent spreadability, and a high drug content of 99.73 ± 1.4%. In vivo, the nanogel improved skin elasticity, resolved psoriatic lesions in oxazolone and imiquimod models, and showed no cytotoxicity or irritation, supporting its effectiveness and safety [132].

Despite the remarkable advantages, their production faces several technological challenges, especially in high-volume manufacturing where reproducibility, stability, and cost-effectiveness are main factors [110,129,133].

Their stability is a critical factor, as they are sensitive to storage conditions such as temperature fluctuations and changes in humidity [110,112,133]. In addition, the desired quality cannot always be ensured in high-volume production [42,110,112].

As these materials are gradually degraded in the body, long-term biocompatibility testing is essential as it is important that they do not cause any toxic side effects [41,133]. In light of this, nanogels have noticeably more advantages than disadvantages, as presented in Figures 3 and 4 [3].

Nevertheless, nanogels represent one of the most promising developments in the field of drug delivery systems and are expected to play a prominent role in the treatment of advanced drug formulations such as psoriasis and other skin diseases in the future [134].

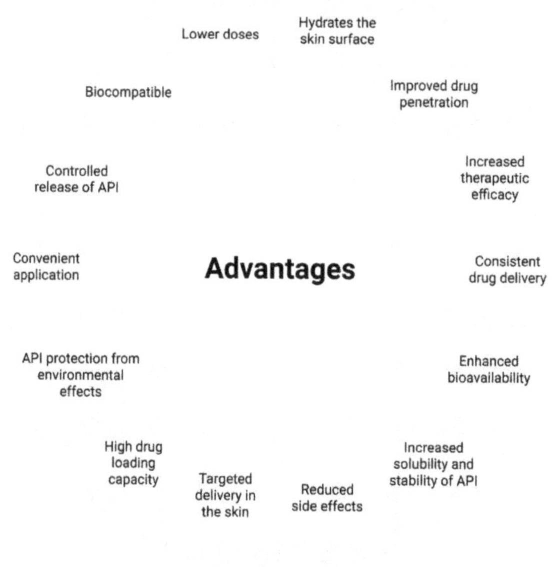

Figure 3. Advantages of nanogels in transdermal drug delivery (created with BioRender.com).

Figure 4. Disadvantages of nanogels in transdermal drug delivery (created with BioRender.com).

11. Tapinarof Delivery Systems

11.1. Patented Tapinarof Formulations

Currently, the only approved and commercially available formulation of tapinarof is VTAMA (tapinarof cream 1%), which was developed and patented by Dermavant Sciences [13,55,88]. This formulation is protected by a series of patents (e.g., US11590088B2, US11938099B2) covering its composition and use for chronic dermatological conditions such as plaque psoriasis [135]. and atopic dermatitis [136]. Other formulations exist at the research level or as patent registrations but not for clinical use. A US patent (US20210000758A1) describes the encapsulation of tapinarof in nanoparticles (e.g., nanomicelles, nanospheres) for ophthalmic applications such as the treatment of uveitis or macular degeneration [137]. Another patent (US20220160650A1) describes various tapinarof-based gel, ointment, and foam formulations designed to improve the stability and skin absorption of the active ingredient [138].

11.2. Challenges in Formulating Tapinarof

One of the major challenges in the dermatological application of tapinarof is its hydrophobic nature and instability, which make it difficult to formulate in conventional systems [139]. The compound is highly sensitive to temperature and humidity, which can negatively impact long-term stability and reduce therapeutic efficacy during storage [79,110,112,133].

Nanogels offer a promising strategy to overcome these challenges, as they are able to encapsulate the active ingredient, thereby protecting it from environmental degradation and improving long-term shelf-life [18,25,41]. Non-ionic surfactants and natural polymers not only support the solubilisation of the drug, but also improve biocompatibility and skin tolerability [10,24].

Furthermore, considering the chronic and relapsing nature of psoriasis, formulations should be designed for safe, prolonged use without causing irritation or sensitisation [10,20]. Patient adherence can be further improved by ensuring favourable application properties (e.g., texture, spreadability) and reducing the dosing frequency. Altogether, these factors highlight the urgent need for effective, steroid-free, and well-tolerated topical therapies that are suitable for long-term monotherapy or combination therapy without the risk of drug–drug interactions [28,33,55].

11.3. Nanogel-Based Tapinarof Formulations

To overcome formulation-related challenges of tapinarof, Balogh et al. developed nanogel-based delivery systems aiming to improve drug solubility, skin penetration, and stability. The nanogels were prepared using Carbopol 940 and 936 polymers combined with excipients such as tween 80, kolliphor, and oleic acid to enhance therapeutic performance. Dynamic light scattering (DLS) confirmed nano-sized particle distributions (151–173 nm), and rheological analyses demonstrated pseudoplastic behaviour with temperature-dependent viscosity, which is very important for spreadability and controlled release at skin temperature. During texture analysis, the formulations exhibited low compressive resistance, ensuring easy topical application. Cytotoxicity was evaluated on HaCaT keratinocyte cells via MTT assays. None of the tested formulations showed cytotoxicity. Notably, nanogel II showed a better release profile in the Franz diffusion assays, as 81% of tapinarof was released after 5 h compared to 52% for nanogel I, with a corresponding increase in drug flow. A wound healing study showed that the incorporation of tapinarof into the nanoformulations enhanced its antiproliferative and antimigratory activity.

Taken together, these findings indicate that nanogel-based tapinarof delivery systems are a promising candidate for topical psoriasis therapy [139].

11.4. Future Directions

Preliminary in vitro studies on tapinarof-loaded nanogels have demonstrated promising results in terms of active ingredient absorption and cellular tolerability, indicating their potential as a novel topical delivery system [139]. However, these findings represent only an early stage of development. To support clinical application, further preclinical and clinical studies are essential to confirm the efficacy and safety of these formulations in vivo.

Preclinical studies in animal models (e.g., IMQ-induced mouse model) can provide information regarding dosing, pharmacokinetics, pharmacodynamics, and long-term stability.

Following successful preclinical testing, phase I clinical trials should be conducted in healthy volunteers to assess safety, local tolerability, and pharmacokinetics. If these results are appropriate, phase II trials involving psoriasis patients can evaluate efficacy and dose optimisation. Following these, phase III randomised, double-blind, placebo-controlled studies are required to confirm therapeutic benefit, long-term safety, and patient adherence. Although this is a lengthy process, it is necessary to make tapinarof nanogels a reliable and effective treatment option for patients with psoriasis.

During development, it is also essential to compare tapinarof nanogels with existing psoriasis treatment modalities, for example, steroids or other topical therapies. These comparative studies may reveal the benefits and potential limitations of nanogels, helping to determine their clinical applicability more accurately [12].

Additionally, efforts must focus on establishing reproducible, cost-effective manufacturing processes, as well as ensuring long-term stability and wide applicability. Addressing these challenges is essential to support the clinical integration of tapinarof nanogels as a safe, patient-friendly alternative for long-term topical management of chronic plaque psoriasis [28,39].

12. Conclusions

The complex and multifactorial pathophysiology of psoriasis continues to make it difficult to develop truly optimal treatments. Despite major advances in the use of systemic biologic agents for the treatment of moderate to severe forms of the disease, a large proportion of psoriasis cases remain untreated, mainly due to side effects of systemic drug treatments or inappropriate drug delivery through the stratum corneum in topical treatments. In this context, nanotechnology offers a revolutionary solution, particularly nanogel formulations that enhance drug penetration and bioavailability at the place of action. With proven potency in the treatment of psoriasis, incorporating tapinarof into nanogels can significantly improve therapeutic outcomes by targeting the skin more effectively and reducing potential side effects. Nanogels can also provide a controlled sustained release of the active ingredient, enhancing the stability and efficacy of tapinarof. This approach can enhance both the convenience and long-term efficacy of topical therapies. Additionally, integrating nanotechnology into psoriasis management could help tailor therapies to the individual needs of patients, improving their adherence to treatment and overall quality of life. Eventually, the incorporation of tapinarof nanogels into clinical practice could revolutionise the way psoriasis is treated, offering a patient-friendly approach to treating this challenging dermatological condition. Future studies should further explore tapinarof-loaded nanogel formulations to support their clinical translation and ensure safety, efficacy, and long-term therapeutic benefits in patients with psoriasis.

Author Contributions: B.B. and I.B. designed and wrote the manuscript. Á.P., P.F. and Z.U. helped review the scientific literature. Á.P., P.F. and Z.U. helped in editing the manuscript and provided their professional opinions and suggestions. All authors have read and agreed to the published version of the manuscript.

Funding: Project no. TKP2021-EGA-18 has been implemented with the support provided by the Ministry of Culture and Innovation of Hungary from the National Research, Development and Innovation Fund, financed under the TKP2021-EGA funding scheme and by 2022-1.2.2-TÉT-IPARI-UZ-2022-00006 Common research and development of different prototypes containing natural herb extract for industrial utilisation. The scientific work/research and/or results publicised in this article were achieved with the sponsorship of Gedeon Richter Talentum Foundation in framework of Gedeon Richter Excellence PhD Scholarship of Gedeon Richter.

Institutional Review Board Statement: Not applicable.

Informed Consent Statement: Not applicable.

Data Availability Statement: Not applicable.

Conflicts of Interest: The authors declare no conflicts of interest.

References

1. Madawi, E.A.; Al Jayoush, A.R.; Rawas-Qalaji, M.; Thu, H.E.; Khan, S.; Sohail, M.; Mahmood, A.; Hussain, Z. Polymeric Nanoparticles as Tunable Nanocarriers for Targeted Delivery of Drugs to Skin Tissues for Treatment of Topical Skin Diseases. *Pharmaceutics* **2023**, *15*, 657. [CrossRef] [PubMed]
2. Singh, S.; Awasthi, R. Breakthroughs and Bottlenecks of Psoriasis Therapy: Emerging Trends and Advances in Lipid Based Nano-Drug Delivery Platforms for Dermal and Transdermal Drug Delivery. *J. Drug Deliv. Sci. Technol.* **2023**, *84*, 104548. [CrossRef]
3. Siafaka, P.I.; Özcan Bülbül, E.; Okur, M.E.; Karantas, I.D.; Üstündağ Okur, N. The Application of Nanogels as Efficient Drug Delivery Platforms for Dermal/Transdermal Delivery. *Gels* **2023**, *9*, 753. [CrossRef]
4. Jurel, P.; Bahadur, S.; Bajpai, M. Treatment of Chronic Plaque Psoriasis: An Overview on Current Update. *Pharmacol. Res.-Rep.* **2024**, *2*, 100004. [CrossRef]
5. Biswasroy, P.; Pradhan, D.; Kar, B.; Ghosh, G.; Rath, G. Recent Advancement in Topical Nanocarriers for the Treatment of Psoriasis. *AAPS PharmSciTech* **2021**, *22*, 164. [CrossRef]
6. Makuch, S.; Dróżdż, M.; Makarec, A.; Ziółkowski, P.; Woźniak, M. An Update on Photodynamic Therapy of Psoriasis—Current Strategies and Nanotechnology as a Future Perspective. *Int. J. Mol. Sci.* **2022**, *23*, 9845. [CrossRef]
7. Dai, P.; Ge, X.; Sun, C.; Jiang, H.; Zuo, W.; Wu, P.; Liu, C.; Deng, S.; Yang, J.; Dai, J.; et al. A Novel Methacryloyl Chitosan Hydrogel Microneedles Patch with Sustainable Drug Release Property for Effective Treatment of Psoriasis. *Macromol. Biosci.* **2023**, *23*, 2300194. [CrossRef]
8. Bodnár, K.; Fehér, P.; Ujhelyi, Z.; Bácskay, I.; Józsa, L. Recent Approaches for the Topical Treatment of Psoriasis Using Nanoparticles. *Pharmaceutics* **2024**, *16*, 449. [CrossRef]
9. Zhu, B.; Jing, M.; Yu, Q.; Ge, X.; Yuan, F.; Shi, L. Treatments in Psoriasis: From Standard Pharmacotherapyto Nanotechnology Therapy. *Adv. Dermatol. Allergol.* **2022**, *39*, 460–471. [CrossRef]
10. Sugumaran, D.; Yong, A.C.H.; Stanslas, J. Advances in Psoriasis Research: From Pathogenesis to Therapeutics. *Life Sci.* **2024**, *355*, 122991. [CrossRef]
11. Igarashi, A.; Tsuji, G.; Fukasawa, S.; Murata, R.; Yamane, S. Tapinarof Cream for the Treatment of Plaque Psoriasis: Efficacy and Safety Results from 2 Japanese Phase 3 Trials. *J. Dermatol.* **2024**, *51*, 1269–1278. [CrossRef] [PubMed]
12. Carmona-Rocha, E.; Rusiñol, L.; Puig, L. New and Emerging Oral/Topical Small-Molecule Treatments for Psoriasis. *Pharmaceutics* **2024**, *16*, 239. [CrossRef] [PubMed]
13. Silverberg, J.I.; Boguniewicz, M.; Quintana, F.J.; Clark, R.A.; Gross, L.; Hirano, I.; Tallman, A.M.; Brown, P.M.; Fredericks, D.; Rubenstein, D.S.; et al. Tapinarof Validates the Aryl Hydrocarbon Receptor as a Therapeutic Target: A Clinical Review. *J. Allergy Clin. Immunol.* **2024**, *154*, 1–10. [CrossRef] [PubMed]
14. Urashima, T.; Katsuda, Y.; Yoshiuchi, H.; Ebihara, S.; Shinozaki, Y.; Kato, T.; Shimazaki, T.; Yasui, Y.; Mera, Y.; Konishi, N. Pharmacological Properties of Tapinarof in Mice as a Novel Topical Agent for Plaque Psoriasis. *BPB Rep.* **2024**, *7*, 116–122. [CrossRef]
15. Saha, S. Transcutaneous Delivery of Disease Specific PI3K/Akt/mTOR Inhibitor Based Hybrid Nanoparticles in Hydrogel System for the Management of Psoriasis: Insights from in Vivo Studies. *Int. J. Pharm.* **2025**, *627*, 125325.

16. Nogueira, S.; Rodrigues, M.A.; Vender, R.; Torres, T. Tapinarof for the Treatment of Psoriasis. *Dermatol. Ther.* **2022**, *35*, e15931. [CrossRef]
17. Mascarenhas-Melo, F.; Carvalho, A.; Gonçalves, M.B.S.; Paiva-Santos, A.C.; Veiga, F. Nanocarriers for the Topical Treatment of Psoriasis—Pathophysiology, Conventional Treatments, Nanotechnology, Regulatory and Toxicology. *Eur. J. Pharm. Biopharm.* **2022**, *176*, 95–107. [CrossRef]
18. Ahmad, M.Z.; Mohammed, A.A.; Algahtani, M.S.; Mishra, A.; Ahmad, J. Nanoscale Topical Pharmacotherapy in Management of Psoriasis: Contemporary Research and Scope. *J. Funct. Biomater.* **2022**, *14*, 19. [CrossRef]
19. Yadav, T.; Yadav, H.; Raizaday, A.; Alam, M.S. The Treatment of Psoriasis via Herbal Formulation and Nanopolyherbal Formulation: A New Approach. *BioImpacts* **2024**, *15*, 30341. [CrossRef]
20. Ortiz-Lopez, L.I.; Choudhary, V.; Bollag, W.B. Updated Perspectives on Keratinocytes and Psoriasis: Keratinocytes Are More Than Innocent Bystanders. *Psoriasis Targets Ther.* **2022**, *12*, 73–87. [CrossRef]
21. Sieminska, I.; Pieniawska, M.; Grzywa, T.M. The Immunology of Psoriasis—Current Concepts in Pathogenesis. *Clin. Rev. Allergy Immunol.* **2024**, *66*, 164–191. [CrossRef] [PubMed]
22. Lé, A.M.; Torres, T. New Topical Therapies for Psoriasis. *Am. J. Clin. Dermatol.* **2022**, *23*, 13–24. [CrossRef] [PubMed]
23. Vasowala, T.; Gharat, S.; Mhase, M.; Momin, M. Advances in Hydrogels Based Cutaneous Drug Delivery System for Management of Psoriasis. *Eur. Polym. J.* **2024**, *202*, 112630. [CrossRef]
24. Aziz Hazari, S.; Kaur, H.; Karwasra, R.; Abourehab, M.A.S.; Ali Khan, A.; Kesharwani, P. An Overview of Topical Lipid-Based and Polymer-Based Nanocarriers for Treatment of Psoriasis. *Int. J. Pharm.* **2023**, *638*, 122938. [CrossRef]
25. Gomes, G.S.; Frank, L.A.; Contri, R.V.; Longhi, M.S.; Pohlmann, A.R.; Guterres, S.S. Nanotechnology-Based Alternatives for the Topical Delivery of Immunosuppressive Agents in Psoriasis. *Int. J. Pharm.* **2023**, *631*, 122535. [CrossRef]
26. Bissonnette, R.; Stein Gold, L.; Rubenstein, D.S.; Tallman, A.M.; Armstrong, A. Tapinarof in the Treatment of Psoriasis: A Review of the Unique Mechanism of Action of a Novel Therapeutic Aryl Hydrocarbon Receptor–Modulating Agent. *J. Am. Acad. Dermatol.* **2021**, *84*, 1059–1067. [CrossRef]
27. Prabath, I.; Subramanian, S.; Rani, J. Tapinarof: A Felicitous Discovery in Psoriasis Treatment. *J. Pharmacol. Pharmacother.* **2022**, *13*, 222–228. [CrossRef]
28. Assaf, J.; Sarkis, J.; Tomb, R. Tapinarof and the Future of Topical Treatments in Plaque Psoriasis. *J. Am. Acad. Dermatol.* **2021**, *84*, e283–e284. [CrossRef]
29. Lin, X.; Meng, X.; Lin, J. The Role of Aryl Hydrocarbon Receptor in the Pathogenesis and Treatment of Psoriasis. *J. Cutan. Med. Surg.* **2024**, *28*, 276–286. [CrossRef]
30. Shen, Q.; Liu, R.; Tan, S.; Xu, X.; Fang, J.; Li, R. Advances in Pathogenesis and Nanoparticles (NPs)-Mediated Treatment of Psoriasis. *Front. Immunol.* **2022**, *13*, 1089262. [CrossRef]
31. Lee, H.-J.; Kim, M. Challenges and Future Trends in the Treatment of Psoriasis. *Int. J. Mol. Sci.* **2023**, *24*, 13313. [CrossRef] [PubMed]
32. Sindrilaru, A.; Filip, A.; Scharffetter-Kochanek, K.; Crisan, D. How Can Nanoparticle-based Technologies Revolutionize the Topical Therapy in Psoriasis? *Exp. Dermatol.* **2020**, *29*, 1097–1103. [CrossRef] [PubMed]
33. Grossmann, M.C.; Pixley, J.N.; Feldman, S.R. A Review of Topical Tapinarof for the Treatment of Plaque Psoriasis. *Ann. Pharmacother.* **2024**, *58*, 76–85. [CrossRef]
34. Chandy, R.J.; Dao, D.-P.D.; Rivis, C.C.; Shan, D.M.; Feldman, S.R. Noncorticosteroid Topical Therapies for the Treatment of Plaque Psoriasis: A Narrative Review. *J. Pharm. Technol.* **2023**, *39*, 247–255. [CrossRef]
35. Wazir, A.; O'Toole, E.A. Itching for Innovation: The Role of Aryl Hydrocarbon Receptor Agonists as a Future Therapy for Atopic Dermatitis. *Clin. Exp. Dermatol.* **2024**, *50*, llae502. [CrossRef]
36. Cai, Z.; Zeng, Y.; Shi, X.; Zhang, X.; Zhu, H.; Wang, W. Benvitimod Inhibits MCM6-Meditated Proliferation of Keratinocytes by Regulating the JAK/STAT3 Pathway. *J. Dermatol. Sci.* **2023**, *109*, 71–79. [CrossRef]
37. Mohd Nordin, U.U.; Ahmad, N.; Salim, N.; Mohd Yusof, N.S. Lipid-Based Nanoparticles for Psoriasis Treatment: A Review on Conventional Treatments, Recent Works, and Future Prospects. *RSC Adv.* **2021**, *11*, 29080–29101. [CrossRef]
38. Armstrong, A.W.; Siegel, M.P.; Bagel, J.; Boh, E.E.; Buell, M.; Cooper, K.D.; Callis Duffin, K.; Eichenfield, L.F.; Garg, A.; Gelfand, J.M.; et al. From the Medical Board of the National Psoriasis Foundation: Treatment Targets for Plaque Psoriasis. *J. Am. Acad. Dermatol.* **2017**, *76*, 290–298. [CrossRef]
39. Peppers, J.; Paller, A.S.; Maeda-Chubachi, T.; Wu, S.; Robbins, K.; Gallagher, K.; Kraus, J.E. A Phase 2, Randomized Dose-Finding Study of Tapinarof (GSK2894512 Cream) for the Treatment of Atopic Dermatitis. *J. Am. Acad. Dermatol.* **2019**, *80*, 89–98.e3. [CrossRef]
40. Pareek, A.; Kumari, L.; Pareek, A.; Chaudhary, S.; Ratan, Y.; Janmeda, P.; Chuturgoon, S.; Chuturgoon, A. Unraveling Atopic Dermatitis: Insights into Pathophysiology, Therapeutic Advances, and Future Perspectives. *Cells* **2024**, *13*, 425. [CrossRef]

41. Murphy, E.C.; Schaffter, S.W.; Friedman, A.J. Nanotechnology for Psoriasis Therapy. *Curr. Dermatol. Rep.* **2019**, *8*, 14–25. [CrossRef]
42. Petit, R.G.; Cano, A.; Ortiz, A.; Espina, M.; Prat, J.; Muñoz, M.; Severino, P.; Souto, E.B.; García, M.L.; Pujol, M.; et al. Psoriasis: From Pathogenesis to Pharmacological and Nano-Technological-Based Therapeutics. *Int. J. Mol. Sci.* **2021**, *22*, 4983. [CrossRef] [PubMed]
43. Valenzuela, F.; Flores, R. Immunogenicity to Biological Drugs in Psoriasis and Psoriatic Arthritis. *Clinics* **2021**, *76*, e3015. [CrossRef]
44. Fitzgerald, T.; Zhdanava, M.; Pilon, D.; Shah, A.; Hilts, A.; Lefebvre, P.; Feldman, S.R. Long-Term Psoriasis Control with Guselkumab, Adalimumab, Secukinumab, or Ixekizumab in the USA. *Dermatol. Ther.* **2023**, *13*, 1053–1068. [CrossRef]
45. Elgaard, C.D.B.; Iversen, L.; Hjuler, K.F. Guselkumab, Tildrakizumab, and Risankizumab in a Real-World Setting: Drug Survival and Effectiveness in the Treatment of Psoriasis and Psoriatic Arthritis. *J. Dermatol. Treat.* **2023**, *34*, 2133531. [CrossRef]
46. Ruggiero, A.; Picone, V.; Martora, F.; Fabbrocini, G.; Megna, M. Guselkumab, Risankizumab, and Tildrakizumab in the Management of Psoriasis: A Review of the Real-World Evidence. *Clin. Cosmet. Investig. Dermatol.* **2022**, *15*, 1649–1658. [CrossRef]
47. Roskoski, R. Deucravacitinib Is an Allosteric TYK2 Protein Kinase Inhibitor FDA-Approved for the Treatment of Psoriasis. *Pharmacol. Res.* **2023**, *189*, 106642. [CrossRef]
48. Ahsan, S.; Degener, R.; Schlamp, M. Non-Invasive Treatments Invade the Psoriasis Pipeline. Available online: https://www.aquestconsulting.com/wp-content/uploads/2023/11/Aquest_Non-Invasive-Psoriasis-Pipeline-White-Paper-FINAL-1123.pdf (accessed on 9 March 2025).
49. Bang, C.-H.; Park, C.-J.; Kim, Y.-S. The Expanding Therapeutic Potential of Deucravacitinib Beyond Psoriasis: A Narrative Review. *J. Clin. Med.* **2025**, *14*, 1745. [CrossRef]
50. Truong, T.M.; Pathak, G.N.; Singal, A.; Taranto, V.; Rao, B.K. Deucravacitinib: The First FDA-Approved Oral TYK2 Inhibitor for Moderate to Severe Plaque Psoriasis. *Ann. Pharmacother.* **2024**, *58*, 416–427. [CrossRef]
51. Gyldenløve, M.; Nissen, C.V.; Stave, S.D.W.; Thomsen, S.F.; Egeberg, A.; Loft, N. Oral Roflumilast in Patients with Psoriasis: A Real-World Cohort Study. *Am. J. Clin. Dermatol.* **2025**, *26*, 147–150. [CrossRef]
52. Lé, A.M.; Yilmaz, O.; Luz, M.; Torres, T. Oral Roflumilast for Psoriasis: A Real-World 24-Week Prospective Cohort Study. *J. Dermatol. Treat.* **2025**, *36*, 2464107. [CrossRef] [PubMed]
53. Ravindran, R.; Muralidharan, P.; Govindarajan, A. Systematizing PDE4 Inhibition in Psoriasis Treatment: Roflumilast Leads the Way. *J. Young Pharm.* **2025**, *17*, 13–18. [CrossRef]
54. De Moraes-Souza, R.; Chahine Chater, R.; Pera Calvi, I.; Mesquita, Y.; Sarto, R.; Lapenda, I.; Figueiredo Pereira, L.; Moury, L.; Herranz-Pinto, P. Efficacy and Safety of Topical Roflumilast for the Treatment of Psoriasis: A Systematic Review and Meta-Analysis of Randomized Controlled Trials. *Clin. Drug Investig.* **2024**, *44*, 655–665. [CrossRef]
55. Armstrong, A.W.; McConaha, J.L. Tapinarof Cream 1% Once Daily for the Treatment of Adults with Mild to Severe Plaque Psoriasis: A Novel Topical Therapy Targeting the Aryl Hydrocarbon Receptor. *J. Manag. Care Spec. Pharm.* **2023**, *29*, S1–S14.
56. Orsini, D.; Assorgi, C.; Bonifati, C.; Cameli, N.; Graceffa, D.; Potestio, L.; Megna, M. Effectiveness and Safety of Risankizumab in VEry Severe Plaque Psoriasis: A Real-Life Retrospective Study (VESPA-Study). *J. Dermatol. Treat.* **2024**, *35*, 2358150. [CrossRef]
57. Gordon, K.B.; Blauvelt, A.; Bachelez, H.; Coates, L.C.; Van Den Bosch, F.E.; Kaplan, B.; Koetse, W.; Ashley, D.G.; Lippe, R.; Sinvhal, R.; et al. Long-Term Safety of Risankizumab in Patients with Psoriatic Disease: A Comprehensive Analysis from Clinical Trials. *Dermatol. Ther.* **2024**, *14*, 2523–2538. [CrossRef]
58. Vender, R.B.; Lynde, C.W. Certolizumab Pegol Use in the Treatment of Moderate-to-Severe Psoriasis: Real-World Data From Two Canadian Centers. *J. Cutan. Med. Surg.* **2022**, *26*, 267–273. [CrossRef]
59. Lee, A.; Scott, L.J. Certolizumab Pegol: A Review in Moderate to Severe Plaque Psoriasis. *BioDrugs* **2020**, *34*, 235–244. [CrossRef]
60. Coto-Segura, P.; Segú-Vergés, C.; Martorell, A.; Moreno-Ramírez, D.; Jorba, G.; Junet, V.; Guerri, F.; Daura, X.; Oliva, B.; Cara, C.; et al. A Quantitative Systems Pharmacology Model for Certolizumab Pegol Treatment in Moderate-to-Severe Psoriasis. *Front. Immunol.* **2023**, *14*, 1212981. [CrossRef]
61. Turkmen, M.; Dogan, S. Certolizumab Pegol in the Treatment of Psoriasis: REAL-LIFE Data. *Dermatol. Ther.* **2021**, *34*, e14929. [CrossRef]
62. Bubna, A.K.; Patil, N. Tildrakizumab in Psoriasis and Beyond. *J. Dermatol. Dermatol. Surg.* **2024**, *28*, 1–10. [CrossRef]
63. Bagel, J. Biologics in Psoriasis: The Next Generation. Practical Dermatology. Available online: https://practicaldermatology.com/topics/psoriasis/biologics-in-psoriasis-the-next-generation/20807/ (accessed on 2 April 2025).
64. Megna, M.; Ruggiero, A.; Tommasino, N.; Brescia, C.; Martora, F.; Cacciapuoti, S.; Potestio, L. Effectiveness and Safety of Tildrakizumab in Psoriasis Patients Who Failed Anti-IL17 Treatment: A 28-Week Real-Life Study. *Clin. Cosmet. Investig. Dermatol.* **2024**, *17*, 1037–1042. [CrossRef] [PubMed]

65. Gebauer, K.; Spelman, L.; Yamauchi, P.S.; Bagel, J.; Nishandar, T.; Crane, M.; Kopeloff, I.; Kothekar, M.; Yao, S.-L.; Sofen, H.L. Efficacy and Safety of Tildrakizumab for the Treatment of Moderate-to-Severe Plaque Psoriasis of the Scalp: A Multicenter, Randomized, Double-Blind, Placebo-Controlled, Phase 3b Study. *J. Am. Acad. Dermatol.* **2024**, *91*, 91–99. [CrossRef]
66. Lamb, Y.N. Guselkumab in Psoriatic Arthritis: A Profile of Its Use. *Drugs Ther. Perspect.* **2021**, *37*, 285–293. [CrossRef]
67. Strober, B.; Coates, L.C.; Lebwohl, M.G.; Deodhar, A.; Leibowitz, E.; Rowland, K.; Kollmeier, A.P.; Miller, M.; Wang, Y.; Li, S.; et al. Long-Term Safety of Guselkumab in Patients with Psoriatic Disease: An Integrated Analysis of Eleven Phase II/III Clinical Studies in Psoriasis and Psoriatic Arthritis. *Drug Saf.* **2024**, *47*, 39–57. [CrossRef]
68. Megna, M.; Ruggiero, A.; Martora, F.; Vallone, Y.; Guerrasio, G.; Potestio, L. Long-Term Efficacy and Safety of Guselkumab in Psoriasis Patients Who Failed Anti-IL17: A Two-Year Real-Life Study. *J. Clin. Med.* **2024**, *13*, 2691. [CrossRef]
69. Blauvelt, A.; Langley, R.G.; Branigan, P.J.; Liu, X.; Chen, Y.; DePrimo, S.; Ma, K.; Scott, B.; Campbell, K.; Muñoz-Elías, E.J.; et al. Guselkumab Reduces Disease- and Mechanism-Related Biomarkers More Than Adalimumab in Patients with Psoriasis: A VOYAGE 1 Substudy. *JID Innov.* **2024**, *4*, 100287. [CrossRef]
70. Puig, L. Brodalumab: The First Anti-IL-17 Receptor Agent for Psoriasis. *Drugs Today* **2017**, *53*, 283. [CrossRef]
71. Lebwohl, M.G.; Armstrong, A.W.; Alexis, A.F.; Lain, E.L.; Jacobson, A.A. Efficacy of Brodalumab in Patients with Psoriasis and Risk Factors for Treatment Failure: A Review of Post Hoc Analyses. *Dermatol. Ther.* **2024**, *14*, 2709–2726. [CrossRef]
72. Menter, A.; Bhutani, T.; Ehst, B.; Elewski, B.; Jacobson, A. Narrative Review of the Emerging Therapeutic Role of Brodalumab in Difficult-to-Treat Psoriasis. *Dermatol. Ther.* **2022**, *12*, 1289–1302. [CrossRef]
73. Syed, Y.Y. Ixekizumab: A Review in Moderate to Severe Plaque Psoriasis. *Am. J. Clin. Dermatol.* **2017**, *18*, 147–158. [CrossRef] [PubMed]
74. Reich, A.; Reed, C.; Schuster, C.; Robert, C.; Treuer, T.; Lubrano, E. Real-World Evidence for Ixekizumab in the Treatment of Psoriasis and Psoriatic Arthritis: Literature Review 2016–2021. *J. Dermatol. Treat.* **2023**, *34*, 2160196. [CrossRef] [PubMed]
75. Miller, J.; Puravath, A.P.; Orbai, A.-M. Ixekizumab for Psoriatic Arthritis: Safety, Efficacy, and Patient Selection. *J. Inflamm. Res.* **2021**, *14*, 6975–6991. [CrossRef]
76. PubChem Tapinarof. Available online: https://pubchem.ncbi.nlm.nih.gov/compound/6439522 (accessed on 9 March 2025).
77. Benvitimod ≥ 98% (HPLC) | Sigma-Aldrich. Available online: https://www.sigmaaldrich.com/HU/en/product/sigma/sml3430 (accessed on 9 March 2025).
78. Tapinarof | 99.98%(HPLC) | AhR Agonist. Available online: https://www.selleckchem.com/products/tapinarof.html (accessed on 9 March 2025).
79. Pecyna, P.; Wargula, J.; Murias, M.; Kucinska, M. More Than Resveratrol: New Insights into Stilbene-Based Compounds. *Biomolecules* **2020**, *10*, 1111. [CrossRef]
80. Smith, S.H.; Jayawickreme, C.; Rickard, D.J.; Nicodeme, E.; Bui, T.; Simmons, C.; Coquery, C.M.; Neil, J.; Pryor, W.M.; Mayhew, D.; et al. Tapinarof Is a Natural AhR Agonist That Resolves Skin Inflammation in Mice and Humans. *J. Investig. Dermatol.* **2017**, *137*, 2110–2119. [CrossRef]
81. Zatloukalova, M.; Hanyk, J.; Papouskova, B.; Kabelac, M.; Vostalova, J.; Vacek, J. Tapinarof and Its Structure-Activity Relationship for Redox Chemistry and Phototoxicity on Human Skin Keratinocytes. *Free Radic. Biol. Med.* **2024**, *223*, 212–223. [CrossRef]
82. Shah, D.K.; Yi, D.S.; Yangxin, D.C.; Ankit, D.S. Treatment Challenges in Atopic Dermatitis with Novel Topical Therapies. *J. Dermatol. Treat.* **2018**, *7*, 136–143.
83. Robbins, K.; Bissonnette, R.; Maeda-Chubachi, T.; Ye, L.; Peppers, J.; Gallagher, K.; Kraus, J.E. Phase 2, Randomized Dose-Finding Study of Tapinarof (GSK2894512 Cream) for the Treatment of Plaque Psoriasis. *J. Am. Acad. Dermatol.* **2019**, *80*, 714–721. [CrossRef]
84. Hu, Y.-Q.; Liu, P.; Mu, Z.; Zhang, J.-Z. Anti-Inflammatory Effects of AhR Agonist Benvitimod in TNFα/IFNγ Stimulated HaCaT Cells and Peripheral Blood Mononuclear Cells from Patients with Atopic Dermatitis. *arXiv* **2022**. [CrossRef]
85. Zhang, J.; Cai, L.; Zheng, M. A Novel Topical Treatment for Plaque Psoriasis: Benvitimod/Tapinarof. *J. Am. Acad. Dermatol.* **2022**, *86*, e137–e138. [CrossRef]
86. Keam, S.J. Tapinarof Cream 1%: First Approval. *Drugs* **2022**, *82*, 1221–1228. [CrossRef] [PubMed]
87. Tian, C.; Zhang, G.; Xia, Z.; Chen, N.; Yang, S.; Li, L. Identification of Triazolopyridine Derivatives as a New Class of AhR Agonists and Evaluation of Anti-Psoriasis Effect in a Mouse Model. *Eur. J. Med. Chem.* **2022**, *231*, 114122. [CrossRef] [PubMed]
88. Bissonnette, R.; Saint-Cyr Proulx, E.; Jack, C.; Maari, C. Tapinarof for Psoriasis and Atopic Dermatitis: 15 Years of Clinical Research. *J. Eur. Acad. Dermatol. Venereol.* **2023**, *37*, 1168–1174. [CrossRef]
89. Ali, S.; Ion, A.; Orzan, O.A.; Bălăceanu-Gurău, B. Emerging Treatments and New Vehicle Formulations for Atopic Dermatitis. *Pharmaceutics* **2024**, *16*, 1425. [CrossRef]
90. Lu, S.T.; Kelly, K.A.; Feldman, S.R. An Overview of Benvitimod for the Treatment of Psoriasis: A Narrative Review. *Expert Opin. Pharmacother.* **2022**, *23*, 327–333. [CrossRef]

91. Strober, B.; Stein Gold, L.; Bissonnette, R.; Armstrong, A.W.; Kircik, L.; Tyring, S.K.; Piscitelli, S.C.; Brown, P.M.; Rubenstein, D.S.; Tallman, A.M.; et al. One-Year Safety and Efficacy of Tapinarof Cream for the Treatment of Plaque Psoriasis: Results from the PSOARING 3 Trial. *J. Am. Acad. Dermatol.* **2022**, *87*, 800–806. [CrossRef]
92. Paller, A.S.; Stein Gold, L.; Soung, J.; Tallman, A.M.; Rubenstein, D.S.; Gooderham, M. Efficacy and Patient-Reported Outcomes from a Phase 2b, Randomized Clinical Trial of Tapinarof Cream for the Treatment of Adolescents and Adults with Atopic Dermatitis. *J. Am. Acad. Dermatol.* **2021**, *84*, 632–638. [CrossRef]
93. Santini, G.; Bonati, L.; Motta, S. From Tapinarof to Novel AhR Modulators: Computational Drug Discovery for Psoriasis Therapeutics. *Res. Sq.* **2025**. *Preprint posted*. [CrossRef]
94. Dutky, S.R. Insect Microbiology. In *Advances in Applied Microbiology*; Umbreit, W.W., Ed.; Academic Press: Cambridge, MA, USA, 1959; Volume 1, pp. 175–200. ISBN 0065-2164.
95. An Open-Label, Long-Term Extension Study to Evaluate the Safety and Efficacy of Tapinarof Cream, 1% in Subjects With Atopic Dermatitis. 2021. Available online: https://clinicaltrials.gov/study/NCT05142774 (accessed on 9 March 2025).
96. Zhao, Y.; Wu, C.; Cai, L.; Mu, Z.; Liu, X.; Zhang, W.; Zhang, J. Tapinarof Inhibits Psoriatic Inflammation through Inducing Tolerogenic Dendritic Cells and Foxp3+ Tregs. *Chin. Med. J.* **2024**, *137*, 2877–2879. [CrossRef]
97. Zhu, X.; Han, R.; Tian, X.; Hochgerner, M.; Li, H.; Wang, J.; Xia, J. The Opposite Effect of Tapinarof between IMQ and IL-23 Induced Psoriasis Mouse Models. *Exp. Dermatol.* **2024**, *33*, e14862. [CrossRef]
98. Jett, J.E.; McLaughlin, M.; Lee, M.S.; Parish, L.C.; DuBois, J.; Raoof, T.J.; Tabolt, G.; Wilson, T.; Somerville, M.C.; DellaMaestra, W.; et al. Tapinarof Cream 1% for Extensive Plaque Psoriasis: A Maximal Use Trial on Safety, Tolerability, and Pharmacokinetics. *Am. J. Clin. Dermatol.* **2022**, *23*, 83–91. [CrossRef] [PubMed]
99. A Phase 3 Study of Tapinarof for the Treatment of Plaque Psoriasis in Pediatric Subjects. 2021. Available online: https://clinicaltrials.gov/study/NCT05172726 (accessed on 9 March 2025).
100. Igarashi, A.; Tsuji, G.; Murata, R.; Fukasawa, S.; Yamane, S. A Phase 2, Randomized, Double-blind, Vehicle-controlled Trial of Tapinarof Cream in Japanese Pediatric Patients with Atopic Dermatitis. *J. Dermatol.* **2025**, *52*, 247–255. [CrossRef] [PubMed]
101. Simpson, E.L.; Hebert, A.A.; Browning, J.; Serrao, R.T.; Sofen, H.; Brown, P.M.; Piscitelli, S.C.; Rubenstein, D.S.; Tallman, A.M. Tapinarof Improved Outcomes and Sleep for Patients and Families in Two Phase 3 Atopic Dermatitis Trials in Adults and Children. *Dermatol. Ther.* **2025**, *15*, 111–124. [CrossRef] [PubMed]
102. A Phase 3 Efficacy and Safety Study of Tapinarof for the Treatment of Moderate to Severe Atopic Dermatitis in Children and Adults. 2021. Available online: https://clinicaltrials.gov/study/NCT05032859 (accessed on 9 March 2025).
103. Silverberg, J.I.; Eichenfield, L.F.; Hebert, A.A.; Simpson, E.L.; Stein Gold, L.; Bissonnette, R.; Papp, K.A.; Browning, J.; Kwong, P.; Korman, N.J.; et al. Tapinarof Cream 1% Once Daily: Significant Efficacy in the Treatment of Moderate to Severe Atopic Dermatitis in Adults and Children down to 2 Years of Age in the Pivotal Phase 3 ADORING Trials. *J. Am. Acad. Dermatol.* **2024**, *91*, 457–465. [CrossRef]
104. Ghani, H.; Podwojniak, A.; Tan, I.J.; Parikh, A.K.; Sanabria, B.; Rao, B. A Comparison of the Safety and Efficacy of Tapinarof and Roflumilast Topical Therapies in the Management of Mild-to-moderate Plaque Psoriasis. *Skin Res. Technol.* **2024**, *30*, e70041. [CrossRef]
105. Teva Pharmaceuticals USA. *A Randomized, Double-Blind, Placebo-Controlled, Parallel-Group, Multiple-Site, Clinical Study to Evaluate the Therapeutic Equivalence of Tapinarof Cream 1% (Teva Pharmaceuticals, Inc.) with VTAMA® Tapinarof (Tapinarof) Cream 1% (Dermavant Sciences, Inc.) in Adult Patients with Plaque Psoriasis*; National Library of Medicine: Bethesda, MD, USA, 2024.
106. A Phase 4, Open-Label, Study to Investigate the Efficacy and Safety of VTAMA® (Tapinarof) Cream 1% in the Treatment of Plaque Psoriasis in Intertriginous Areas. 2022. Available online: https://clinicaltrials.gov/study/NCT05680740 (accessed on 9 March 2025).
107. Gul, R.; Ahmed, N.; Ullah, N.; Khan, M.I.; Elaissari, A.; Rehman, A.U. Biodegradable Ingredient-Based Emulgel Loaded with Ketoprofen Nanoparticles. *AAPS PharmSciTech* **2018**, *19*, 1869–1881. [CrossRef]
108. Abdellatif, A.A.H.; Alsowinea, A.F. Approved and Marketed Nanoparticles for Disease Targeting and Applications in COVID-19. *Nanotechnol. Rev.* **2021**, *10*, 1941–1977. [CrossRef]
109. Yin, Y.; Hu, B.; Yuan, X.; Cai, L.; Gao, H.; Yang, Q. Nanogel: A Versatile Nano-Delivery System for Biomedical Applications. *Pharmaceutics* **2020**, *12*, 290. [CrossRef]
110. Mastella, P.; Todaro, B.; Luin, S. Nanogels: Recent Advances in Synthesis and Biomedical Applications. *Nanomaterials* **2024**, *14*, 1300. [CrossRef]
111. Botha, N.L.; Mushonga, P.; Onani, M.O. Review on Nanogels and Their Applications on Dermal Therapy. *Polym. Polym. Compos.* **2023**, *31*, 09673911231192816. [CrossRef]
112. Gupta, V.; Mohapatra, S.; Mishra, H.; Farooq, U.; Kumar, K.; Ansari, M.; Aldawsari, M.; Alalaiwe, A.; Mirza, M.; Iqbal, Z. Nanotechnology in Cosmetics and Cosmeceuticals—A Review of Latest Advancements. *Gels* **2022**, *8*, 173. [CrossRef] [PubMed]

113. Fytianos, G.; Rahdar, A.; Kyzas, G.Z. Nanomaterials in Cosmetics: Recent Updates. *Nanomaterials* **2020**, *10*, 979. [CrossRef] [PubMed]
114. Sainaga Jyothi, V.G.S.; Ghouse, S.M.; Khatri, D.K.; Nanduri, S.; Singh, S.B.; Madan, J. Lipid Nanoparticles in Topical Dermal Drug Delivery: Does Chemistry of Lipid Persuade Skin Penetration? *J. Drug Deliv. Sci. Technol.* **2022**, *69*, 103176. [CrossRef]
115. Aggarwal, G.; Dhawan, B.; Harikumar, S. Enhanced Transdermal Permeability of Piroxicam through Novel Nanoemulgel Formulation. *Int. J. Pharm. Investig.* **2014**, *4*, 65. [CrossRef]
116. Algahtani, M.S.; Ahmad, M.Z.; Nourein, I.H.; Albarqi, H.A.; Alyami, H.S.; Alyami, M.H.; Alqahtani, A.A.; Alasiri, A.; Algahtani, T.S.; Mohammed, A.A.; et al. Preparation and Characterization of Curcumin Nanoemulgel Utilizing Ultrasonication Technique for Wound Healing: In Vitro, Ex Vivo, and In Vivo Evaluation. *Gels* **2021**, *7*, 213. [CrossRef]
117. Li, N.; Qin, Y.; Dai, D.; Wang, P.; Shi, M.; Gao, J.; Yang, J.; Xiao, W.; Song, P.; Xu, R. Transdermal Delivery of Therapeutic Compounds With Nanotechnological Approaches in Psoriasis. *Front. Bioeng. Biotechnol.* **2022**, *9*, 804415. [CrossRef]
118. Walunj, M.; Doppalapudi, S.; Bulbake, U.; Khan, W. Preparation, Characterization, and in Vivo Evaluation of Cyclosporine Cationic Liposomes for the Treatment of Psoriasis. *J. Liposome Res.* **2020**, *30*, 68–79. [CrossRef]
119. Abu Hashim, I.; Abo El-Magd, N.; El-Sheakh, A.; Hamed, M.; Abd El-Gawad, A.E.-G. Pivotal Role of Acitretin Nanovesicular Gel for Effective Treatment of Psoriasis: Ex Vivo–in Vivo Evaluation Study. *Int. J. Nanomed.* **2018**, *13*, 1059–1079. [CrossRef]
120. Vohra, P.; Varekar, S.; Shah, V. Nano-Transferosomes of Aloe-Vera and Vitamin-E for Management of Psoriasis: An Archetype in Herbal Drug Technology. *Int. J. Innov. Res. Technol.* **2023**, *9*, 876–882.
121. Dadwal, N.; Amisha; Singh, D.; Singh, A. Quality-by-Design Approach for Investigating the Efficacy of Tacrolimus and Hyaluronic Acid–Loaded Ethosomal Gel in Dermal Management of Psoriasis: In Vitro, Ex Vivo, and In Vivo Evaluation. *AAPS PharmSciTech* **2023**, *24*, 220. [CrossRef]
122. Serini, S.; Trombino, S.; Cassano, R.; Marino, M.; Calviello, G. Anti-Inflammatory Effects of Curcumin-Based Nanoparticles Containing α-Linolenic Acid in a Model of Psoriasis In Vitro. *Nutrients* **2025**, *17*, 692. [CrossRef] [PubMed]
123. Llorente, X.; Esteruelas, G.; Bonilla, L.; Agudelo, M.G.; Filgaira, I.; Lopez-Ramajo, D.; Gong, R.C.; Soler, C.; Espina, M.; García, M.L.; et al. Riluzole-Loaded Nanostructured Lipid Carriers for Hyperproliferative Skin Diseases. *Int. J. Mol. Sci.* **2023**, *24*, 8053. [CrossRef] [PubMed]
124. Tomar, Y.; Pandit, N.; Priya, S.; Singhvi, G. Evolving Trends in Nanofibers for Topical Delivery of Therapeutics in Skin Disorders. *ACS Omega* **2023**, *8*, 18340–18357. [CrossRef] [PubMed]
125. Kang, N.-W.; Kim, M.-H.; Sohn, S.-Y.; Kim, K.-T.; Park, J.-H.; Lee, S.-Y.; Lee, J.-Y.; Kim, D.-D. Curcumin-Loaded Lipid-Hybridized Cellulose Nanofiber Film Ameliorates Imiquimod-Induced Psoriasis-like Dermatitis in Mice. *Biomaterials* **2018**, *182*, 245–258. [CrossRef]
126. Du, H.; Liu, P.; Zhu, J.; Lan, J.; Li, Y.; Zhang, L.; Zhu, J.; Tao, J. Hyaluronic Acid-Based Dissolving Microneedle Patch Loaded with Methotrexate for Improved Treatment of Psoriasis. *ACS Appl. Mater. Interfaces* **2019**, *11*, 43588–43598. [CrossRef]
127. Nguyen, T.-T.-L.; Duong, V.-A. A Review on Nanosystem-Based Delivery of Tofacitinib for Enhanced Treatment of Autoimmune Diseases and Inflammation. *BioNanoScience* **2024**, *14*, 2048–2064. [CrossRef]
128. Manimaran, V.; Nivetha, R.P.; Tamilanban, T.; Narayanan, J.; Vetriselvan, S.; Fuloria, N.K.; Chinni, S.V.; Sekar, M.; Fuloria, S.; Wong, L.S.; et al. Nanogels as Novel Drug Nanocarriers for CNS Drug Delivery. *Front. Mol. Biosci.* **2023**, *10*, 1232109. [CrossRef]
129. Suhail, M.; Rosenholm, J.M.; Minhas, M.U.; Badshah, S.F.; Naeem, A.; Khan, K.U.; Fahad, M. Nanogels As Drug-Delivery Systems: A Comprehensive Overview. *Ther. Deliv.* **2019**, *10*, 697–717. [CrossRef]
130. Chandrashekhar, B.S. Tretinoin Nanogel 0.025% Versus Conventional Gel 0.025% in Patients with Acne Vulgaris: A Randomized, Active Controlled, Multicentre, Parallel Group, Phase IV Clinical Trial. *J. Clin. Diagn. Res.* **2015**, *9*, WC04. [CrossRef]
131. Avasatthi, V.; Pawar, H.; Dora, C.P.; Bansod, P.; Gill, M.S.; Suresh, S. A Novel Nanogel Formulation of Methotrexate for Topical Treatment of Psoriasis: Optimization, in Vitro and in Vivo Evaluation. *Pharm. Dev. Technol.* **2016**, *21*, 554–562. [CrossRef]
132. Kakade, P.; Patravale, V.; Patil, A.; Disouza, J. Formulation Development of Nanostructured Lipid Carrier-Based Nanogels Encapsulating Tacrolimus for Sustained Therapy of Psoriasis. *Int. J. Pharm.* **2024**, *660*, 124172. [CrossRef] [PubMed]
133. Vashist, A.; Perez Alvarez, G.; Andion Camargo, V.; Raymond, A.D.; Arias, A.Y.; Kolishetti, N.; Vashist, A.; Manickam, P.; Aggarwal, S.; Nair, M. Recent Advances in Nanogels for Drug Delivery and Biomedical Applications. *Biomater. Sci.* **2024**, *12*, 6006–6018. [CrossRef] [PubMed]
134. Tariq, L.; Arafah, A.; Ali, S.; Beigh, S.; Dar, M.A.; Dar, T.U.H.; Dar, A.I.; Alsaffar, R.M.; Masoodi, M.H.; Rehman, M.U. Nanogel-Based Transdermal Drug Delivery System: A Therapeutic Strategywith Under Discussed Potential. *Curr. Top. Med. Chem.* **2023**, *23*, 44–61. [CrossRef] [PubMed]
135. Kraus, J.E.; Lee, J. Use of Tapinarof for the Treatment of Chronic Plaque Psoriasis. 2023. Available online: https://patents.google.com/patent/US11590088B2/en (accessed on 19 May 2025).

136. Kraus, J.E.; Lee, J. Use of Tapinarof for the Treatment of Atopic Dermatitis. 2024. Available online: https://patents.google.com/patent/US11938099B2/en?oq=US11938099B2 (accessed on 19 May 2025).
137. Arkin, M.; Zighelboim, M. Treatment of Ocular Diseases with Ophthalmic Tapinarof Compositions. 2021. Available online: https://patents.google.com/patent/US20210000758A1/en?oq=US20210000758A1 (accessed on 19 May 2025).
138. Jain, P.; Tabolt, G.; Rubenstein, D.S.; Kalluri, H.; Catubig, R.; Roughan, B.; Buchta, R.; Bonjour, F.M.G.; Brady, B.P.; Sutcliffe, L.H. Gel, Ointment, and Foam Formulations of Tapinarof and Methods of Use 2022. U.S. Patent Application 17/533,885, 19 May 2025.
139. Balogh, B.; Pető, Á.; Haimhoffer, Á.; Sinka, D.; Kósa, D.; Fehér, P.; Ujhelyi, Z.; Argenziano, M.; Cavalli, R.; Bácskay, I. Formulation and Evaluation of Different Nanogels of Tapinarof for Treatment of Psoriasis. *Gels* **2024**, *10*, 675. [CrossRef]

pharmaceutics MDPI

Article

Voriconazole-Loaded Nanohydrogels Towards Optimized Antifungal Therapy for Cystic Fibrosis Patients

Shaul D. Cemal [1,†], María F. Ladetto [2,3,†], Katherine Hermida Alava [2], Gila Kazimirsky [1], Marcela Cucher [2], Romina J. Glisoni [4], María L. Cuestas [2,*] and Gerardo Byk [1,*]

1 Department of Chemistry, Bar-Ilan University, Ramat Gan 5290002, Israel; shaulcemal11@gmail.com (S.D.C.); gilakazimirsky@gmail.com (G.K.)

2 Instituto de Investigaciones en Microbiología y Parasitología Médica (IMPaM), CONICET, Universidad de Buenos Aires, Buenos Aires 1121, Argentina; mflorencialadetto@gmail.com (M.F.L.); katherinehermidaalava@gmail.com (K.H.A.); marcecucher@gmail.com (M.C.)

3 Centro de Investigación y Desarrollo en Fermentaciones Industriales (CINDEFI), Laboratorio de Nanobiomateriales, Departamento de Química, Facultad de Ciencias Exactas, Universidad Nacional de La Plata (UNLP)-CONICET (CCT La Plata), Buenos Aires 1121, Argentina

4 Instituto de Nanobiotecnología (NANOBIOTEC), CONICET, Universidad de Buenos Aires, Buenos Aires 1121, Argentina; romy.glisoni@gmail.com

* Correspondence: marilucuestas@gmail.com (M.L.C.); gerardo.byk@biu.ac.il (G.B.)

† These authors contributed equally to this work.

Abstract: Background/Objectives: Filamentous fungi, in particular the species *Aspergillus*, *Scedosporium*, and *Exophiala*, frequently colonize the lungs of cystic fibrosis (CF) patients. Chronic colonization is linked to hypersensitivity reactions and persistent infections leading to a significant long-term decline in lung function. Azole antifungal therapy such as voriconazole (VRC) slows disease progression, particularly in patients with advanced CF; however, excessive mucus production in CF lungs poses a diffusional barrier to effective treatment. **Methods**: Here, biodegradable nanohydrogels (NHGs) recently developed as nanocarriers were evaluated for formulating VRC as a platform for treating fungal infections in CF lungs. The NHGs entrapped up to about 30 μg/mg of VRC, and physicochemical properties were investigated via dynamic laser light scattering and nanoparticle tracking analysis. Diameters were 100–400 nm, and excellent colloidal stability was demonstrated in interstitial fluids, indicating potential for pulmonary delivery. Nano-formulations exhibited high in vitro cytocompatibility in A549 and HEK293T cells and were tested for the release of VRC under two different sink conditions. **Results**: Notably, the antifungal activity of VRC-loaded nanohydrogels was up to eight-fold greater than an aqueous suspension drug against different fungal species isolated from CF sputum, regardless of the presence of a CF artificial mucus layer. **Conclusions**: These findings support the development of potent VRC nano-formulations for treating fungal disorders in CF lungs.

Keywords: voriconazole; biodegradable nanohydrogels; cystic fibrosis; CF mucus model; drug delivery; drug release

1. Introduction

Cystic fibrosis (CF) is the most common inherited life-limiting disease among Caucasian people, affecting 90,000 people worldwide [1]. This single-gene disease, with autosomal recessive inheritance is caused by a mutation in the CF gene encoding the transmembrane conductance regulator (CFTR) [2], which forms a chloride channel that is critical to efficient mucus transport. Mutations in CFTR disrupt chloride secretion, sodium reabsorption, and water transport, leading to mucus hyper concentration and decreased mucociliary clearance [3].

The resultant dehydrated form of mucus is characterized by a reduced mesh size (60−300 nm) [4] in comparison to physiological mucus (497−503 nm) [5]. Dehydrated mucus secretions lead to endobronchial infection with a narrow spectrum of distinctive bacteria and fungi and an exaggerated inflammatory response, resulting in the development of severe bronchiectasis rather than fibrosis and, eventually, respiratory failure [6]. Therefore, in mucus-related pathologies such as CF, an overexpression of mucins, accumulation of extracellular DNA, and cellular debris, as well as the persistent presence of bacteria and fungi, confer mucus stasis, leading to a vicious cycle of infection and inflammation that can be chronically sustained [7,8]. This pathological mucus strongly limits the absorption of antibiotics, antifungals, and anti-inflammatory drugs, preventing them from reaching target sites [9–13].

Despite great advancements in disease management in recent decades, pulmonary failure remains the main cause of morbidity and mortality in CF patients [2]. Recurrent lung infections that lead to chronic airway inflammation and respiratory failure remain the major prognostic problem accounting for over 90% of mortality [14]. Clinical research and therapy mostly focus on the role of bacterial species with little attention to the role of fungi in the pathogenesis of lung function decline, despite their frequent isolation in respiratory samples [15].

Filamentous fungi, especially *Aspergillus*, *Scedosporium*, and *Exophiala* species, are frequent lung colonizers in CF patients [16]. Chronic filamentous fungal colonization is particularly observed at a late stage of the disease and is associated with allergic bronchopulmonary mycosis and chronic infections related to a significant decline in lung function that persists over time [16–23]. Chronic antifungal therapy with azole drugs such as voriconazole (VRC) slows down this progression, mainly in patients with advanced disease [18].

However, the use of VRC is often limited by its poor solubility and bioavailability. VRC is a relatively lipophilic molecule with a Log D = 1.8 at pH 7.4, with an epithelial lining fluid/plasma ratio between 6 and 11 when administered through a systemic route in humans [24]. The low aqueous solubility of VRC leads to challenges in formulation and reduced absorption in the gastrointestinal tract. This poor bioavailability necessitates higher doses to achieve therapeutic levels, increasing the risk of adverse effects and drug interactions, particularly CF patients who receive multiple concomitant medications [25]. Overcoming these limitations is crucial for improving the efficacy and safety of VRC in chronic therapy.

Currently, there are commercially available VRC dosage forms for oral and parenteral delivery. However, the oral administration of VRC may induce dose-related side effects (hepatotoxicity, cutaneous malignancies, visual disturbances, and neuro-psychiatric symptoms) [26]. The usefulness of VRC may also be limited by the inability to reach therapeutic levels in the main focus of infection: the lung using the oral route [27]. No inhaling dosage form of VRC is commercially available nowadays. However, inhaled VRC has been used in clinical and investigational settings through compounded nano-formulations or extemporaneous preparations such as VRC suspension [28]. In addition, the pathological CF mucus strongly limits the absorption of drugs and their ability to reach target sites in the lung (bronchi, alveoli).

To date, no VRC formulation has been specifically designed for CF. Existing formulations or drug delivery systems for VRC have been developed with a focus on invasive fungal infections [29–33], which do not involve the pathological mucus environment characteristic of CF patients.

Therefore, more effective drug delivery to mucosal surfaces is needed to achieve optimized antifungal therapy in CF patients. For this purpose, drugs and drug carriers must bypass the adhesive mucus barrier [5].

Nanoparticles (NPs), densely coated with poly(ethylene glycol) (PEG), a muco-inert material, can rapidly diffuse through human mucus secretions [34]. These "mucus-penetrating particles (MPPs)" demonstrated improved mucosal distribution, enhanced pharmacokinetics, and superior treatment efficacy in applications such as CF lung gene therapy [35].

This study involves the synthesis of series of NHGs based on our established methodology as a platform for loading VRC as a delivery system for treating broncho-pulmonary fungal infections in CF patients. As the mucus presents a diffusional barrier, the NHGs used here were composed of an outer shell of poly(ethylene oxide) (PEO), which was recently shown to cross lung-generated mucus [34]. An in vitro model of pulmonary pathological mucus that mimics the chemical composition, structural features, and viscoelastic properties of CF mucus [36] allowed us to study their penetration across this diffusional barrier and to assess their activity against the three major molds affecting CF patients worldwide. The biocompatible and biodegradable monodisperse NHG libraries, which range in size from 100 to 400 nm, are disclosed in Figure 1. NHGs are generated through the polymerization of mixtures of N-isopropylacrylamide and di-block (hydrophilic-hydrophobic) and tri-block (hydrophobic-hydrophilic-hydrophobic) copolymer acrylamide macro-monomers made of Jeffamine using recently developed biodegradable cross-linkers [37]. The schematic structure of the NHG is presented in Figure 1, notably a radial distribution of hydrophilic PEO at a high ratio in the outer shell of the NHGs (Figure 1a in blue) and poly N-isopropyl acrylamide (PNIPAM) at a high ratio in the center of the NHGs (Figure 1b in yellow). NHGs possess two significant advantages: mono-dispersity and biocompatibility. Their unique design allows for the fine-tuning and optimization of biological applications by adjusting the particle size within the 100–400 nm range, while maintaining the same particle composition. This tunability offers potential for diverse applications, including targeted drug delivery, imaging, and tissue engineering, where particle size plays a critical role in determining the efficacy and specificity of the intended biological interaction.

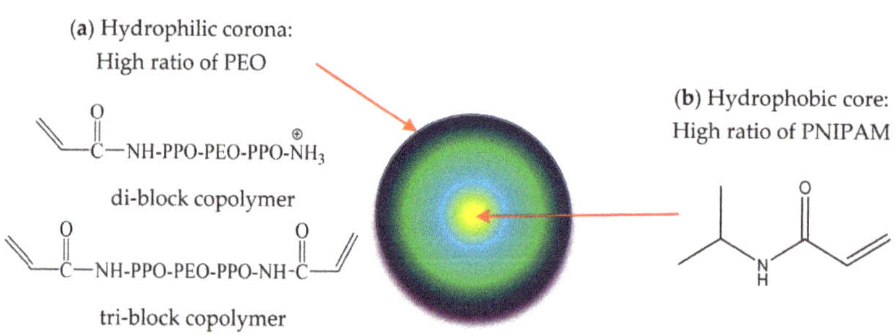

Figure 1. Schematic representation of all NHGs. (**a**) Hydrophilic corona: High ratio of PEO, (**b**) Hydrophobic core: High ratio of PNIPAM.

2. Materials and Methods

Polymerizations were performed in a DAIHAN scientific water bath shaker polymerizer. Jeffamine® ED-2003 (Mn = 1900), acryloyl chloride, polyvinyl-pyrrolidone (PVP_{360000}), potassium persulfate (KPS), ethylene glycol, triethanol amine, and Amicon ultrafilter tube (100 kDa cutoff) were purchased from Merck (Jerusalem, Israel); N-isopropylacrylamide

(NIPAM, 99.0%) from Acros Organics Co. Ltd. (St. Louis, MI, USA); and Cellulose Ester Spectra/Pore (1000 kDa cutoff) dialysis membrane from Spectrum Laboratories, Inc. (Rancho Dominguez, CA, USA).

Hydrodynamic sizes and size distributions were determined using a Zetasizer 3000 HSA (Malvern Instruments Ltd., Malvern, UK). NP tracking analysis (NTA) was conducted to determine the zeta potential (ζ) and NP number (concentration) using a ZetaView PMX-230 instrument (Particle Metrix GmbH, Inning am Ammersee, Germany). An XTT assay kit (Biological Industries, Beit Haemeq, Israel) was used according to the manufacturer's instructions. A humidified water-jacketed automatic CO_2 incubator (US Autoflow, NuAire Inc., Plymouth, MA, USA) was used for cell culture. Dimethyl sulfoxide (DMSO) and VRC were purchased from Merck-Sigma-Aldrich (Buenos Aires, Argentina), and mucin from porcine stomach (type II, powder), sodium alginate, calcium carbonate ($CaCO_3$), D-(+)-gluconic acid δ-lactone (GDL), and sodium chloride (NaCl) were used to develop the artificial airway mucus model. All solvents used were of analytical- or high-performance liquid chromatography (HPLC)-grade.

2.1. Synthesis of NHGs

The novel NHGs of PNIPAAM-co-PPO-PEO-PPO (Poly(N-isopropylacrylamide)-co-Poly(propylene oxide)-Poly(ethylene oxide)-Poly(propylene oxide)) were obtained via free-radical polymerization using monomeric mixtures of (Acr)1.1Jeffamine1900 and NIPAM, at differing ratios for obtaining different particle sizes, as previously shown [37].

In addition, for obtaining two different families of biodegradable NHGs, the ester-functionalized cross-linkers (Figure 2), Ethane-1,2-diyl diacrylate (EDDA), and Nitrilotris(ethane-2,1-diyl) triacrylate (NTEDTA) were introduced in separate reactions in the presence of potassium persulfate (KPS) as an initiator together with the surfactant polyvinylpyrrolidone (PVP360000) [37]. These cross-linkers were selected based on their hydrolytically labile ester bonds, which introduce controlled degradability into the polymer backbone under physiological conditions.

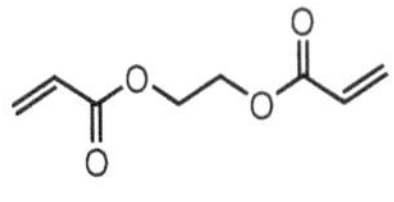

(**a**) EDDA

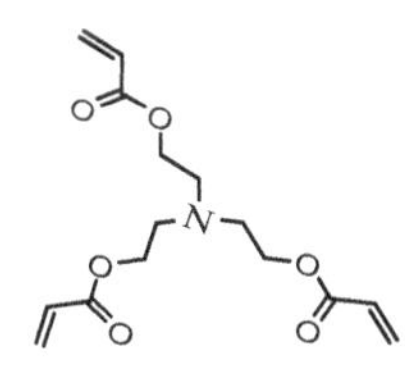

(**b**) NTEDTA

Figure 2. Structure of cross-linkers: (**a**): EDDA; (**b**): NTEDTA.

E_{1-3}/E_{4-6} series present EDDA/NTEDTA cross-linked NHGs, respectively (Table 1). Polymerization was performed at 73 °C in H_2O (10 mL) with PVP and KPS (5 mg each).

Table 1. Synthesis of E_{1-3} and E_{4-6} NHG series (mg) [37].

#	$(Acr)_{1.1}$Jeffamine$_{1900}$	NIPAM	EDDA	NTEDTA
E_1	130	120	15	-
E_2	260	120	15	-
E_3	320	120	15	-
E_4	200	188	-	19
E_5	300	188	-	19
E_6	400	188	-	19

2.2. DLS Analysis

Hydrodynamic sizes and size distributions were measured using diluted NHGs (50 μg/3 mL) with a Zetasizer 3000 HSA (Malvern Instruments Ltd., UK) operating with a 4 mW HeNe laser (632.8 nm), a detector positioned at a scattering angle of 90°, and a temperature-controlled cuvette. Three measurements consisting of 10 sub runs were performed for each sample.

2.3. Atomic Force Microscopy Characterization

Samples were prepared using the simple spin-coating method. NHGs (1 mg/mL, 50 μL) were coated on a silicon wafer substrate. The coating timer was adjusted for 30 s at 2000 RPM. After spin-coating, samples were thoroughly dried with nitrogen and transferred for AFM imaging (Bruker, GmbH—Karlsruhe, Germany).

2.4. Concentration and Zeta Potential (ζ) by NTA

NTA was conducted to determine the ζ and particle number (concentration) of the NPs using the ZetaView instrument described above. Data were collected and analyzed using ZetaView software (version 8.05.16 SP7). Reported particle numbers correspond to the median.

All NHG samples were stored at 4 °C until measurement and were diluted to 1:50,000 in particle-free sterile Milli-Q (MQ) water immediately prior to measurement. The samples were analyzed in light scattering mode using a 488 nm laser. Particles were counted and size-distributed at 25 °C. Instrumental parameters were adjusted as follows: minimum brightness of 30 AU, frame rate of 30 per second (10 per cycle), sensitivity of 85 AU, and shutter of 100 AU. Data for two exposures at 11 measurement positions were collected for each sample. The zeta potential was measured twice at 25 °C with sensitivity of 80 AU and shutter of 150.

2.5. Mucus Penetration Assay for NHGs Monitored by NTA

To further investigate the interaction of nano-formulations with mucus, an assay was set to track the movement of single particles through simulated mucus. For this purpose, we used the previously described (Section 2.3) NTA instrument for quantifying the number of NHGs of different sizes that were able to cross a mucus layer prepared as described by Pacheco et al. [36]. To evaluate the NP penetration, an experiment using a dual-chamber model was designed. In the upper compartment, a quantified equal number of NHGs (see Table S1) was placed onto a homogeneous film formed by 100 μL of simulated mucus, supported in a Transwell® insert with a 1 μm microporous membrane. Meanwhile, the lower receiving compartment was filled with 2 mL of PBS. The same setup without the mucus layer was taken as the control. In both cases, the entire setup was incubated at 37 °C for 24 h. At predetermined time intervals (0, 4, and 24 h), 200 μL samples were withdrawn from the lower compartment for analysis. To keep the lower-well volume constant, an appropriate refill of medium (PBS) was performed at each sampling point. For each withdrawn PBS sample, the number of NHGs that managed to pass through the mucus layer was measured with the NTA instrument. Until quantification, samples were preserved at 4 °C, and prior to measurement, dilution in particle-free sterile MQ water was performed, i.e., 1:25 v/v for samples up to 8 h and 1:2000 v/v otherwise (up to 24 h).

2.6. Loading of VRC

VRC-loaded NHGs (VRC-NHGs) were prepared using the solvent evaporation method [38]. Briefly, VRC (10 mg) was dissolved in chloroform (1 mL), and 200 μL aliquots of the stock solution (10 mg/mL) were dropped into 1 mL of NHGs aqueous solution

(10 mg/mL) and mixed overnight at room temperature (RT) to volatilize the chloroform. The obtained NHGs were isolated using an Amicon® ultrafilter tube by centrifugation at 6000 rpm for 20 min and repeated washings with DDW until no signal from the filtrate was detected by UV-visible spectroscopy. The loaded amount of drug was determined using a systematic method. A calibration plot (see Figure S1) was obtained using a set of standard solutions prepared with known concentrations of VRC in acetonitrile. The linear portion of this plot served as a reference to calculate the concentration of drug in NHGs.

The VRC-NHGs (1 mL, 10 mg/mL) were lyophilized for loading calculations. The freeze-dried powder was treated with 5 mL MeCN and filtered to obtain a clear solution that was analyzed by UV spectroscopy. The absorbance was measured at 255 nm, enabling the determination of the loaded amount of drug through the application of the calibration equation derived from the linear portion. This method provides a reliable approach for quantifying encapsulated drug in NHGs [39].

The drug loading efficiency (DLE) and drug loading capacity (DLC) were calculated as:

$$\text{DLE (\%)} = (\text{weight of loaded drug}/\text{weight of drug used in formulation}) \times 100 \quad (1)$$

$$\text{DLC (\%)} = (\text{weight of encapsulated drug}/\text{total weight}) \times 100 \quad (2)$$

2.7. In Vitro Release of VRC

In vitro drug release studies were performed under sink conditions as previously reported [40]. Briefly, VRC-NHGs containing equivalent amounts of VRC were transferred in a suspended dialysis bag to a container for urine analysis containing 100 mL phosphate-buffered saline (PBS) at pH 7.4 and 37 °C with continuous orbital agitation. Samples (1 mL) from the PBS bath were taken at specific intervals (1, 2, 4, 6, 24, 48, 72 h) for quantification of VRC, and an equal amount of fresh PBS was replaced to maintain the volume constant. A solution of free VRC at the same concentration placed in a dialysis bag using the same procedure was used as the control. The amount of VRC released at each interval was analyzed with a fluorimeter (λ_{ex} = 255 nm, λ_{em} = 372 nm). Assays were performed in duplicates, and the results were expressed as the average ± SD.

A number of studies also reported the use of simulated interstitial lung fluid (SILF, pH 7.4) to investigate the release properties of drug formulations due to its close replication of the mucosal fluids in the respiratory system [41–43]. A release experiment was also conducted in SILF (pH 7.4) supplemented with 1% *w/v* Tween® 80 as previously reported [42]. Dialysis in SILF (pH 7.4) with the addition of 1% *w/v* Tween® 80 as external medium was carried out to investigate the in vitro release profiles from the polymeric matrices of the NHGs. SILF was prepared in 1 L of distilled water (pH 7.4): $MgCl_2$ (98.5 mg), NaCl (6.032 g), KCl (404 mg), Na_2HPO_4 (436 mg), $NaHCO_3$ (2.59 g), Na_2SO_4 (83 mg), $CaCl_2$ (359 mg), sodium acetate (607 mg), and trisodium citrate (115 mg).

2.8. Cell Culture

Human lung cancer cell line A549 and embryonic kidney cell line HEK293T were cultured in DMEM medium supplemented with 10% fetal bovine serum (FBS, Sigma-Aldrich, New York, NY, USA), 100 µg/mL streptomycin, and 100 U/mL penicillin at 37 °C in a 5% CO_2 atmosphere.

2.9. Cytotoxicity

The in vitro cytotoxicity of VRC-NHGs was determined in the A549 and HEK293T cell lines using an XTT assay and compared to free VRC. For this purpose, 8×10^3 cells/well were pre-incubated in a 96-well plate for 24 h and incubated for 48 h at 37 °C with 5% CO_2 and different concentrations of non-loaded NHGs, free VRC, or VRC-NHGs (60 μM VRC). Cells treated with 2% triton were used as the positive control and fresh medium as the negative control. Subsequently, 50 μL of XTT reagent (with initiator) was added, and the cells were incubated for 4 h. The absorbance was read at 570 nm using a TECAN microplate reader.

Cell viability (%) = $(I_{sample}/I_{control}) \times 100$, where I_{sample} is the absorbance of NHG treated wells and $I_{control}$ is the absorbance of control wells without NHG treatment. Experiments were repeated twice. Assays were carried out according to known procedures [44].

2.10. Antifungal Susceptibility Testing

Antifungal susceptibility testing was performed in the mold isolates most frequently recovered from respiratory secretions of CF patients (*A. fumigatus, A. flavus, A. terreus, A. niger, A. nidulans, S. aurantiacum*, and *E. dermatitidis*) against non-loaded NHGs, VRC-NHGs, VRC suspension, and VRC/DMSO following the clinical laboratory standard institute (CLSI) M38 3rd ed. Document [36]. All clinical fungal isolates included in this study were wild-type with respect to azole antifungals and were collected from spontaneously expectorated sputum samples of CF patients as part of a surveillance study conducted in Argentina [45]. All these isolates are available in the culture collection of the Mycology bank of the Center of Mycology of the School of Medicine (University of Buenos Aires, Argentina). *A. fumigatus* ATCC 204305 (obtained from American Type Culture Collection) was used as the quality control strain for susceptibility testing.

VRC suspension, VRC/DMSO (5 mg/mL), and VRC-loaded NHGs were prepared in sterile saline buffer. Two-fold dilutions of each drug formulation (8 to 0.015 μg/mL) were prepared in 100 μL of RPMI−MOPS in 96-well plates. In total, 5×10^4 CFU/mL were added to each well. The results were recorded 48 h post incubation at 37 °C. The MIC was defined as the lowest drug concentration causing the complete inhibition of visible growth.

Antifungal susceptibilities were interpreted using the available epidemiological cutoff values (ECVs) for VRC and the VRC clinical breakpoint for A. fumigatus published in the CLSI documents M57S-Ed4 and M38M51S-Ed3, respectively, as previously reported by Brito Devoto et al. [45], who characterized the antifungal susceptibility profile of all the clinical strains included in this study.

Aiming at developing an optimized method for the administration of VRC through the mucus barrier, the antifungal activity of VRC-NHGs was evaluated using the CF mucus composition previously described by Pacheco et al. for drug diffusion assays [36]. A 16.3 mg/mL NaCl solution was prepared, and a solution of 21 mg/mL alginate sodium salt was dissolved in it using gentle magnetic agitation. Concurrently, a suspension containing 43.7 mg/mL mucin was meticulously prepared in DDW and left to gently agitate overnight. Subsequently, the alginate and mucin solutions were combined at a 1:4 ratio utilizing interconnected luer-lock syringes, and a suspension of 7 mg/mL $CaCO_3$ in the 16.3 mg/mL NaCl solution was intricately blended at 1:5 proportions. A freshly concocted 70 mg/mL GDL solution in 16.3 mg/mL NaCl was merged with the previously created suspension (comprising alginate, mucin, and $CaCO_3$) at 1:6 proportions, always under aseptic conditions. Lastly, 100 μL of the prepared mucus was accurately pipetted onto the 96-well surface, previously inoculated with fungal conidia. The different VRC formulations (VRC-NHGs, VRC suspension, and VRC/DMSO) were added onto the surface of the mucus layer, finalizing the experimental procedure. This antifungal susceptibility procedure was performed according to the abovementioned CLSI document against *A. fumigatus* as the most common filamentous fungus isolated from the airways of CF patients [46].

3. Results and Discussion

3.1. Physico-Chemical Characterization of NHGs and VRC-NHGs

Chronic pulmonary infections caused by opportunistic pathogens affect more than 90% of CF patients. In addition to bacterial colonization, these patients are predisposed to fungal colonization due to the capacity of environmental fungi to colonize the lower respiratory tract and to the frequent cycling of antibiotics required to control the disease.

One of the main goals of the present study was the synthesis of a series of NHGs based on our established methodology as a platform for loading the antifungal drug VRC as a delivery system for an optimized antifungal therapy of broncho-pulmonary infections caused by molds in CF patients. As the mucus presents a diffusional barrier to the antifungal treatment, the NHGs used in this study were composed of a shell of PEG, recently shown to be capable of crossing lung-generated mucus [34].

With this in mind, we developed novel polymeric monodispersed NHGs with a large range of sizes starting from 100 nm up to 400 nm, which according to previous results exhibited high biocompatibility both in vitro and in vivo [37,39,47,48]. This non-toxic profile may support its potential use for treating fungal infections especially in CF patients.

In the present study, monomeric mixtures in water at RT spontaneously formed star/flower-like self-assemblies with uncontrollable poly-dispersed sizes around 100–800 nm. Upon heating at 75 °C, NIPAM and the hydrophobic elements of the macro-monomers (PPO) collapse together into well-defined self-assemblies probably with a high concentration of the hydrophobic elements in the core (NIPAM and PPO chains) and a high concentration of the hydrophilic elements of the macro-monomeric chains (PEO and NH2 groups) at the outer shell. These self-assemblies can be polymerized, leading to particles with sizes that correlate with those of the intermediate self-assemblies. The size of the assemblies can be tuned/calibrated by combining different ratios of the starting monomeric mixtures at high temperature, as previously reported [37,39,47,48]. Larger amounts of (Acr)1.1Jeffamine1900 correspond to smaller NHG diameters due to increased hydrophobic interactions during self-assembly.

Dh represents the particle size of pristine and VRC-NHGs as measured by DLS at 25 °C, and the polydispersity index (PDI) represents the relative variance in the particle size distribution [49,50]. Each value is the average of three measurements by DLS. The results shown in Table 2 indicate that both EDDA cross-linked (series E_{1-3}) and NTEDTA cross-linked NHGs (series E_{4-6}) are highly monodisperse and thermo-responsive [37].

Table 2. Physical properties of NHGs and VRC-NHG series at 25 °C. D_h [nm] results represent the mean ± SD, (n = 3).

#	NHGs				VRC-NHGs			
	D_h [nm]	PDI	ζ [mV]	NPs/mL	D_h [nm]	PDI	ζ [mV]	NPs/mL
E_1	410 ± 4	0.17	−27.9	5.7×10^{12}	424 ± 2	0.12	−21.7	1.4×10^{12}
E_2	242 ± 8	0.28	−28.8	7.4×10^{12}	251 ± 6	0.19	−9.2	4.8×10^{12}
E_3	136 ± 5	0.17	−23.3	3.7×10^{12}	145 ± 5	0.22	−27.5	5.2×10^{12}
E_4	392 ± 2	0.08	−27.1	1.3×10^{12}	407 ± 4	0.14	−26.7	1.6×10^{12}
E_5	198 ± 6	0.15	−27.3	2.4×10^{12}	220 ± 5	0.23	−23.5	6.7×10^{12}
E_6	121 ± 7	0.24	−11.9	6.4×10^{12}	134 ± 3	0.19	−20.0	5.6×10^{12}

The D_H of the VRC-NHGs is a key parameter for their potential clinical use as nanocarriers in CF patients. Carriers in the nano scale have a greater potential to penetrate the airway mucus of CF patients than microparticles, avoiding steric interactions [51]. Thus, VRC-NHGs can overcome the mucus barrier for a better treatment outcome. The data summarized in Table 2 show that all VRC-NHGs maintain their size upon encapsulation. For instance, the E1 batch of NHGs and VRC-NHGs span 410 and 424 nm, respectively. In both cases, there is a unimodal narrow size distribution according to the PDI (0.17/0.12 for NHGs/VRC-NHGs).

The zeta potential measures the electrostatic charge on the surface of particles in a colloidal dispersion (at the shear plane of the particle) [49,50]. The potentials of NHGs and VRC-NHGs were examined by NTA at 25 °C, pH 7.4. All ζ measured at neutral pH were clearly negative (Table 2). This negativity is attributed to the presence of sulfate groups generated at the starting polymeric chains by KPS initiator that are accommodated at the shear plane of the NHGs. A negative zeta potential offers several benefits for NPs. It enhances stability by repelling charges, preventing aggregation, and ensuring a more uniform dispersion, crucial for colloidal stability [52,53]. In biomedical applications, NPs with higher negative potential typically show improved bioavailability. Cellular uptake and intracellular trafficking can also be influenced, providing better control over drug release kinetics and targeting specificity, which is particularly advantageous for drug delivery systems [52]. Moreover, the reduced interaction with cells and tissues due to a high negative ζ can potentially lower the toxicity of NPs, enhancing their biocompatibility, particularly in biomedical imaging, diagnostics, and therapeutics [53].

AFM analysis characterization of E_{1-3} and E_{4-6} NHGs (Figure 3) showed spherical and monodispersed particles. Line profiles from the marked region in the AFM image indicate smaller sizes on a Si wafer compared to DLS (Table 2), as we expected due to the NHGs dryness during AFM measurement.

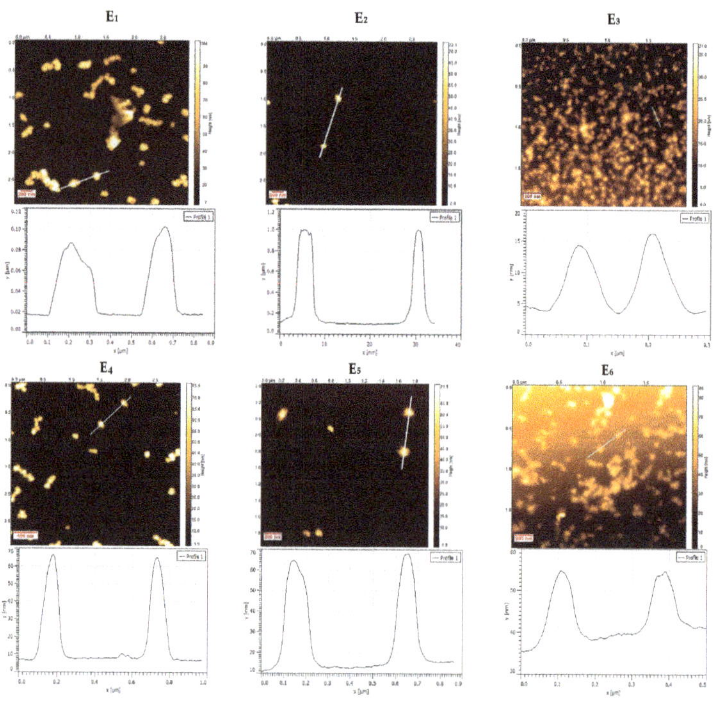

Figure 3. AFM topography images of E_{1-3} (**top**) and E_{4-6} (**bottom**) NHGs.

3.2. Mucus Penetration Assay of NHGs Monitored by NTA

The goal of this experiment was to evaluate NHG diffusion through artificial mucus using a dual-chamber setup (Figure 4a). PBS samples taken from the lower compartment at specific intervals (0, 4, and 24 h) were measured by NTA to quantify the particles reaching that compartment after crossing the mucus layer.

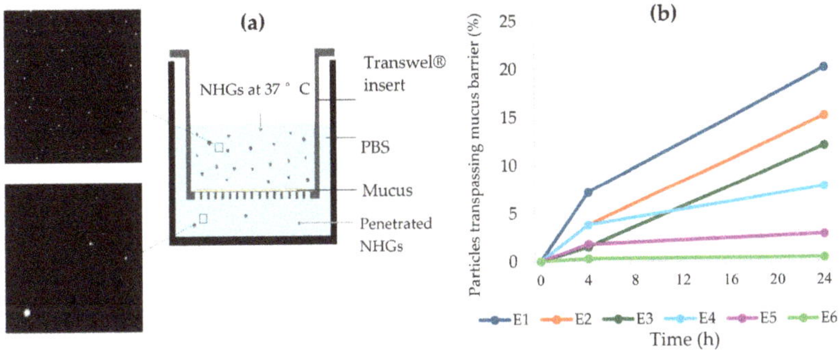

Figure 4. (**a**) Illustration of double-chamber setup and NTA images. (**b**) Percentage of NHG particles trespassing mucus over time.

The entire setup was incubated at 37 °C for 24 h. As our NHGs are based on thermo-responsive poly(N-isopropylacrylamide) (PNIPAM), exposure to temperatures above the low critical solution temperature (LCST) value of 37 °C induces a reversible phase transition in the polymer chains—from a swollen coil state to a collapsed globule. This thermally driven transition can be utilized in an on–off delivery system [54–56]. In other words, the NHGs decrease in size when the temperature exceeds the LCST. Consequently, under the experimental conditions, all NHGs were smaller than the mesh size of the mucus layer [37].

Percentages (see Table 3 and Table S2) were calculated relative to the total number of particles on the receptor media of the same setup without mucus (Figure 4b). The results are shown in Table 3 and Figure 4b as percentages of an equally added number of particles crossing the mucus over time.

Table 3. Percentage of NHGs crossing the mucus barrier over time.

	% Particles Trespassing Mucus Barrier					
Time [h]	E_1	E_2	E_3	E_4	E_5	E_6
4	7.2	3.8	1.5	3.8	1.8	0.3
24	20.1	15.1	12	7.8	2.8	0.4

Series $E_{1\text{–}3}$ shows, in general, better penetration than series $E_{4\text{–}6}$. As for the size, larger NPs display more penetration for both series. We speculate that smaller particles seem to exhibit more interaction with the mucus film interstices and therefore spend more time within this layer compared to larger ones, in contrast to previous reports [57,58].

Our findings support recent results [59], which showed that larger PEGylated nanoparticles (200 and 500 nm) penetrate mucus more effectively than smaller counterparts. This observation suggests two possible explanations: (1) the mucus mesh spacing may be significantly larger and more heterogeneous than previously believed, and/or (2) mucus may exhibit an "exclusion" mechanism similar to exclusion chromatography. In this model, particles navigate through two distinct pathways—one consisting of tightly condensed regions with small mesh pores that trap smaller particles, and another comprising larger mesh channels that allow rapid transit of larger particles. Regardless of the mechanism, PEG-coated large nanoparticles appear to diffuse through low-viscosity channels without substantially disrupting mucus structure.

In addition to enhanced transport, larger nanoparticles offer other benefits, including higher drug-loading capacity, increased stability during formulation and storage, and more sustained drug release profiles. However, while larger sizes confer these advantages, excessively large particles may suffer from limited diffusion and decreased cellular uptake. Therefore, optimizing nanoparticle size is essential to maximize therapeutic efficacy across various mucosal tissues [59].

3.3. VRC Loading into NHGs

In all loading experiments, the applied amount of VRC was 2 mg in 10 mg of NHGs. The DLE and DLC were calculated according to these parameters. The results in Table 4 indicate that larger NHGs have a better drug loading capacity that is directly proportional to their increased hydrophobicity due to their higher content of PNIPAM in their structure as measured by NMR in a previous report [37]. Both series display a similar loading capacity.

Table 4. Drug loading efficiency and capacity of NHG series. Data represent the mean ± SD, (n = 3).

#	Loaded VRC [μg]	DLE [%]	* DLC [%]
E_1	325 ± 8	16.2	3.2
E_2	276 ± 9	13.8	2.8
E_3	208 ± 5	10.4	2.1
E_4	312 ± 10	15.6	3.1
E_5	245 ± 7	12.2	2.4
E_6	198 ± 6	9.9	2.0

* In all cases, 2 mg of drug was applied into 10 mg of NHGs. DLE/DLC was calculated accordingly.

Overall, both NP hydrophobicity and size, as well as the hydrophobic nature of the drug, contribute to the drug loading capacity [60].

3.4. In Vitro Drug Release Study of VRC-NHGs

To investigate the drug release from the NHGs, the fluorescence of released VRC was measured at predetermined time intervals (λ_{ex} = 255 nm, λ_{em} = 372 nm). VRC-NHGs and free VRC as the control were dialyzed in PBS (50 mL), as largely studied by others under sink conditions [40,61]. VRC-NHGs from both series (Figure 5a,b) showed similar release profiles with approximately 50% of encapsulated VRC released within 24 h and around 80% released after 72 h. In contrast, the free VRC control demonstrated a more rapid release, exceeding 80% after just 8 h, emphasizing the role of NHGs in facilitating a controlled release mechanism.

A release experiment was also conducted in SILF (pH 7.4), supplemented with 1% *w*/*v* Tween® 80 as previously reported (Figure 5c,d) [42]. This external medium was utilized to explore the in vitro release profiles of VRC. Here, we dialyzed VRC-NHGs and free VRC as the control in SILF (50 mL), as largely studied by others under sink conditions [41–43].

In the SILF environment, both series of VRC-NHGs exhibited similar release profiles (Figure 5c,d) with over 60% of encapsulated VRC released within 24 h and ~85% after 72 h. The free VRC control exhibited a release of 85% after 8 h. This further highlights the role of NHGs in enabling a sustained release of VRC.

The slightly faster release of VRC observed in the SILF environment compared to the PBS environment can be attributed to its ionic composition, which closely mimics lung interstitial conditions, enhancing drug solubility and diffusion.

VRC exhibits very low solubility in phosphate-buffered saline (PBS), which can limit its release from nanohydrogel formulations in such environments. In PBS, the release of VRC from nanohydrogels is primarily governed by the drug's dissolution rate, which is constrained due to its poor solubility in aqueous media. This limitation affects the diffusion of VRC from the nanohydrogel matrix into the surrounding medium.

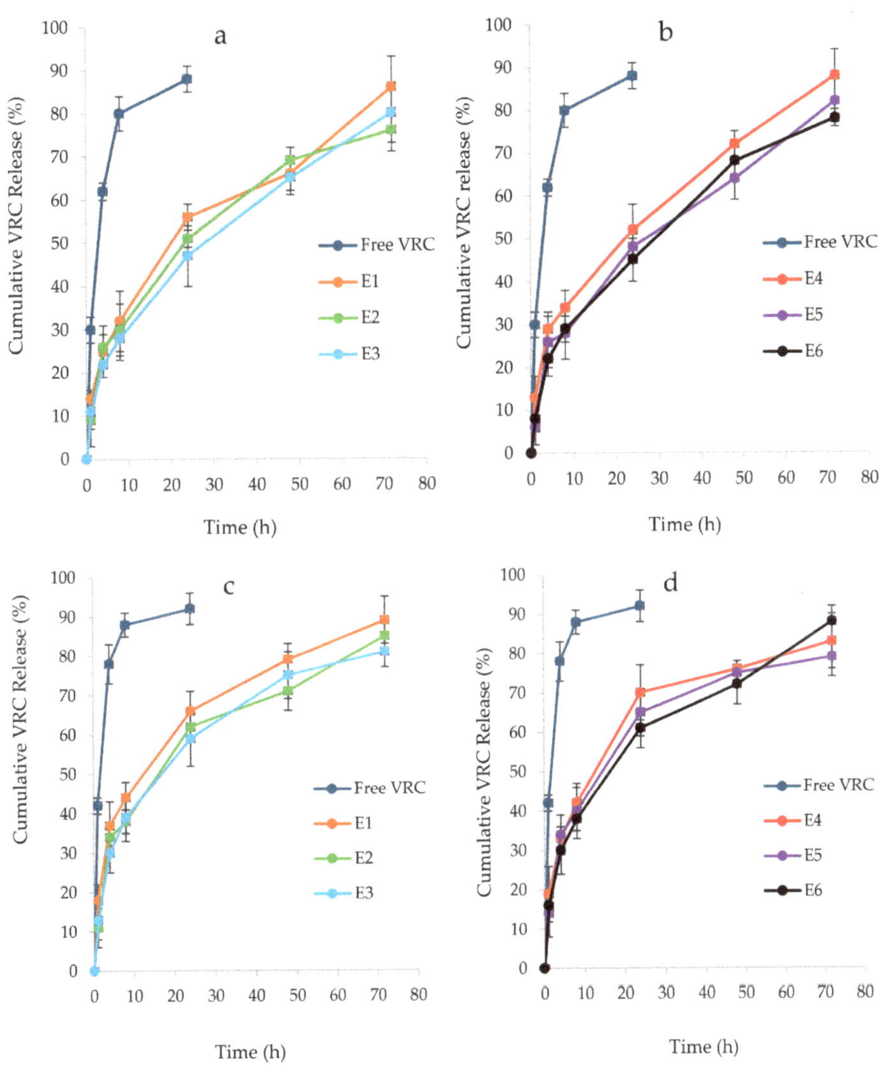

Figure 5. In vitro release profiles of VRC from NHGs: (**a**) (E_{1-3}) and (**b**) (E_{4-6}) in PBS; (**c**) (E_{1-3}) and (**d**) (E_{4-6}) in SILF environment. Data represents the mean ± SD, (n = 2).

In contrast, when evaluating VRC release in SILF, the addition of 1% w/v Tween® 80 to the release medium was employed. Tween® 80, a non-ionic surfactant, enhances the solubility of hydrophobic drugs like VRC by forming micelles that encapsulate the drug molecules, thereby increasing their apparent solubility in aqueous environments. This micellar solubilization facilitates a more efficient release of VRC from the nanohydrogel into the SILF medium.

Moreover, the ionic composition of the release medium can influence the swelling behavior of hydrophilic polymers within the nanohydrogel, potentially affecting drug release dynamics. However, the primary role of Tween® 80 in this context is to improve the solubility and release profile of VRC by mitigating its inherent hydrophobicity, rather than directly altering the ionic characteristics of the medium.

In summary, the incorporation of Tween® 80 into the SILF release medium serves to enhance the solubility and subsequent release of VRC from nanohydrogel formulations, addressing the solubility challenges posed by the drug's hydrophobic nature.

3.5. Cytotoxicity Studies of NHGs and VRC-NHGs

Ensuring the safety and compatibility of the formulation with lung cells is essential for developing a pulmonary drug delivery system, as inhalation of toxic substances may deplete lung surfactant and trigger phagocytic cell activation [62]. Both NHGs and VRC-NHGs were tested for cell toxicity using XTT viability assay and were compared to the relevant controls on HEK-293T and A549 cell lines. The results demonstrate that NHGs (Figure 6a) are devoid of toxicity at concentrations of 2 mg/mL on A549 and HEK293T cell lines after incubating for 48 h. Both cell lines show beyond 90% viability. This concentration is by far higher than those reported for most of the NPs reported as having low toxicity [28]. VRC-NHGs are non-toxic with 20 μg/mL of VRC on A549 and HEK-293T after incubating for 48 h (Figure 6b).

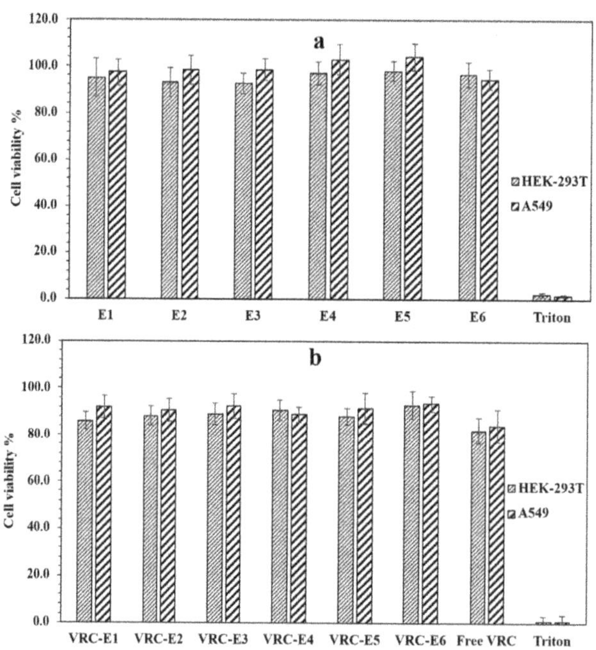

Figure 6. In vitro cytotoxicity of (**a**) NHGs and (**b**) VRC-NHGs (20 μg/mL) at 48 h incubation in HEK293T and A549 cell lines. Each column represents the mean ± SD (n = 3).

Both cell lines show beyond 80% viability. The viability of the control group (free VRC: 20 μg/mL) was 80–84%.

These results show that VRC-NHGs could be considered a "safe nano-formulation" where cell viability is not compromised. Cell viability above 70% is considered safe in vitro assays according to the ISO 10993-5 guideline [63].

3.6. Antifungal Effect of VRC-NHGs

The antifungal studies showed that only 400 nm VRC-NHGs (see E_1 and E_4 in Table 5) maintained high activity, comparable to free VRC in DMSO and up to eight-fold higher than the aqueous suspension. For instance, 400 nm VRC-NHGs exhibited lower MIC values against *A. fumigatus* (four-fold), *A. flavus* (four-fold), *A. terreus* (four-fold), *A. niger* (four-fold), *A. nidulans* (four-fold), *S. aurantiacum* (four-fold), and *E. dermatitidis* (eight-fold) than those MIC values observed for the drug aqueous suspension (Table 5). These data might be related to the capacity of the NHGs to increase VRC aqueous solubility and sustain its release compared to the drug suspension where VRC is homogenously dispersed (but not solubilized) in aqueous medium. The non-loaded NHGs did not display any antifungal activity against any of the studied strains. Activity decreases proportionally with size, resulting in a MIC similar to that of VRC suspension (see E_2, E_3, E_5, and E_6 in Table 5).

Table 5. MIC (μg/mL) values of VRC-NHGs (no mucus layer). Each value is the mean of two measurements.

Fungal Species	VRC-DMSO	VRC Susp.	VRC-E_1	VRC-E_2	VRC-E_3	VRC-E_4	VRC-E_5	VRC-E_6
A. fumigatus	0.25	2	0.5	1	1	0.5	1	2
* *A. fumigatus*	0.25	2	0.5	1	1	0.5	1	2
A. terreus	1	2	0.5	1	2	0.5	1	2
A. flavus	0.5	2	0.5	1	2	0.5	1	2
A. niger	0.5	2	0.5	1	2	0.5	1	2
A. nidulans	0.5	2	0.5	1	1	0.25	1	2
E. dermatitidis	0.5	2	0.25	1	1	0.25	1	2
S. auarantiacum	0.5	2	0.5	1	2	0.5	1	1

* The same activity was found in the presence of a mucus layer (the second row).

By using an in vitro model of pulmonary pathological mucus that mimics the chemical composition, structural features, and viscoelastic properties of CF mucus [36], we studied their penetration across this diffusional barrier and assessed the antimicrobial capacity of the developed NHGs against A. fumigatus, which is the most frequent fungal agent affecting CF patients worldwide. The results indicate that the introduction of a mucus barrier for A. fumigatus (denoted by a star) did not significantly alter the antifungal effect of all VRC-NHGs, which remained comparable to that observed under conditions without the mucus barrier. These results are in line with chitosan-based polymeric micelles for the pulmonary delivery of itraconazole and with Soluplus-based polymeric micelles for the pulmonary delivery of VRC, where the micellar system and a drug solution in DMSO were compared [64,65].

Furthermore, these data suggest that the encapsulation of VRC within the 400 nm NHGs have an improved antimicrobial potential to the aqueous suspension itself, considering that nowadays only an aqueous VRC suspension and a dug parenteral solution represent the commercially available dosage forms.

Regarding the mucus penetrating assay, the results suggest that the 400 nm NHGs (E_1 and E_4) can penetrate through the mucus better, as shown also for the non-loaded

NHGs in Figure 4, serving as a sustained release depot, improving the antifungal efficacy of VRC against the tested major molds. It should be further explored whether these mucus-penetrating NHGs, containing different hydrophobic therapeutics effectively bypass the mucus barrier, allowing better drug distribution and absorption at mucosal surfaces.

The ability of these novel 400 nm NHGs to bypass the mucosal barrier offers a significant advantage for the treatment of CF, where highly viscoelastic mucus presents a major diffusional barrier relevant to drug delivery challenges in CF.

These results are promising, as our VRC nano-formulation demonstrated improved in vitro antifungal activity, overcoming a physical barrier like the artificial CF mucus. The data denote the potential of our 400 nm NHG system for enhanced clinical management of CF patients employing a VRC respirable nanocarrier. In particular, the MIC results observed for the VRC-NHG formulation were both lower than the VRC clinical breakpoint for A. fumigatus (1 μg/mL), regardless of the presence of the mucus layer, and lower than the VRC ECV described for the other fungal species (1 μg/mL for *A. flavus* and *A. terreus*, and 2 μg/mL for *A. niger* and *A. nidulans*). Currently, there are no established ECVs for VRC specifically for *S. aurantiacum* and *E. dermatitidis*. In a study involving clinical isolates of various Scedosporium [66] species, including *S. aurantiacum*, VRC demonstrated significant activity, with an MIC of $\leq$2 μg/mL for most isolates. Another study evaluating 51 E. dermatitidis strains isolated from CF patients found that the MIC range for VRC was 0.03 to 0.5 μg/mL, suggesting good in vitro activity for the drug [67].

Then, VRC-loaded NHGs could be effective in reducing pulmonary fungal load and improving clinical outcome in CF patients.

4. Conclusions

Ester-functionalized cross-linked NHGs with tunable sizes (100–400 nm), and devoid of toxicity, were prepared by free radical polymerization. The present study demonstrates that the newly developed NHGs are effective for controlled drug delivery, as evidenced by the in vitro controlled release of VRC-NHGs under different sink conditions.

The 400 nm VRC-NHGs (E_1 and E_4) exhibit high antifungal activity, comparable to free VRC in DMSO and superior to an aqueous VRC suspension. Notably, even in the presence of a mucus barrier, the antifungal effect remained consistent. This research showcases the significant advancements in the field of biodegradable NHGs and their potential for diverse biological applications, such as the therapy of lung disorders caused by fungi in CF patients. These findings are at the basis of ongoing in vivo investigations aimed at exploring the applications and advantages of these promising NHGs.

Supplementary Materials: The following supporting information can be downloaded at https://www.mdpi.com/article/10.3390/pharmaceutics17060725/s1. Figure S1: Calibration curve of VRC/MeCN for quantification of loading VRC in NHGs; Table S1: Initial number of NHGs for mucus penetration assay; Table S2: Percentage of NHGs crossing the mucus barrier over time.

Author Contributions: Conceptualization, S.D.C., M.C., R.J.G., M.L.C. and G.B.; Methodology, S.D.C., M.F.L., K.H.A., G.K. and M.L.C.; Validation, S.D.C., K.H.A. and G.K.; Investigation, S.D.C., M.F.L., K.H.A., M.C., R.J.G. and M.L.C.; Writing—original draft, S.D.C., M.L.C. and G.B.; Supervision, M.L.C. and G.B.; Funding acquisition, M.L.C. and G.B. All authors have read and agreed to the published version of the manuscript.

Funding: Israel Science Foundation: 1482/22; Ministry of Science and Technology: 1001563914.

Institutional Review Board Statement: Not applicable.

Informed Consent Statement: Not applicable.

Data Availability Statement: The data presented in this study are available on request from the corresponding author due to patent registration.

Acknowledgments: This work was supported by grants from the Ministry of Science, Technology and Innovation of the Argentine Republic (MINCyT) and the Ministry of Innovation, Science and Technology of the State of Israel (MoST) within the framework of the first Argentine–Israel Scientific Research Program. For S.D.C., this work is part of his PhD dissertation. M.C., R.J.G., and M.L.C. are staff members of CONICET. M.F.L. and K.H.A. thank CONICET for their doctoral scholarships. S.D.C. is indebted to Bar Ilan University for a PhD scholarship. We are indebted for the support of the Marcus Center for Medicinal Chemistry at Bar Ilan University and to the Israel Science Foundation grant N°1482/22. Finally, we would like to thank Alejandro David Nusblat for comprehensive editing of the manuscript.

Conflicts of Interest: The authors declare no conflicts of interest.

References

1. Warris, A.; Bercusson, A.; Armstrong-James, D. *Aspergillus* colonization and antifungal immunity in cystic fibrosis patients. *Med. Mycol.* **2019**, *57*, S118–S126. [CrossRef] [PubMed]
2. Farinha, C.M.; Callebaut, I. Molecular mechanisms of cystic fibrosis–how mutations lead to misfunction and guide therapy. *Biosci. Rep.* **2022**, *42*, BSR20212006. [CrossRef] [PubMed]
3. Turcios, N.L. Cystic fibrosis lung disease: An overview. *Respir. Care* **2020**, *65*, 233–251. [CrossRef] [PubMed]
4. Suk, J.S.; Lai, S.K.; Wang, Y.Y.; Ensign, L.M.; Zeitlin, P.L.; Boyle, M.P.; Hanes, J. The penetration of fresh undiluted sputum expectorated by cystic fibrosis patients by non-adhesive polymer nanoparticles. *Biomaterials* **2009**, *30*, 2591–2597. [CrossRef]
5. Yu, T.; Chisholm, J.; Choi, W.J.; Anonuevo, A.; Pulicare, S.; Zhong, W.; Chen, M.; Fridley, C.; Lai, S.K.; Ensign, L.M.; et al. Mucus-Penetrating Nanosuspensions for Enhanced Delivery of Poorly Soluble Drugs to Mucosal Surfaces. *Adv. Healthc. Mater.* **2016**, *5*, 2745–2750. [CrossRef]
6. Hansson, G.C. Mucus and mucins in diseases of the intestinal and respiratory tracts. *J. Intern. Med.* **2019**, *285*, 479–490. [CrossRef]
7. Fahy, J.V.; Dickey, B.F. Airway Mucus Function and Dysfunction. *N. Engl. J. Med.* **2010**, *363*, 2233–2247. [CrossRef]
8. Meldrum, O.W.; Chotirmall, S.H. Mucus, microbiomes and pulmonary disease. *Biomedicines* **2021**, *9*, 675. [CrossRef]
9. Cone, R.A. Barrier properties of mucus. *Adv. Drug Deliv. Rev.* **2009**, *61*, 75–85. [CrossRef]
10. Pearson, J.P.; Chater, P.I.; Wilcox, M.D. The properties of the mucus barrier, a unique gel-how can nanoparticles cross it? *Ther. Deliv.* **2016**, *7*, 229–244. [CrossRef]
11. Murgia, X.; Loretz, B.; Hartwig, O.; Hittinger, M.; Lehr, C.M. The role of mucus on drug transport and its potential to affect therapeutic outcomes. *Adv. Drug Deliv. Rev.* **2018**, *124*, 82–97. [CrossRef] [PubMed]
12. Sigurdsson, H.H.; Kirch, J.; Lehr, C.M. Mucus as a barrier to lipophilic drugs. *Int. J. Pharm.* **2013**, *453*, 56–64. [CrossRef] [PubMed]
13. Boegh, M.; Nielsen, H.M. Mucus as a barrier to drug delivery-Understanding and mimicking the barrier properties. *Basic Clin. Pharmacol. Toxicol.* **2015**, *116*, 179–186. [CrossRef]
14. King, J.; Brunel, S.F.; Warris, A. *Aspergillus* infections in cystic fibrosis. *J. Infect.* **2016**, *72*, S50–S55. [CrossRef]
15. Chotirmall, S.H.; McElvaney, N.G. Fungi in the cystic fibrosis lung: Bystanders or pathogens? *Int. J. Biochem. Cell Biol.* **2014**, *52*, 161–173. [CrossRef]
16. Martín-Gómez, M.T. Taking a look on fungi in cystic fibrosis: More questions than answers. *Rev. Iberoam. Micol.* **2020**, *37*, 17–23. [CrossRef]
17. Montagna, M.T.; Barbuti, G.; Paglionico, F.; Lovero, G.; Iatta, R.; De Giglio, O.; Cuna, T.; Coretti, C.; Santostasi, T.; Polizzi, A.; et al. Retrospective analysis of microorganisms isolated from cystic fibrosis patients in Southern Italy, 2002–2010. *J. Prev. Med. Hyg.* **2011**, *52*, 209–214.
18. Viñado, C.; Girón, R.M.; Ibáñez, E.; García-Ortega, A.; Pérez, I.; Polanco, D.; Pemán, J.; Solé, A. Filamentous fungi in the airway of patients with cystic fibrosis: Just spectators? *Rev. Iberoam. Micol.* **2021**, *38*, 168–174. [CrossRef]
19. Seidel, D.; Meißner, A.; Lackner, M.; Piepenbrock, E.; Salmanton-García, J.; Stecher, M.; Mellinghoff, S.; Hamprecht, A.; Graeff, L.D.; Köhler, P.; et al. Prognostic factors in 264 adults with invasive *Scedosporium* spp. and *Lomentospora prolificans* infection reported in the literature and FungiScope®. *Crit. Rev. Microbiol.* **2019**, *45*, 1–21. [CrossRef]
20. Pihet, M.; Carrere, J.; Cimon, B.; Chabasse, D.; Delhaes, L.; Symoens, F.; Bouchara, J.P. Occurrence and relevance of filamentous fungi in respiratory secretions of patients with cystic fibrosis—A review. *Med. Mycol.* **2009**, *47*, 387–397. [CrossRef]
21. Cimon, B.; Carrère, J.; Vinatier, J.F.; Chazalette, J.P.; Chabasse, D.; Bouchara, J.P. Clinical significance of *Scedosporium apiospermum* in patients with cystic fibrosis. *Eur. J. Clin. Microbiol. Infect. Dis.* **2000**, *19*, 53–56. [CrossRef] [PubMed]

22. Güngör, Ö.; Tamay, Z.; Güler, N.; Erturan, Z. Frequency of fungi in respiratory samples from Turkish cystic fibrosis patients. *Mycoses* **2013**, *56*, 123–129. [CrossRef]
23. Ziesing, S.; Suerbaum, S.; Sedlacek, L. Fungal epidemiology and diversity in cystic fibrosis patients over a 5-year period in a national reference center. *Med. Mycol.* **2016**, *54*, 781–786. [CrossRef]
24. Felton, T.; Troke, P.F.; Hope, W.W. Tissue penetration of antifungal agents. *Clin. Microbiol. Rev.* **2014**, *27*, 68–88. [CrossRef]
25. Lakhani, P.; Patil, A.; Majumdar, S. Challenges in the Polyene- and Azole-Based Pharmacotherapy of Ocular Fungal Infections. *J. Ocul. Pharmacol. Ther.* **2019**, *35*, 6–22. [CrossRef]
26. Levine, M.T.; Chandrasekar, P.H. Adverse effects of voriconazole: Over a decade of use. *Clin. Transplant.* **2016**, *30*, 1377–1386. [CrossRef]
27. Bentley, S.; Gupta, A.; Balfour-Lynn, I.M. Subtherapeutic itraconazole and voriconazole levels in children with cystic fibrosis. *J. Cyst. Fibros.* **2013**, *12*, 418–419. [CrossRef]
28. Brunet, K.; Martellosio, J.P.; Tewes, F.; Marchand, S.; Rammaert, B. Inhaled Antifungal Agents for Treatment and Prophylaxis of Bronchopulmonary Invasive Mold Infections. *Pharmaceutics* **2022**, *14*, 641. [CrossRef]
29. Sinha, B.; Mukherjee, B.; Pattnaik, G. Poly-lactide-co-glycolide nanoparticles containing voriconazole for pulmonary delivery: In vitro and in vivo study. *Nanomed. Nanotechnol. Biol. Med.* **2013**, *9*, 94–104. [CrossRef]
30. Beinborn, N.A.; Du, J.; Wiederhold, N.P.; Smyth, H.D.C.; Williams, R.O. Dry powder insufflation of crystalline and amorphous voriconazole formulations produced by thin film freezing to mice. *Eur. J. Pharm. Biopharm.* **2012**, *81*, 600–608. [CrossRef]
31. Das, P.J.; Paul, P.; Mukherjee, B.; Mazumder, B.; Mondal, L.; Baishya, R.; Debnath, M.C.; Dey, K.S. Pulmonary Delivery of Voriconazole Loaded Nanoparticles Providing a Prolonged Drug Level in Lungs: A Promise for Treating Fungal Infection. *Mol. Pharm.* **2015**, *12*, 2651–2664. [CrossRef] [PubMed]
32. Paul, P.; Sengupta, S.; Mukherjee, B.; Shaw, T.K.; Gaonkar, R.H.; Debnath, M.C. Chitosan-coated nanoparticles enhanced lung pharmacokinetic profile of voriconazole upon pulmonary delivery in mice. *Nanomedicine* **2018**, *13*, 501–520. [CrossRef] [PubMed]
33. Liao, Q.; Yip, L.; Chow, M.Y.T.; Chow, S.F.; Chan, H.K.; Kwok, P.C.L.; Lam, J.K.W. Porous and highly dispersible voriconazole dry powders produced by spray freeze drying for pulmonary delivery with efficient lung deposition. *Int. J. Pharm.* **2019**, *560*, 144–154. [CrossRef]
34. Wan, F.; Bohr, S.S.R.; Kłodzińska, S.N.; Jumaa, H.; Huang, Z.; Nylander, T.; Thygesen, M.B.; Sørensen, K.K.; Jensen, K.J.; Sternberg, C.; et al. Ultrasmall TPGS-PLGA Hybrid Nanoparticles for Site-Specific Delivery of Antibiotics into *Pseudomonas aeruginosa* Biofilms in Lungs. *ACS Appl. Mater. Interfaces* **2020**, *12*, 380–389. [CrossRef]
35. Suk, J.S.; Kim, A.J.; Trehan, K.; Schneider, C.S.; Cebotaru, L.; Woodward, O.M.; Boylan, N.J.; Boyle, M.P.; Lai, S.K.; Guggino, W.B.; et al. Lung gene therapy with highly compacted DNA nanoparticles that overcome the mucus barrier. *J. Control. Release* **2014**, *178*, 8–17. [CrossRef]
36. Pacheco, D.P.; Butnarasu, C.S.; Vangosa, F.B.; Pastorino, L.; Visai, L.; Visentin, S.; Petrini, P. Disassembling the complexity of mucus barriers to develop a fast screening tool for early drug discovery. *J. Mater. Chem. B* **2019**, *7*, 4940–4952. [CrossRef]
37. Cemal, S.D.; Kazimirsky, G.; Shadkchan, Y.; Eswaran, L.; Abramovitch, R.; Abudi, N.; Cuestas, M.L.; Osherov, N.; Byk, G. Biocompatible narrow size nanohydrogels for drug delivery. *Nanomed. Nanotechnol. Biol. Med.* **2025**, *66*, 102824. [CrossRef]
38. Yang, H.; Wang, Q.; Chen, W.; Zhao, Y.; Yong, T.; Gan, L.; Xu, H.; Yang, X. Hydrophilicity/hydrophobicity reversable and redox-sensitive nanogels for anticancer drug delivery. *Mol. Pharm.* **2015**, *12*, 1636–1647. [CrossRef]
39. Palakkal, S.; Logviniuk, D.; Byk, G. Tuning the size and hydrophobicity of nanohydrogels exploiting a self-assembly assisted polymerization mechanism for controlled drug delivery. *J. Nanoparticle Res.* **2020**, *22*, 1–16. [CrossRef]
40. Shi, F.; Ding, J.; Xiao, C.; Zhuang, X.; He, C.; Chen, L.; Chen, X. Intracellular microenvironment responsive PEGylated polypeptide nanogels with ionizable cores for efficient doxorubicin loading and triggered release. *J. Mater. Chem.* **2012**, *22*, 14168–14179. [CrossRef]
41. Gupta, V.; Ahsan, F. Influence of PEI as a core modifying agent on PLGA microspheres of PGE 1, a pulmonary selective vasodilator. *Int. J. Pharm.* **2011**, *413*, 51–62. [CrossRef] [PubMed]
42. Costabile, G.; D'Angelo, I.; Rampioni, G.; Bondì, R.; Pompili, B.; Ascenzioni, F.; Mitidieri, E.; D'Emmanuele Di Villa Bianca, R.; Sorrentino, R.; Miro, A.; et al. Toward Repositioning Niclosamide for Antivirulence Therapy of *Pseudomonas aeruginosa* Lung Infections: Development of Inhalable Formulations through Nanosuspension Technology. *Mol. Pharm.* **2015**, *12*, 2604–2617. [CrossRef]
43. Pellosi, D.S.; d'Angelo, I.; Maiolino, S.; Mitidieri, E.; d'Emmanuele di Villa Bianca, R.; Sorrentino, R.; Quaglia, F.; Ungaro, F. In vitro/in vivo investigation on the potential of Pluronic® mixed micelles for pulmonary drug delivery. *Eur. J. Pharm. Biopharm.* **2018**, *130*, 30–38. [CrossRef]
44. Gulati, N.; Rastogi, R.; Dinda, A.K.; Saxena, R.; Koul, V. Characterization and cell material interactions of PEGylated PNIPAAM nanoparticles. *Colloids Surf. B Biointerfaces* **2010**, *79*, 164–173. [CrossRef]
45. Devoto, T.B.; Hermida-Alva, K.; Posse, G.; Finquelievich, J.L.; García-Effrón, G.; Cuestas, M.L. Antifungal susceptibility patterns for *Aspergillus*, *Scedosporium*, and *Exophiala* isolates recovered from cystic fibrosis patients against amphotericin B, and three triazoles and their impact after long-term therapies. *Med. Mycol.* **2023**, *61*, myad089. [CrossRef]

46. *CLSI M38*; Clinical Laboratory Standard Institute (CLSI). Reference Method for Broth Dilution Antifungal Susceptibility Testing of Filamentous Fungi. Clinical Laboratory Standard Institute: Wayne, PA, USA, 2017.
47. Khandadash, R.; Machtey, V.; Shainer, I.; Gottlieb, H.E.; Gothilf, Y.; Ebenstein, Y.; Weiss, A.; Byk, G. Novel biocompatible hydrogel nanoparticles: Generation and size-tuning of nanoparticles by the formation of micelle templates obtained from thermo-responsive monomers mixtures. *J. Nanoparticle Res.* **2014**, *16*, 2796. [CrossRef]
48. Eswaran, L.; Kazimirsky, G.; Byk, G. New Biocompatible Nanohydrogels of Predefined Sizes for Complexing Nucleic Acids. *Pharmaceutics* **2023**, *15*, 332. [CrossRef]
49. Yang, S.C.; Paik, S.Y.R.; Ryu, J.; Choi, K.O.; Kang, T.S.; Lee, J.K.; Song, C.W.; Ko, S. Dynamic light scattering-based method to determine primary particle size of iron oxide nanoparticles in simulated gastrointestinal fluid. *Food Chem.* **2014**, *161*, 185–191. [CrossRef]
50. Yang, F.H.; Zhang, Q.; Liang, Q.Y.; Wang, S.Q.; Zhao, B.X.; Wang, Y.T.; Cai, Y.; Li, G.F. Bioavailability enhancement of paclitaxel via a novel oral drug delivery system: Paclitaxel-loaded glycyrrhizic acid micelles. *Molecules* **2015**, *20*, 4337–4356. [CrossRef]
51. Ensign, L.M.; Schneider, C.; Suk, J.S.; Cone, R.; Hanes, J. Mucus penetrating nanoparticles: Biophysical tool and method of drug and gene delivery. *Adv. Mater.* **2012**, *24*, 3887–3894. [CrossRef]
52. Li, L.; Wang, H.; Ye, J.; Chen, Y.; Wang, R.; Jin, D.; Liu, Y. Mechanism Study on Nanoparticle Negative Surface Charge Modification by Ascorbyl Palmitate and Its Improvement of Tumor Targeting Ability. *Molecules* **2022**, *27*, 4408. [CrossRef] [PubMed]
53. Gumustas, M.; Sengel-Turk, C.T.; Gumustas, A.; Ozkan, S.A.; Uslu, B. Effect of Polymer-Based Nanoparticles on the Assay of Antimicrobial Drug Delivery Systems. In *Multifunctional Systems for Combined Delivery, Biosensing and Diagnostics*; Elsevier: Amsterdam, The Netherlands, 2017; pp. 67–108. [CrossRef]
54. Blackburn, W.H.; Lyon, L.A. Size-controlled synthesis of monodisperse core/shell nanogels. *Colloid Polym. Sci.* **2008**, *286*, 563–569. [CrossRef] [PubMed]
55. Shekhar, S.; Mukherjee, M.; Sen, A.K. Swelling, thermal and mechanical properties of NIPAM-based terpolymeric hydrogel. *Polym. Bull.* **2016**, *73*, 125–145. [CrossRef]
56. Costa, M.C.M.; Silva, S.M.C.; Antunes, F.E. Adjusting the low critical solution temperature of poly(N-isopropyl acrylamide) solutions by salts, ionic surfactants and solvents: A rheological study. *J. Mol. Liq.* **2015**, *210*, 113–118. [CrossRef]
57. Cu, Y.; Saltzman, W.M. Controlled surface modification with poly(ethylene)glycol enhances diffusion of PLGA nanoparticles in human cervical Mucus. *Mol. Pharm.* **2009**, *6*, 173–181. [CrossRef]
58. Murgia, X.; Pawelzyk, P.; Schaefer, U.F.; Wagner, C.; Willenbacher, N.; Lehr, C.M. Size-Limited Penetration of Nanoparticles into Porcine Respiratory Mucus after Aerosol Deposition. *Biomacromolecules* **2016**, *17*, 1536–1542. [CrossRef]
59. Lai, S.K.; Wang, Y.Y.; Hanes, J. Mucus-penetrating nanoparticles for drug and gene delivery to mucosal tissues. *Adv. Drug Deliv. Rev.* **2009**, *61*, 158–171. [CrossRef]
60. Shen, S.; Wu, Y.; Liu, Y.; Wu, D. High drug-loading nanomedicines: Progress; current status; prospects. *Int. J. Nanomed.* **2017**, *12*, 4085–4109. [CrossRef]
61. Abouelmagd, S.A.; Sun, B.; Chang, A.C.; Ku, Y.J.; Yeo, Y. Release kinetics study of poorly water-soluble drugs from nanoparticles: Are we doing it right? *Mol. Pharm.* **2015**, *12*, 997–1003. [CrossRef]
62. Patil, J.S.; Sarasija, S. Pulmonary drug delivery strategies: A concise, systematic review. *Lung India* **2012**, *29*, 44–49. [CrossRef]
63. *ISO 10993-5:2009*; Biological Evaluation of Medical Devices—Part 5: Tests for In Vitro Cytotoxicity. International Organization for Standardization: Geneva, Switzerland, 2009.
64. Moazeni, E.; Gilani, K.; Najafabadi, A.R.; Rouini, M.R.; Mohajel, N.; Amini, M.; Barghi, M.A. Preparation and evaluation of inhalable itraconazole chitosan based polymeric micelles. *DARU J. Pharm. Sci.* **2012**, *20*, 85. [CrossRef] [PubMed]
65. Fernández, V.B.; Alava, K.H.; Bernabeu, E.; Fuentes, P.; Devoto, T.B.; Höcht, C.; Chiappetta, D.A.; Cuestas, M.L.; Moretton, M.A. Highly effective inhalable voriconazole-loaded nanomicelles for fungal infections in cystic fibrosis patients: A promising therapeutic strategy for allergic bronchopulmonary aspergillosis. *J. Drug Deliv. Sci. Technol.* **2024**, *100*, 106126. [CrossRef]
66. Czech, M.; Stock, F.; Aneke, C.; Lionakis, M.; Cuellar-Rodriguez, J.; Seyedmousavi, A. 2758. Clinical Significance and Antifungal Susceptibility Profile of 103 Clinical *Scedosporium* Species Complex and *Lomentospora prolificans* Isolated from NIH Hospitalized Patients. *Open Forum Infect. Dis.* **2023**, *10*, ofad500.2369. [CrossRef]
67. Gülmez, D.; Doğan, Ö.; Boral, B.; Döğen, A.; İlkit, M.; de Hoog, G.S.; Arikan-Akdagli, S. In vitro activities of antifungal drugs against environmental *Exophiala* isolates and review of the literature. *Mycoses* **2018**, *61*, 561–569. [CrossRef]

MDPI AG
Grosspeteranlage 5
4052 Basel
Switzerland
Tel.: +41 61 683 77 34

Pharmaceutics Editorial Office
E-mail: pharmaceutics@mdpi.com
www.mdpi.com/journal/pharmaceutics

www.ingramcontent.com/pod-product-compliance
Lightning Source LLC
Chambersburg PA
CBHW040816250926
51850CB00007BA/1186
9783725864706